D0217752

DANGEROUS EARTH

DANGEROUS EARTH

An Introduction to Geologic Hazards

BARBARA W. MURCK

University of Toronto

BRIAN J. SKINNER

Yale University

STEPHEN C. PORTER

University of Washington

JOHN WILEY & SONS, INC.

NEW YORK • CHICHESTER • BRISBANE • TORONTO • SINGAPORE • WEINHEIM

Acquisition Editor *Clifford Mills*
Developmental Editor *Rachel Nelson*
Marketing Manager *Cathy Faduska*
Senior Production Editor *Jeanie Berke*
Text Designer *Karin Gerdes Kincheloe*
Cover Designer *Laurie Ierardi*
Cover Photo *©Eric Meola/The Image Bank*
Manufacturing Manager *Dorothy Sinclair*
Senior Photo Editor *Mary Ann Price*
Photo Editor *Alexandra Truitt*
Photo Assistant *Kim Khatchatourian*
Illustration Coordinator *Anna Melhorn*

This book was set in 10/12 Adobe Garamond by Ruttle, Shaw & Wetherill
and printed and bound by Von Hoffmann Press.
The cover was printed by Lehigh Press.

Recognizing the importance of preserving what has been written, it is a
policy of John Wiley & Sons, Inc. to have books of enduring value published
in the United States printed on acid-free paper, and we exert our best
efforts to that end.

John Wiley & Sons, Inc., places great value on the environment and is actively involved in efforts to
preserve it. Currently, paper of high enough quality to reproduce full-color art effectively contains a
maximum of 10% recovered and recycled post-consumer fiber. Wherever possible, Wiley uses paper
containing the maximum amount of recycled fibers. In addition, the paper in this book was manufac-
tured by a mill whose forest management programs include sustained yield harvesting of its timber-
lands. Sustained yield harvesting principles ensure that the number of trees cut each year does not
exceed the amount of new growth.

**TOTAL 10% RECYCLED PAPER
ALL POST-CONSUMER WASTE**

Copyright © 1997, by John Wiley & Sons, Inc.

All rights reserved. Published simultaneously in Canada.

Reproduction or translation of any part of
this work beyond that permitted by Sections
107 and 108 of the 1976 United States Copyright
Act without the permission of the copyright
owner is unlawful. Requests for permission
or further information should be addressed to
the Permissions Department, John Wiley & Sons, Inc.

Library of Congress Cataloging-in-Publication Data:
Murck, Barbara Winifred, 1954–
     Dangerous Earth : an introduction to geologic hazards / Barbara W.
  Murck, Brian J. Skinner, Stephen C. Porter.
         p.   cm.
     Includes bibliographical references.
     ISBN 0-471-13565-8 (paper : alk. paper)
     1. Natural Disasters.   I. Skinner, Brian J., 1928–   .
  II. Porter, Stephen C.   III. Title.
  GB5014.M87   1996
  550–dc20   96-23720
                                                          CIP

Printed in the United States of America

10 9 8 7 6 5 4 3 2 1

ABOUT THE AUTHORS

The authors of this book bring a wealth of professional and personal knowledge, training, and experience to the project. Among them, they have carried out geologic fieldwork on all of the Earth's continents. The diversity demonstrated in their own careers reflects the broad range of challenges that characterize geology today.

As an undergraduate *Barbara Murck* was a confirmed nonscientist, until an introductory geology course changed her plans. Since then her professional focus has ranged from igneous geochemistry and ore-deposit petrography to alternative energy sources and state-of-the-environment reporting. Her current work focuses primarily on environmental management training for decision makers in developing countries.

Throughout his career as a geologist, *Brian Skinner's* research has focused on the physical properties of minerals and on the genesis of base-metal deposits. He has worked extensively in Australia, Africa, and North America and with students in Asia and Europe. With Yale University colleagues, he has had the opportunity to explore a diversity of Earth science topics, including oceanography and climatic change, volcanic gases, economic models of resource depletion, and the geologic aspects of the space program.

Stephen Porter's professional career has largely been concerned with studies of glaciation in many of the world's major mountain systems and with the history of the climatic changes their deposits record. He has also studied the evolution of midocean and continental volcanoes and the products of their prehistoric eruptions and how volcanic eruptions may have influenced the Earth's climate. With colleagues from around the world, he has studied the hazards of large rockfalls in the Alps and the thick, extensive deposits of windblown dust in China that provide one of the longest continuous records of climatic change during the past several million years.

The authors' global perspective is reflected in this book by examples and illustrations from numerous foreign areas, for it is important to emphasize that geology is a global science—a science that recognizes no political boundaries.

Only by studying the Earth in its entirety can we hope to understand how our amazing planet works, how geologic hazards affect our lives, and how, in turn, human activities affect the functioning of the Earth system.

PREFACE

HOW *DANGEROUS EARTH* CAME TO BE

Geologic hazards are always with us. Even when we are not personally threatened by events such as floods, earthquakes, tornadoes, and volcanic eruptions, we are exposed to a constant stream of news reports of the impacts of hazardous events elsewhere. Recognizing the magnitude of these impacts on modern society, the National Academy of Sciences, supported by the United Nations and other institutions around the world, declared the 1990s to be the International Decade of Natural Hazard Reduction. In light of so much attention, departments of geology and environmental science have found it increasingly helpful to develop undergraduate courses devoted to the study of hazards. The concept for *Dangerous Earth* grew out of the need for a comprehensive, full-color hazards text to serve such courses. As explained on page xi, *Dangerous Earth* comprises those parts of a more comprehensive text, *Environmental Geology*, that are specifically devoted to hazards and the basic geologic concepts needed to understand them.

Students respond best when they understand the reasons for learning; hence, frequent explanations are offered. They ask, for example, why it is important to learn about the interior of the Earth? The answer: because internal Earth processes are fundamental in forming the landscape and in causing hazardous Earth processes like earthquakes and volcanoes. They wonder, why is it necessary to study the properties of rocks, minerals, and soils? The response: because these properties can affect human interests in a wide variety of ways—in their ability to resist mass wasting, or in the way they transmit earthquake waves, for example. The basic geology topics covered in this book were selected to provide the foundation of concepts and terminology needed to understand the impacts of geologic hazards on human interests. We hope that *Dangerous Earth* will help readers become more aware of the geologic nature of our environment and the role of geoscientists in events

of public concern. And we hope our readers will emerge better prepared to make informed decisions about the natural processes that affect our lives on a daily basis.

Organization and Special Features

In *Part I: Geologic Framework* we provide a brief background of Earth system science and physical geology. In this part of the book you will find basic concepts and terminology concerning the structure and materials of the Earth and the functioning of Earth systems and cycles. *Part II: Hazardous Geologic Processes* covers the broad range of geologic events that are damaging to human interests, such as earthquakes, volcanic eruptions, landslides, floods, meteorite impacts, and others. We look at both the human impacts of such events and the geologic processes that underlie them. Each of the two main parts opens with an essay designed to set the context and provide an overview of fundamental concepts in the ensuing chapters. The essay for Part I, entitled *The Home Planet*, puts the Earth in perspective by placing it in the context of the solar system as a whole and examining some of the Earth-forming processes that have made this planet hospitable to life. *Assessing Geologic Hazards and Risks*, the essay for Part II, introduces natural, geologic, technologic, and anthropogenic hazards as well as some of the approaches used to assess human vulnerability to these hazards.

Each chapter opens with a short vignette or anecdote. The purpose of these vignettes is to provide a glimpse into some aspect of the geologic environment and to show how that environment affects humans (and vice versa). Each chapter also includes a *Summary* of key points and a list of *Important Terms to Remember*. A page reference is given for each of the important terms, and a complete *Glossary* is provided at the end of the book. The *Questions and Activities* that end each chapter are meant to stimulate independent thought and study and critical thinking. Throughout each chapter you will also find material set aside in shaded boxes. The boxes perform two distinct functions: (1) Boxes entitled *The Human Perspective* are intended to highlight particular aspects of the human–planet relationship, such as human impacts on the environment, the impacts of geologic processes on human interests, or human institutions (projects, programs, etc.) devoted to some aspect of geology. (2) Boxes entitled *Focus On . . .* are intended to provide an in-depth look at some of the more technical aspects of geology and related sciences. The *Appendices* at the end of the book contain useful reference information for students (and instructors) on units and conversions, the chemical elements, and the geologic time scale.

Supplements

A full range of supplementary material is available to assist both instructors and students using either *Environmental Geology* or *Dangerous Earth*. The **Environmental Geology Study Guide** provides an inexpensive way for students to get the most out of these textbooks. It includes chapter summaries, brief discussions of the most important terms and key points in each chapter, study pointers and guidelines, and practice questions to help students review and apply concepts and prepare for tests. The **Environmental Geology Instructor's Manual and Test Bank** includes chapter synopses and lecture lead-ins, sample syllabi and options for course organization, suggestions for further reading, a full description of supplementary materials, and additional written, audiovisual, and computer resources. The *Test Bank* is also available in computerized format.

The **Wiley Geology Transparency Set** includes 150 full-color textbook illustrations, resized and edited for maximum effectiveness in large lecture halls. The **Wiley Geology Slide Set** comprises the 150 images provided as transparency acetates in 35-mm slide form. The **Environmental Geology Overhead Transparency Set** consists of full-color line drawings, also available as 35-mm slides in the **Environmental Geology Slide Set**. The **Wiley Geosciences CD-ROM** provides animations of key concepts and many images from *Environmental Geology* and *Dangerous Earth*,

as well as from *Dynamic Earth* and *The Blue Planet*, by Brian J. Skinner and Stephen C. Porter.

ACKNOWLEDGMENTS

As always, it has been a pleasure to work with the talented, efficient, and ever-patient professionals at John Wiley & Sons and the freelance experts associated with them. The concept of producing the *Environmental Geology* text originated with Barry Harmon, then Earth Sciences editor, and publisher Kaye Pace, and continued under the editorial guidance of Chris Rogers. Developmental editor Rachel Nelson guided *Environmental Geology* to a successful finish. Others who contributed their considerable talents to the success of *Environmental Geology* include (in no particular order): Bonnie Cabot, Stella Kupferberg, Alexandra Truitt, Kim Khatchatourian, Michelle Orlans, Anna Melhorn, Karin Kincheloe, John Woolsey and his staff, Eric Stano, Carolyn Smith, Cathy Faduska, Diane Kraut, Beth Balch, and Pui Szeto. Special mention is due to Judith Peatross, who contributed editorial advice and authored a number of the boxes that appear throughout the book.

The idea of building on the foundation of *Environmental Geology* to produce a full-color geologic hazards text was proposed by editor Cliff Mills, assisted by Cathy Donovan. The project was approved by publisher Kaye Pace and executive editor Nedah Rose. The production coordinators for *Dangerous Earth* were Elizabeth Swain and Jeanie Berke, who deftly handled many last-minute changes. Many of those mentioned above also contributed further efforts to *Dangerous Earth*.

Careful reading and extensive commentary by many colleagues improved both *Environmental Geology* and *Dangerous Earth* immeasurably. The thoughtful suggestions of these reviewers touched on every aspect, from overall organization to the tiniest details. They helped keep us up-to-date in a science that is constantly changing. Through their comments the reviewers made available to us their many years of collective experience in conveying this material to students. Thank you to those who assisted us by reviewing all or part of these manuscripts. They include:

REVIEWERS

Yemane Asmerom
University of New Mexico

James Bell
Linn Benton Community College

Jane Boger
State University of New York at Geneseo

Phillip Boger
State University of New York at Geneseo

Marsha S. Bollinger
Winthrop University

James Bugh
State University of New York at Cortland

John Busby
Hardin-Simmons University

F. William Cambray
Michigan State University

Ellen Cowan
Appalachian State University

Stanley Dart
University of Nebraska at Kearney

Pascal de Caprariis
Indiana University–Purdue University

Mitra Fattahipour
San Diego State University

Jeremy Fein
McGill University

Robert Furlong
Wayne State University

Henry C. Halls
University of Toronto

Stephen B. Harper
East Carolina University

Judith R. Hepburn
Boston College

Malcolm Hill
Northeastern University

Jerry Horne
San Bernadino Valley College

Larry Kodosky
Oakland Community College

Joan Licari
Cerritos College

Barbara Manner
Duquesne University

George Meyer
College of the Desert

Kevin Mickus
Southwest Missouri State University

Michael Nielson
University of Alabama at Birmingham

Nathaniel Ostrom
Michigan State University

Donald Palmer
Kent State University

Eugene Perry
Northern Illinois University

Frank Revetta
Potsdam College

J. Donald Rimstidt
Virginia Polytechnic Institute and State University

Glenn Roquemore
Irvine Valley Community College

Peter M. Sadler
University of California at Riverside

Laura Sanders
Northern Illinois University

Catherine Shrady
St. Lawrence University

Frank Simpson
University of Windsor

Lonnie Thompson
Ohio State University

John Vitek
Oklahoma State University

Daniel Zarin
University of Pennsylvania

BRIEF TABLE OF CONTENTS

PUBLISHER'S NOTE

Dangerous Earth is the first ten chapters of a more comprehensive 18 chapter text entitled Environmental Geology by Barbara W. Murck, Brian J. Skinner and Stephen C. Porter. These ten chapters are devoted to basic geologic concepts (Part I) and geologic hazards (Part II). The Preface and Introduction have been rewritten to reflect the content of the new book. In the text of Dangerous Earth, you will occasionally run across references to the material contained in the omitted chapters, so please overlook these references. The glossary also contains material from the chapters of Environmental Geology, and some of this material is not relevant to Dangerous Earth. It is our hope that this concise and focused text will meet the needs of courses devoted to the study of geologic and natural hazards.

The chapters contained in the text Environmental Geology that are not contained in Dangerous Earth are as follows: Part III, Using and Caring for Earth Resources; Chapter 11, Energy from Fossil Fuels; Chapter 12, Energy Alternatives; Chapter 13, Mineral Resources; Chapter 14, Soil Resources; Chapter 15, Water Resources. Part IV, Human Impacts on the Environment; Chapter 16, Waste Disposal; Chapter 17, Contaminants in the Geologic Environment; Chapter 18, Atmospheric Change.

CONTENTS

CHAPTER 1
WEB LINKS
EARTH SYSTEMS AND CYCLES

Earth Viewer URL:
http://www.fourmilab.ch/earthview/vplanet.html

View either a map of the Earth showing the day and night regions at this moment, or view the Earth from the Sun, the Moon, the night side of the Earth, above any location on the planet specified by latitude, longitude, and altitude, or from a satellite in Earth orbit. Images can be generated based on a topographical map of the Earth, up-to-date weather satellite imagery, or a composite image of cloud cover superimposed on a map of the Earth.

Earth and Environmental Science URI.:
http://info.er.usgs.gov/network/science/earth/earth.html

CHAPTER 2
WEB LINKS
EARTH STRUCTURE AND MATERIALS

Petrographic Workshop URL:
http://pong.igpp.ucla.edu/pet/pet_intro.html

An interactive database on petrography. The program is designed as a source of mineralogical information used in the identification process of rocks and minerals.

Rock "U" URL:
http://www.ucs.usl.edu/~amg6262/rocku.html

Explains various rock groups with images and text.

The Active Tectonics Web Site URL:
http://www.muohio.edu/tectonics/activetectonics.html

Plate Tectonics and the Dynamic Earth URL:
http://zebu.uoregon.edu/geol.html

Structural Geology Home Page URL:
http://www-geol.unine.ch/STRUCTURAL/Structural_HomePage.html

CHAPTER 3
WEB LINKS
EARTHQUAKES

Earthquakes and Plate Tectonics URL:
http://gldfs.cr.usgs.gov/neis/general/handouts/rift_man.html

An introduction via text and a few diagrams by the USGS.

Access to the Current Seismicity URL:
http://quake.wr.usgs.gov/QUAKES/CURRENT/current.html

The April 22, 1991 Valle de la Estrella Costa Rica Earthquake URL:
http://www.eqe.com/publications/costaric/costaric.htm

Oklahoma Earthquake Catalog URL:
gopher://wealake.oksurvey1.gov:70/11/okeqcat

The Southern Arizona Seismological Observatory URL:
http://www.geo.arizona.edu/saso/

An online seismograph.

CHAPTER 4
WEB LINKS
VOLCANIC ERUPTIONS

The Electronic Volcano URL:
http://mmm.dartmouth.edu/pages/stoiber/elecvolc.html

Eruptions of Mount Spurr Volcano. Alaska URL:
ftp://mojave.wr.usgs.gov/pub/spurr/Spurr.htm

Photographs of the 1992 eruptions.

Volcano Watch URL:
http://www.soest.hawaii.edu/hvo/

Volcano World URL:
http://volcano.und.nodak.edu/

World-Wide Volcanism URL:
http://skye.gsfc.nasa.gov/wwvolcano.html

Natural Disaster Info URL:
http://web66.coled.umn.edu/hillside/franklin/disaster/Project.html

This site, created by students, is an excellent introduction to all kinds of natural disasters such as earthquakes and volcanoes.

CHAPTER 5
WEB LINKS
TSUNAMIS

Tsunami URL:
http://tsunami.ce.washington.edu/tsumani/intro.html

Hawaii Tsunami Page URL:
http://lumahai.soest.hawaii.edu/tsunami.html

CHAPTER 7
WEB LINKS
SUBSIDENCE

The Edwards Aquifer Home Page URL:
http://www.txdirect.net/users/eckhardt/

CHAPTER 8

WEB LINKS

FLOODS

Mississippi River URL:
http://www.jpl.nasa.gov/sircxsar/mississippi.html

This image shows regions of the southern United States that are prone to flooding.

CHAPTER 9

WEB LINKS

HAZARDS OF OCEAN AND WEATHER

Sea-Level Increase URL:
ftp://ftp.hmc.edu/pub/science/sci.answers/.mirror.OLD/sea-level

Weather Maps & Movies URL:
http://rs560.cl.msu.edu/weather/

Hurricanes

Hurricane Dynamics URL:
http://www.gfdl.gov/hurricane.html

Hurricane Tracking Map URL:
http://lumahai.soest.hawaii.edu/Tropical_Weather/atlantic_track.gif

Scanned image of a blank tracking map.

Hurricane Watch URL:
http://www.netcreations.com/hurricane/

Hurricane tracking resources.

Hurricane.com URL:
http://www.huricane.com/

Will accept information, pictures, or data about any present or future hurricane or tropical storm.

Hurricane Tropical Data URL:
http://thunder.atms.purdue.edu/hurricane.html

Hurricane—Living With Tropical Weather Systems URL:
http://www.flinet.com/~reiter/

Information and graphics packed resource on hurricane and tropical weather. Practical information on preparing for, surviving, and recovering from a hurricane.

Tracking the Eye URL:
http://www.cyberspy.com/~gencode/trackeye.html

Hurricane tracking software for Windows or Windows 95.

This page also provides links to current hurricane coordinate information.

USA Today-Guide to Hurricanes URL:
http://www.yahoo.com/?http://www.usatoday.com/weather-whur0.htm

VNO: All About Hurricanes URL:
http://www.yatcom.com/neworl/weather/hurricane.html

Worldwide Hurricane/Typhoon Tracks and Forecasts URL:
http://www.solar.ifa.hawaii.edu/Tropical/tropical.html

Tornadoes

Bloom, Steve - Stock Photography URL:
http://ourworld.compuserve.com/homepages/S_Bloom/

Digital photographic artist specialising in wildlife and "impossible" photography. See his tornado, running cheetah, charging rhino, leaping dolphins, and exploding prism.

Bears Cage URL:
http://www.ionet.net/~tornado1/index.shtml

Storm chaser information with links to other chasers and pictures of tornadoes and other weather images.

The Tornado Page URL:
http://cc.usu.edu/~kforsyth/Tornado.html

An introduction to tornadoes.

CHAPTER 10

WEB LINKS

METEORITE IMPACTS

Near Earth Asteroid Rendezvous URL:
http://utopia.eps.jhu.edu/near.html

Planet Earth Home Page URL:
http://www.nosc.mil/planet_earth/info_modern.html

Virtual library text version.

Planetary Geophysics Home Page URL:
http://www.wdcb.rssi.ru/

Asteroid and Comet Impact Hazard URL:
http://ccf.arc.nasa.gov/sst/

Fireball Reporting Form URL:
http://www_dsa.uqac.uquebec.ca/Divers/Higgins/fireball.htm

Earth Science and Solar System Exploration Division URL:
http://www-sn.jsc.nasa.gov/

DANGEROUS EARTH

LIVING WITH DANGER

Geology controls our lives.
• Jack Oliver

*E*gypt has always been dominated by its geology. Upper Egypt consists of a long valley carved by the Nile River into the rocky desert plateau of the eastern Sahara. Lower Egypt is a vast delta built outward into the Mediterranean Sea by the deposition of sediment from the Nile. Ancient Egyptians called the Nile Valley *Kemit,* "the Black Land," referring to the dark, fertile soil along the banks of the river. The deserts bordering the valley were known as "the Red Land," also named for their distinctive soil. This region was harsh and forbidding, yet it was the source of the great mineral wealth that made Egypt a powerful center of civilization.

The role of geology in Egyptian civilization is reflected in mythology, especially in the figure of Osiris. Osiris was killed by his jealous brother Set, who cut his body into fourteen pieces and threw the pieces into the Nile. Isis, the wife of Osiris, found the pieces and, with the help of the gods, restored Osiris to life. The power of resurrection vested in Osiris was associated with the fertilizing effects of the annual flooding of the Nile, which watered and enriched the valley. Each year for a period of about 100 days the river swelled as if by magic, since Egypt experienced no great rains. The flooding left behind deposits rich in

◄ ━━━━━━━━━━━━━

Egypt and the River Nile viewed from space. This LANDSAT image was made in the infrared so vegetation shows up red. The red ribbon is the floodplain of the Nile and it has the color it does because it is farmed intensively. The river is the thin, wavy blue line inside the red ribbon. The direction of river flow is northward, from the bottom of the page to the top; the triangular region at the top is the delta, where the river empties into the Mediterranean Sea. Cairo is the blue-grey area at the apex of the delta. The Aswân Dam and Lake Nasser are visible at the bottom of the image.

potash and soft silts that were easily tilled. These rich muds formed a natural fertilizer that, when warmed by the spring sun, allowed crops to flourish. Yet as soon as one crop was harvested, the nutrients in the sediment were essentially exhausted. Only the next flooding by the Nile would render the field suitable for cultivation again.

The Nile was also a destructive force for the ancient Egyptians. Only through the coordination of human efforts could its power be brought under control. The success of Egyptian agriculture required not only local control of the river but a united effort throughout the two lands. Regions were divided into political districts and taxes were based on the height of flood waters. The river thus became the impetus for agricultural innovation, technological advancement, and the development of governmental and legal systems.

For modern Egyptians the Nile remains both a provider and a destroyer. The heritage of technological control over geologic processes survives in the form of the Aswân High Dam, which generates more than half the electricity produced in Egypt. The construction of the dam and the creation of Lake Nasser added significantly to the amount of arable land in Egypt. At the same time, however, the dam has interfered with the distribution of sediment to agricultural areas downstream, forcing them to become dependent on chemical fertilizers. The lack of nutrient-laden sediments reaching the Mediterranean has affected the offshore fishing industry. Other unexpected effects include salinization of irrigated areas, erosion of sediment-starved coastlines, saltwater intrusion of coastal freshwater supplies, and the spread of water-borne diseases such as schistosomiasis. The Egyptian people have exerted a massive amount of effort to control the force of the Nile and to draw resources from it, but these activities have had some unanticipated consequences.

HAZARDOUS PROCESSES

Geologic processes affect every inhabitant of the Earth every day. The impacts of some of these processes, such as earthquakes, landslides, and floods, are obvious. Others are more subtle. They include the role of mountains in controlling the weather and shaping climatic zones; the influence of volcanism on the chemical evolution of the atmosphere; and the contribution of floodwaters to the creation of fertile agricultural soils, as in the example of the Nile Valley. Because the Earth is a dynamic planet, many of these geologic processes have risks associated with them; that is, they may negatively affect human interests, activities, or health. In order to understand these processes and assess the risks and impacts associated with them, we turn to **geology,** the scientific study of the Earth.

The Role of Geology

Traditionally, geology has been divided into two broad areas with related but differing aims. **Physical geology** is concerned with understanding the *processes* that operate at or beneath the surface of the Earth and the *materials* on which those processes operate. The causes of volcanic eruptions, earthquakes, landslides, and floods are processes. Materials include soils, sediments, rocks, air, and seawater. **Historical geology** is concerned with the chronology of

events, both physical and biological, that have occurred in the past. It seeks to resolve questions such as when the oceans formed, when dinosaurs first appeared, when the Rocky Mountains rose, and when and where the first trees appeared.

To these two traditional branches of geology we may now add a third, **environmental geology,** which focuses on the ways in which Earth systems and geologic processes affect and are affected by human activities. One component of environmental geology is the study of **geologic hazards,** that is, the wide range of geologic circumstances, materials, processes, and occurrences that are harmful, hazardous, or costly to humans, such as earthquakes, volcanic eruptions, floods, and landslides. In addition to geologic hazards, environmental geology is concerned with Earth resources—such as minerals, soil, water, and fossil fuels—and the behavior of wastes and contaminants in the natural environment.

Environmental geology in general, and the study of geologic hazards in particular, necessarily encompass some aspects of both historical *and* physical geology. The physical characteristics and chemical composition of the Earth, inside and out, affect our lives in many different ways. For example, the internal structure of the Earth plays a significant role in shaping our landscape and causing geologic events that may be hazardous for humans. Physical geology emphasizes the study of processes such as the release of heat from the interior of the Earth (Fig. In.1), the movement of

▲ F I G U R E In.1
Lava dome at Soufrière Hills on the island of Montserrat, West Indies. At the beginning of 1996, nearby residents were evacuated because of increasing eruptive activity from this volcano. The release of heat from the Earth's interior is directly or indirectly responsible for many natural hazards, including volcanic eruptions.

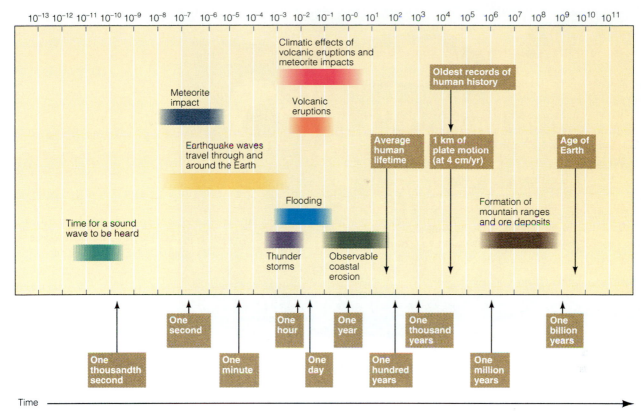

▲ F I G U R E In.2

Human beings and geologic processes operate on different time scales. The things that are important to us socially and politically are measured on a scale of days to years. Geologic processes, meanwhile, operate on scales ranging from a few seconds (e.g., an earthquake) to a few millennia (e.g., soil formation) to a few hundred million years (e.g., the uplifting of a mountain range).

material in the hydrologic cycle, and circulation patterns in the atmosphere, all of which are responsible for causing different natural hazards. The materials of the Earth also have distinct physical and chemical properties that may affect their role in hazardous processes, such as their tendency to flow or to fail, or their ability to transmit earthquake waves.

Historical geology also contributes to the study of hazards. The human–planet relationship is a dynamic one, characterized by change. Humans are affected by change in the natural environment, but we also contribute to such changes on a massive scale. Historical geology helps us understand the past and establishes a context within which to understand the magnitude, scope, and direction of planetary changes. Historical geology can also provide some perspective on the time scale of environmental change. Geologic processes and human beings operate on different time scales (Fig. In.2). The things that are important to us socially and historically are measured on time scales of years, decades, or centuries. Yet the processes that affect our lives on a daily basis operate on time scales ranging from a few seconds (e.g., an earthquake) to a few millennia (e.g., soil formation) to millions, even billions of years (e.g., the for-

mation of mountain ranges and the evolution of the atmosphere).

Risk and the Human–Planet Relationship

Hazardous geologic processes have always existed. Earth processes that we term "hazardous"—floods, earthquakes, volcanic eruptions, drought—are natural geologic processes (Fig. In.3). They have operated, for the most part, since early in Earth history, although in some cases the magnitude or timing of the events may have changed. We are *anthropocentric* (human-centered) in that we define Earth processes as hazardous only when they have a direct negative impact on human interests. Some hazardous processes, such as massive volcanic eruptions, are beyond human influence. Many natural processes, however, can be influenced—positively or negatively—by human activities (Fig. In.4). In fact, human actions have generated a whole new category of hazardous processes—*technological hazards*—that encompasses hazards involved in the use of Earth materials in the built environment.

Risk is characteristic of the human–planet relationship.

▲ F I G U R E In.3

In April, 1996, volunteers scrambled to get furniture and clothing out of a residence in West Frankfort, Illinois, as rising floodwaters from the Big Muddy River crept higher and higher. This family was among dozens forced to leave their homes due to flash flooding, the worst in this area in over three decades. Flooding is a natural part of the hydrologic cycle, but it can have a devastating impact on families and communities.

Geologic processes challenge us ruthlessly and continuously with events that seem, to us, damaging or even catastrophic. Floods and volcanic eruptions routinely strike the most fertile land; earthquakes and hurricanes level cities; and the inexorable processes of climate and weathering wear away at the solid Earth around us. One of the main goals of environmental geology and the study of geologic hazards is to understand such processes so that we can predict and control them if possible, or at least mitigate their effects. This involves finding answers to questions such as: Where are geologic hazards most likely to strike, and how often? What are the physical effects of hazards, and who is vulnerable to them? And, how can hazards be predicted or mitigated?

◀ F I G U R E In.4

Attempting to hold desertification at bay in Southern Morocco, on the edge of the Sahara. Drought and resulting advance and retreat of deserts are natural climatic processes. However, human activities such as overly intensive farming can greatly accelerate the process of desertification. In this view, palm fronds are woven into fences in an effort to slow the drift of sand.

International Decade for Natural Disaster Reduction

Some of the answers to these questions will be found through research currently being carried out. In 1987, the United Nations General Assembly passed a unanimous declaration supporting the National Academy of Sciences in the establishment of an *International Decade for Natural Disaster Reduction (IDNDR)*, beginning in 1990. The IDNDR is devoted to reducing the loss of life and property damage caused by natural disasters worldwide (Table In.1). Approximately three million people have been killed by natural disasters in the past two decades, with about 800 million more adversely affected. Worldwide, economic losses over the same period may be as high as $300 billion (U.S.). The risk to life and property from natural hazards is increasing steadily, partly because of population growth and the increasing concentration of population in urban areas.

Through the IDNDR, a global community of scientists, engineers, technicians, and decision makers will work cooperatively to understand natural hazards more fully and to apply this knowledge as effectively as possible. More than 150 nations are participating in a wide range of interdisciplinary research projects associated with Decade activities. Projects are focused on the identification and intensive study of potentially damaging hazards, the development of strategies and guidelines for predicting and responding to hazardous events, and the transfer of information and technology to developing countries, where vulnerability to natural disasters is disproportionately high.

GOALS FOR THE FUTURE

Geology has always been an interdisciplinary science because Earth processes involve interacting biological, physi-

T A B L E In.1A •

The Number of Major Disasters Around the World During the Period 1963–1992, as Defined by Type, Damage, Persons Affected, and Number of Deaths

Number of significant disasters, as defined by type	>1% of total annual GNP lost	>1% of country's population affected	>100 deaths caused
Floods	76	162	202
Tropical Storms	73	100	153
Drought	53	167	21
Earthquakes	24	20	102
Landslides	1	2	54
Volcanoes	2	9	12
Tsunamis	1	1	9

T A B L E In.1B •

The Number of Major Disasters Around the World During the Period 1963–1992, in Five-year Periods, as Defined by Damage, Number of Persons Affected, and Number of Deaths. The Table Shows a Steady Increase in the Number of Disasters in Each Category Over the Time Period.

Number of significant disasters, as defined by five-year periods	>1% of total annual GNP lost	>1% of country's population affected	>100 deaths caused
1963–1967	16	39	89
1968–1972	15	54	98
1973–1977	31	56	95
1978–1982	55	99	138
1983–1987	58	116	162
1988–1992	66	139	205

Source: Disasters Around the World—A Global and Regional View, World Conference on Natural Disaster Reduction, Yokohama, Japan, May, 1994, U.N. Information Paper DHA/94/132.

▲ F I G U R E In.5

This infrared satellite image of the Northern Hemisphere shows (from left to right) Tropical Storm Luis, Tropical Storm Karen, Hurricane Humberto, and Hurricane Iris. On the far left, near the Florida coast, are the remnants of Tropical Storm Jerry. This parade of storms was captured by the Geostationary Operational Environmental Satellite (GOES-8) on August 30, 1995. Hurricanes and tropical storms are manifestations of complex interactions among the Earth's atmosphere, oceans, and land masses. Scientists are only just beginning to understand the factors that control these interactions.

cal, and chemical processes. Yet we are discovering that the interactions are more complex and dynamic than we would have believed as recently as a few decades ago. We now know more about how the Earth's climate is intimately connected with oceanic and atmospheric circulation; how volcanoes contribute to processes as diverse as the evolution of the atmosphere and the shifting of continents; how certain activities can trigger or worsen natural events such as landslides, floods, and even earthquakes. We have a more profound appreciation of the role of humans in geologic change and the need to study the Earth system as a whole rather than in individual fragments.

We still have much to learn about the functioning of Earth systems and the factors that control and initiate haz-

ardous processes. We are only beginning to understand the complexities and interrelationships of systems like the Earth's climate, oceans, and shifting continents (Fig. In.5). Sophisticated modeling tools and advanced technologies for detecting and monitoring small changes in the environment have helped our understanding considerably. As stated in the goals of the IDNDR, we need to improve the capacity of each country to mitigate the effects of natural disasters; foster scientific and engineering endeavors aimed at closing critical gaps in knowledge in order to reduce loss of life and property; and disseminate existing and new information related to measures for the assessment, prediction, prevention, and mitigation of natural disasters.

Only by refining and enhancing our understanding of

Earth systems and processes can we really begin to assess the risks and benefits of living in this dynamic system. As geologist Léo LaPorte once wrote:

The geologic hazards that threaten humans are, usually, only as dangerous as we make them by settling directly within their midst. For us to live in active areas of floods, landslides, and earthquakes is indeed dangerous to our health. By knowing something of the location, magnitude, and frequency of large-scale geologic phenomena, we can avoid in large measure the dangers to life and limb posed by them. (Encounter with the Earth: Wastes and Hazards, Harper & Row, New York, 1975, p. v).

IMPORTANT TERMS TO REMEMBER

environmental geology

geologic hazards

geology

historical geology

physical geology

GEOLOGIC FRAMEWORK

For those who have seen the Earth from space, and for the hundreds and perhaps thousands more who will, the experience most certainly changes your perspective. The things that we share in our world are far more valuable than those which divide us.

• Donald Williams (U.S. Astronaut)

★ E • S • S • A • Y ★

THE HOME PLANET

Perhaps the single most powerful image of all times is a photograph of the Earth taken from space (Fig. I.1). We first saw the image in the 1960s, at the dawn of the "space age." There it was, the entire planet in one sweeping view. We could see everything at a glance—the clouds, the oceans, polar ice caps, and continents—all at the same time and in their proper scale. As never before it was possible to see and understand that the Earth really is just a small planetary body—"Spaceship Earth"—in orbit around an ordinary, medium-sized star. This understanding has fueled a global outpouring of concern for the environment. It has been fundamental in shaping societal views about how the resources available on Spaceship Earth should be managed and utilized for the expanding population and at the same time nurtured and protected for generations to come. To fully appreciate Spaceship Earth, to comprehend its limits and vulnerabilities, it is helpful to look first at the Earth's place among its neighbors in space.

THE SOLAR SYSTEM

Studies of the Earth in the context of the **solar system**— the group of objects in orbit around the Sun—reveal a planet that is at once akin to and distinct from other planets in the system. The Earth is one of nine planets in the solar system, which also includes the Sun, at least 61 moons, a vast number of asteroids, millions of comets, and innumerable small fragments of rock and dust called *mete-*

oroids. All the objects in the solar system move through space in smooth, regular orbits, held in place by gravitational attraction. The planets, asteroids, and meteoroids all circle the Sun, whereas the moons circle the planets.

The planets can be separated into two groups on the basis of their physical characteristics, especially density, and

▲ F I G U R E I.1
The Earth from space. The blue ocean, the thin, oxygen rich layer of atmosphere, and the green, plant-covered land areas make it clear that the Earth is a unique planet.

The Earth rising over the Moon.

▲ FIGURE I.2

The planets can be divided into two groups—the terrestrial planets, which are the small, rocky planets closest to the Sun—and the jovian planets, which are the large gaseous bodies distant from the Sun.

their closeness to the Sun (Fig. I.2). The innermost planets—Mercury, Venus, Earth, and Mars—are small, rocky, and relatively dense. Each has an overall density of 3 g/cm^3 or more (which is about the density of the average rock). They are similar in size and composition and are called **terrestrial planets** because as a group they resemble *Terra* (the Latin word for the Earth). With the exception of Pluto, the planets farther from the Sun than Mars are much larger than the terrestrial planets yet much less dense. The masses of Jupiter and Saturn, for example, are 317 and 95 times the mass of the Earth, but their densities are only 1.3 and 0.7 g/cm^3, respectively. These **jovian planets**—Jupiter, Saturn, Uranus, Neptune, and Pluto—take their name from *Jove,* an alternate designation for the Roman god Jupiter. They all probably have small solid centers that resemble terrestrial planets, but (again with the exception of Pluto) most of their planetary mass is contained in thick atmospheres of hydrogen, helium, and other gases. These atmospheres (which are what we actually see when we observe these planets) are the reason for the low densities of the jovian planets (Fig. I.3).

*T*HE ORIGIN OF THE SOLAR SYSTEM

How did the solar system form? We may never know the precise answer to this question, but we can discern the outlines of the process from evidence obtained by astronomers and from the laws of physics and chemistry.

The process began with space that was not entirely empty because earlier suns had exploded in what as-

▲ FIGURE I.3

Jupiter, the largest planet, is a gas-shrouded giant that conceals its presumably solid interior. This image of Jupiter was taken February 5, 1979 from a distance of 29 million km by the spacecraft *Voyager 2.* The banded pattern is due to turbulence and Jupiter's rotation. A particularly violent storm is visible in the lower left-hand corner.

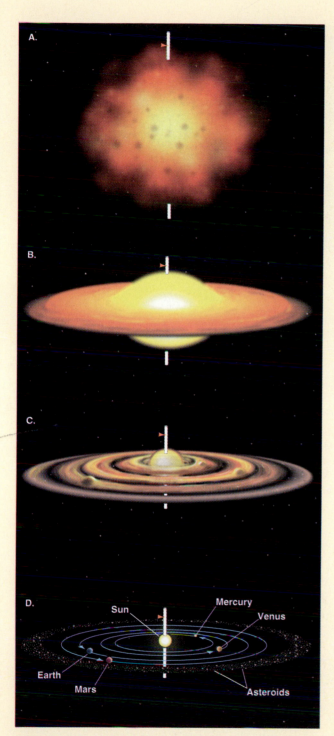

tronomers call supernovas. The explosions scattered atoms of various elements throughout a huge volume of space. Most of the atoms were hydrogen and helium, but small percentages of all the other chemical elements were also present. It is from these thinly spread atoms that everything in the solar system was eventually constructed. In a sense, the Earth and everything on it is made of star dust. Even though they were thinly spread, the atoms of star dust formed a turbulent, swirling cloud of cosmic gas. Over a very long period the gas thickened as a result of a slow gathering of the atoms in response to mutual gravitational attraction. As the atoms moved closer together, the gas became hotter and denser. Near the center of the cloud of gas, hydrogen and helium atoms eventually became so tightly pressed and so hot that they began to fuse and form heavier elements. At that point, estimated at about 6 billion years ago, the Sun was born.

At some stage the cool outer portions of the cosmic gas cloud became compacted enough to allow solid objects to condense, in the same way that ice condenses from water vapor to form snow (Fig. I.4). The solid condensates eventually formed the planets, moons, and other solid objects of the solar system.

Condensation of the gas cloud is only the first part of the planetary birth story. Condensation formed innumerable small rocky fragments, but the fragments still had to be joined together to form a planet. This happened as a result of impacts between fragments drawn together by gravitational attraction. The largest masses slowly swept up more and more of the condensed rocky fragments, grew larger, and became the planets. Meteorites, such as the ones in Fig. I.5, still fall on the Earth, proving that even now

▲ **F I G U R E I.4**
Birth of a solar system. A. The gathering of atoms in space created a rotating cloud of dense gas. B. The center of the gas cloud eventually became the Sun; the planets formed by condensation of the outer portions of the gas cloud (C, D).

▲ **F I G U R E I.5**
Messengers from space carrying some of the history of the earliest days of the solar system. These stony meteorites fell to the Earth at Pueblo de Allende, Mexico, in 1969.

▲ FIGURE I.6

Meteor Crater, near Flagstaff, Arizona. The crater was created by the impact of a meteorite about 50,000 years ago. It is 1.2 km in diameter and 200-m deep. Note the raised rim and the blanket of broken rock debris thrown out of the crater. Many impacts larger than the Meteor Crater event have occurred during the Earth's long history, and some have caused major disruptions to living creatures.

some ancient rocky fragments still exist in space. Meteorites and the scars of ancient impacts (Fig. I.6) provide evidence of the way the terrestrial planets grew to their present sizes. The growth process—the gathering of more and more bits of solid matter from surrounding space—is called *planetary accretion.* The formation of the Earth, the terrestrial planets, and the other large planetary bodies through the processes of condensation and planetary accretion was essentially complete approximately 4.6 billion years ago.

Brothers and Sisters: A Comparison of the Terrestrial Planets

The nine planets and the other planetary bodies are like siblings; all were born of the same processes that gave rise to the whole solar system family. Some of the siblings are more alike than others. The Earth and its close neighbor Venus, for example, are so much alike in size, density, and overall composition that they are often called "sister" planets.

As a group, the terrestrial planets have many things in common beyond their small size and rocky composition. They have all been hot and, indeed, partially melted at some time early in their histories. All of them appear to have dense metallic iron cores, which probably separated from the rocky outer layers during this period of partial melt. The planets have all experienced volcanic activity, dominated by one type of rock, *basalt,* which is common to all the terrestrial planets. They have all passed through a period of intense cratering and surface modification by meteorite impacts, a process that continues today. Unlike the jovian planets, all the terrestrial planets have lost their envelopes of gaseous material, which were swept away by the solar wind. The terrestrial planets that ended up with atmospheres (Earth, Mars, and Venus) had to evolve their own secondary gaseous envelopes.

In sum, when we look at the solar system we see a group of planets and other objects that are related by birth and by their association with our Sun. Within this system is a smaller group, the terrestrial planets, linked even more closely by their similar geologic histories and planetary characteristics. The Earth is what it is and the environment works the way it does because of all the things that have happened during its long history. The Earth's history is just sufficiently different that planet Earth is habitable but the other terrestrial planets are not.

THE HABITABLE PLANET

Although the terrestrial planets are alike in composition and density, they differ greatly in the compositions of their atmospheres and the presence or absence of water and life. The space image of our blue planet reveals just what makes the Earth unique and different from the other planetary bodies.

Spheres of the Earth

Atmosphere and Hydrosphere

The Earth has an overall blue and white hue because it is surrounded by an **atmosphere** of gases, predominantly nitrogen, oxygen, carbon dioxide, and water vapor. No other planet in the solar system has such an atmosphere. The Earth's atmosphere contains white clouds of condensed water vapor. The clouds form because water evaporates from the **hydrosphere** ("water sphere"), which consists of the oceans, lakes, and streams; underground water; and snow and ice, including glaciers. The hydrosphere is another unique characteristic of the Earth. Planets farther from the Sun are too cold for water to exist as both a liquid and a gas (water vapor), and planets closer to the Sun are too hot. Other planets have hydrospheres, but only the Earth has a hydrosphere consisting of water, ice, and water vapor.

Biosphere and Regolith

Another unique feature of the Earth is the **biosphere** ("life sphere")—the totality of the Earth's living matter. When the Earth is viewed from space, the biosphere is most dramatically revealed by blankets of green plants on some of the land masses (Fig. I.7). The biosphere embraces innumerable living things, large and small, which are grouped into millions of different species. It also includes dead plants and animals that have not yet been completely decomposed.

A fourth reason the Earth is special concerns its solid surface. Regions that are not green because of dense plant cover appear brown and weather-beaten because of *weathering*—the chemical alteration and mechanical breakdown of rock during exposure to the atmosphere, hydrosphere, and biosphere. As a result of weathering, the Earth is covered by an irregular blanket of loose rock debris, or **regolith.** Soils, muds in river valleys, sands in the desert, and all the other friable rock debris are part of the regolith. Other planets and planetary bodies with rocky surfaces have regolith too, but in those cases the regolith has formed primarily as a result of endless meteorite impact cratering. The Earth's regolith is unique because it is formed by complex interactions among physical, chemical, and biologic processes, usually involving water, and because it is teeming with life forms. Most of the world's plants and animals live on or in the regolith or in the hydrosphere.

▲ F I G U R E I.7
The biosphere determines the color of the land surface viewed from space. A composite of numerous satellite images shows densely vegetated regions in green, dry deserts in yellow or brown, and ice-covered regions in white. The variations in color reflect climatic variations and their influence on the biosphere.

Lithosphere

Finally, the Earth differs from all other planets we know of in the unique relationship that exists between the outermost rocky layer, the **lithosphere,** and the hotter, more plastic material in the planet's interior. The Earth is the largest of the terrestrial planets; for this reason, it has cooled off more slowly than the others. Because the Earth's lithosphere is a thin, cold, brittle shell, it has broken up into a series of enormous rocky plates. These plates move around and jostle one another as a result of movements in the hot, mobile material underneath.

Plate Tectonics

The result of the movement of lithospheric plates is a set of processes that we refer to as **plate tectonics.** Plate tectonic activity—that is, the motion and consequent interactions of lithospheric plates—has been a force throughout much of geologic history (at least 2 billion years) and has shaped the landscape. It is responsible for the creation of mountains (Fig. I.8), volcanoes, and deep ocean basins, and has influenced the formation of the atmosphere, the development of climatic zones, and the evolution of life. As we will see in later chapters, plate tectonics has also provided geologists with a concept that enables many years' worth of observations of natural processes to be understood in a unified context.

We know of no other planet where plate tectonics has played, and continues to play, such an important role in forming the environment. We know of no other planet, anywhere, where the temperature permits water to exist near the surface in solid, liquid, and gaseous forms. No other planet that we now know of would have been hospitable to the origin and evolution of life as we know it. There are billions upon billions of suns in the universe, so it is almost inevitable that there are billions of planets too; surely a few of these planets must be Earthlike and therefore may support life. However, if a relatively advanced civilization does exist on a planet somewhere out in space, so far we haven't heard or seen any sign of it. The Earth might well be unique in the universe.

THE RELATIONSHIP BETWEEN LIFE AND THE EARTH

A fundamental principle of environmental geology is that the relationship between life and the Earth is one of mutual influences. Nowhere is this delicately balanced relationship better portrayed than in the story of the atmosphere.

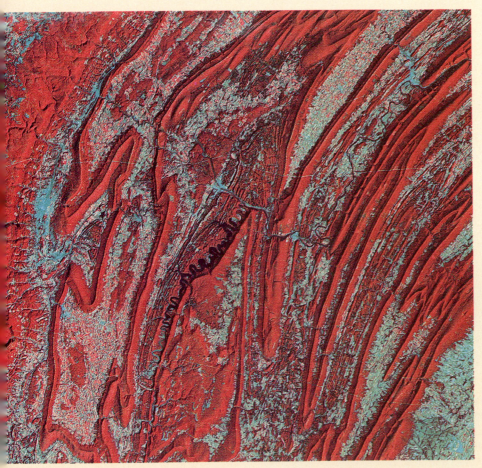

◄ F I G U R E I.8
The Appalachian mountain range in Pennsylvania viewed from space. Layers of rock, once horizontal, were twisted and contorted as a result of a collision between two plates of lithosphere several hundred million years ago. The hills are the eroded roots of a once much grander mountain range.

Life on the Earth

Life originated on the Earth because of this planet's special characteristics, especially the presence of liquid water. In turn, life itself has had a profound influence on the chemical evolution of the Earth's atmosphere. For example, carbonate rocks called *limestones,* which are formed from discarded shells of billions of tiny sea creatures, provide a home, or "sink," for carbon dioxide (CO_2) (Fig. I.9). Since relatively early in the Earth's history, these limestones—which would never have existed if life itself had not first evolved—have kept huge quantities of carbon dioxide bound up at the bottoms of shallow seas where oceans meet continents. If all the carbon dioxide in limestones were released, we would have a carbon dioxide–dominated atmosphere like that of Venus. The concentrations of carbon dioxide in Venus' atmosphere trap so much of the Sun's warmth that the planet's surface temperature is hot enough to melt lead! Carbon dioxide performs the same function in the Earth's atmosphere, but because so much carbon dioxide is bound up in limestone there is only a small amount of it in the atmosphere, and the planet's surface stays just warm enough for life to continue to exist.

▲ **FIGURE I.9**
The Dolomites, a mountain range in northern Italy, are an example of the huge carbonate masses in which carbon dioxide is locked up in rocks.

The process of *photosynthesis,* whereby green plants utilize the energy of the Sun to combine water and carbon dioxide to make carbohydrates and oxygen, represents another fundamental interaction between life and the Earth. The Earth's early atmosphere was probably much more reducing (i.e., oxygen-poor) than the present atmosphere. We can trace the development of an oxygen-bearing atmosphere, as well as the oxygenation of the hydrosphere, by looking at weathered materials preserved in the regolith of 1 billion years ago and more. These materials are a record of the interaction between the oxygen-bearing atmosphere, the hydrosphere, and the lithosphere. In them we can actually observe the transition from oxygen-poor to oxygen-rich minerals. This transition became possible only after the advent of photosynthetic plants, especially algae.

Expanding Horizons for a Shrinking Earth

We, who belong to the present generation, have inherited the fruits of countless thousands of inquiring minds. We are heirs to a vast legacy of knowledge about the Earth—our environment, our home. Anyone can, if he or she will study this legacy, understand how the Earth works, how the natural activities that shape it are interwoven in complex ways, how the environment of which each of us is a part was created. Anyone can understand the threat that the environment can be changed, distorted, and even shattered by human activities.

Being heirs to a body of knowledge is nothing new. Each generation enjoys that privilege. This generation, however, is not merely another in the long line of generations; it is unique because it is the first to face the reality that humans are now so powerful and so numerous that collectively we rival the other forces of nature that nurture, shape, and change our fragile environment. Therefore, we have a crisis on our hands—one that came to a head so recently and so quickly that earlier generations, confused by long-accustomed ways of doing and thinking, have failed to join battle with it. But we also have something that prior generations lacked; we have a point of view toward the human place in the natural world. We must now use our collective understanding of the Earth to impel this generation, and future ones, to make the hard decisions, to sort out and choose a list of priorities by which the quality of life can be maintained and enhanced, and to accept the discipline that our decisions and priorities will require.

IMPORTANT TERMS TO REMEMBER

atmosphere (p. 15)
biosphere (p. 15)
hydrosphere (p. 15)
jovian planets (p. 12)
lithosphere (p. 16)

plate tectonics (p. 16)
regolith (p. 15)
solar system (p. 11)
terrestrial planets (p. 12)

EARTH SYSTEMS AND CYCLES

We shall not cease from exploration.
And the end of all our exploring
Will be to arrive where we started
And know the place for the first time.

• **T.S. Eliot** *Four Quartets*

A farmer in Indonesia tends his rice fields. The soil is rich, full of minerals and nutrients. But the slopes are steep, and the farmer must terrace his fields carefully to avoid losing the soil through erosion. Further up the mountainside, near a steaming vent by the shore of a volcanic lake, a worker carries a basket full of yellow sulfur, the day's harvest. The sulfur will be sold in town, to be processed and combined with phosphate to make fertilizer. Downslope, women are busy scooping baskets of volcanic sand from the bottom of a river, to be used in making cement for road construction.

Volcanoes are a fundamental part of life in Indonesia, a chain of more than 400 volcanic islands along the southwestern edge of the Pacific Ocean. Even the earliest human inhabitants of the region were affected by volcanic eruptions as far back as 70,000 years ago; the tools and weapons of prehistoric hunters have been found buried in volcanic ash nearby. Tales of catastrophic eruptions are common in Indonesian folklore, such as the legend of the eruption of Tangkubanprahu. Many years ago, according to the legend, there lived a prince who sought to marry his own mother. But the mother, a beautiful princess, feared the wrath of the gods over such a union and demanded of her son a seemingly impossible task: to create a lake in the mountains nearby and build a suitable nuptial canoe, all before the sun went down that day. The prince set to with a will, damming a river gorge and hewing a canoe from an enormous tree. The wedding took place, and the gods were not pleased. They sent down a holocaust of fire and thunder that overturned the canoe and drowned the participants. To this day, it is said, the upturned hull may be seen

in the elongated shape of Tangkubanprahu, which means "capsized canoe."

Indonesians depend on the mineral-rich soils and other resources provided by the volcanoes. But the impacts of volcanic eruptions on Indonesian history and civilization represent only a small part of the story. The geologic turmoil that characterizes the Pacific Ocean rim causes the volcanoes to erupt. The largest eruptions cause changes in regional and global climate, sometimes lasting for years. The eruptions also form mountains, change the landscape, replenish the soils, and deposit sulfur and volcanic ash. While volcanic activity replenishes the land, the rains come and wash away the minerals, eventually carrying them to the sea. Volcanoes erupt, rain falls, water flows, and the cycle continues. In Indonesia, the interconnections among various parts of the Earth system and their impacts on human activities are very apparent.

THE EARTH SYSTEM

A new approach is taking hold in the Earth sciences. The traditional way to study the Earth has been to focus on separate units—the atmosphere, the oceans, or even a single mountain range—in isolation from the others. In the new approach, the Earth is studied as a whole and viewed as a unified system. In particular, Earth scientists are now focusing on interactions and interrelationships among the various parts of the Earth system.

It is sometimes said that the Earth is a *closed system*. This is not just a catchy saying; it has a specific scientific meaning and is rich in connotations that are particularly applicable to the study of environmental geology. In this chapter we will take a close look at the meaning of the word *system*, at some of the characteristics of the Earth system and its component parts, at the interactions that characterize the system, and at how those interactions combine to produce

Terraced rice field on the steep volcanic slopes of the island of Bali, Indonesia.

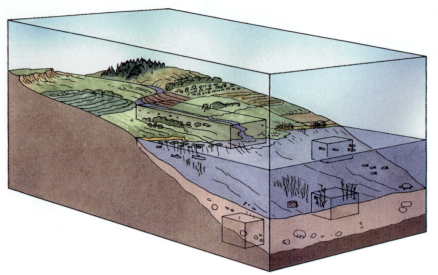

<figcaption>◀ **F I G U R E   1.1**
The system concept. The river is a system, as is the lake it flows into. Together they form a larger system—the watershed. The small volumes of water and sediment indicated by boxes are examples of smaller systems.</figcaption>

the environment in which humans have evolved. It will be useful to begin our discussion by considering the system concept in greater detail.

The System Concept

The system concept is a helpful way to break down a large, complex problem into smaller, more easily studied pieces. A **system** can be defined as any portion of the universe that can be isolated from the rest of the universe for the purpose of observing changes. By saying that *a system is any portion of the universe,* we mean that the system can be whatever

the observer defines it to be. That's why a system is only a concept; you choose its limits for the convenience of your study. It can be large or small, simple or complex (Fig. 1.1). You could choose to observe the contents of a beaker in a laboratory experiment. Or you might study a lake, a hand sample of rock, an ocean, a volcano, a mountain range, a continent, or even an entire planet. A leaf is a system, but it is also part of a larger system (a tree), which in turn is part of an even larger system (a forest).

The first step in viewing the Earth as a system is to identify the smaller systems that are its component parts. There are four principal systems within the larger Earth system:

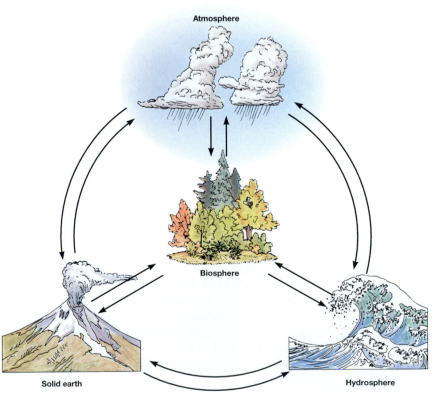

<figcaption>◀ **F I G U R E   1.2**
The four parts of the Earth system that most directly concern environmental geology: lithosphere, biosphere, atmosphere, and hydrosphere.</figcaption>

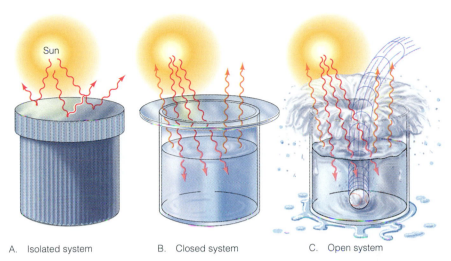

◄ F I G U R E 1.3
**The three basic types of systems:
A. An isolated system. B. A closed
system. C. An open system.**

A. Isolated system   B. Closed system   C. Open system

the atmosphere, the hydrosphere, the biosphere, and the lithosphere (Fig. 1.2). Each of these can be further divided into smaller, more manageable study units. We can divide the hydrosphere into the oceans, glacier ice, streams, and groundwater, for example.

The fact that a system has been *isolated from the rest of the universe* means that it must have boundaries that set it apart from its surroundings. The nature of those boundaries is one of the most important defining characteristics of a system, leading to three basic kinds of systems, as shown in Fig. 1.3. The simplest type of system to understand is an *isolated system;* in this case the boundaries are such that they prevent the system from exchanging either matter or energy with its surroundings. The concept of an isolated system is easy to understand, but such a system is

imaginary because although it is possible to have boundaries that prevent the passage of matter, in the real world it is impossible for any boundary to be so perfectly insulating that energy can neither enter nor escape.

The nearest thing to an isolated system in the real world is a **closed system;** such a system has boundaries that permit the exchange of energy, but not matter, with its surroundings. An example of a closed system is an oven, which allows the material inside to be heated but does not allow any of that material to escape. The third kind of system, an *open system,* is one that can exchange *both* matter and energy across its boundaries. Rain falling on an island is a simple example of an open system: some of the water runs off via streams and groundwater while some evaporates back to the atmosphere (Fig. 1.4).

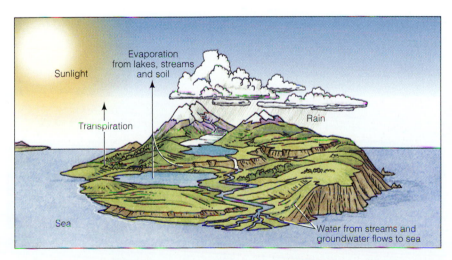

◄ F I G U R E 1.4
Example of an open system. Energy (sunlight) and water (rainfall) reach an island from external sources. The energy leaves the island as long-wavelength radiation; the water either evaporates or drains into the sea.

Box Models

Systems are commonly depicted in the form of box models, as shown in Fig. 1.3. The advantages of doing so are simplicity and convenience. A box model can be used to show the following essential features of a system:

1. The rates at which material and/or energy enter and leave the system.

2. The amount of matter or energy in the system.

The island system mentioned above can be depicted in this way (Fig. 1.5). Water stays on the island for some time before it flows off or is evaporated. The island thus is a **reservoir,** or storage tank, for water in this system. The average length of time water spends in the reservoir is called the **residence time.** The essential features of a box model of the island's water budget are, therefore, the rate at which water falls as rain (*input*), the rate at which water leaves the island (*output*), and the average amount of water on the island at any given time (the size of the reservoir). The size of the reservoir is a function of input and output. If input increases or output decreases, the size of the reservoir must increase to compensate. If input decreases or output increases, the reservoir's size must decrease. The main challenge in using box models to study systems lies in identifying and measuring all the inputs and outputs in order to understand how the system reaches a balance.

Living in a Closed System

As you may have realized, the Earth is a natural closed system—or at least a very close approach to such a system (Fig. 1.6). Energy reaches the Earth in abundance in the form of solar radiation. Energy also leaves the system in the form of longer wavelength infrared radiation. It is not quite correct to say that no matter crosses the boundaries of the Earth system, because we lose a small but steady stream of hydrogen atoms from the upper part of the atmosphere and we gain some extraterrestrial material in the form of meteorites. However, the amount of matter that enters or leaves the Earth system is so minuscule compared with the mass of the system as a whole that for all practical purposes the Earth is a closed system.

The fact that the Earth is a closed system has two important implications for environmental geology:

1. *The amount of matter in a closed system is fixed and finite.* This means that the mineral resources on this planet are all we have and, for the foreseeable future, all we will ever have. Someday it may be possible to visit an asteroid for the purpose of mining nickel and iron; there may even be a mining space station on the Moon or Mars at some time in the future. But for now it is realistic to think of the Earth's resources as being finite and therefore limited. This means that we must treat them with respect and use them wisely and cautiously.

Another consequence of living in a closed system is that material wastes must remain within the confines of the Earth system. As environmentalists are fond of saying, "There is no *away* to throw things to."

2. *When changes are made in one part of a closed system, the results of those changes will eventually affect other parts of the system.* Even though the Earth system is closed, its innumerable smaller parts are open systems. These systems are in a dynamic and sometimes delicate state of balance. When something disturbs one of the smaller systems, the rest also change as they seek to reestablish a state of balance or *equilibrium.* Sometimes an entire chain of events may ensue; for example, a volcanic eruption in Indonesia could throw so much dust into the atmosphere that it could initiate climatic changes leading to floods in South America and droughts in California, eventually affecting the price of grain in west Africa.

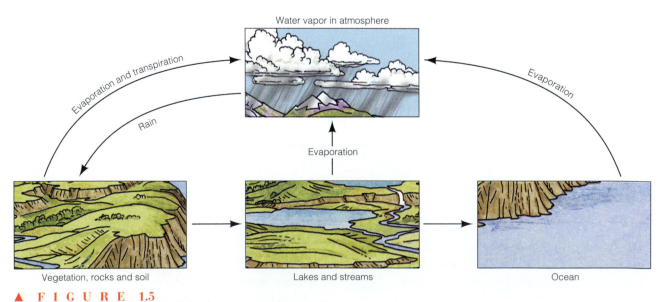

▲ **F I G U R E   1.5**
Depiction of the open system in Figure 1.4 by a box model.

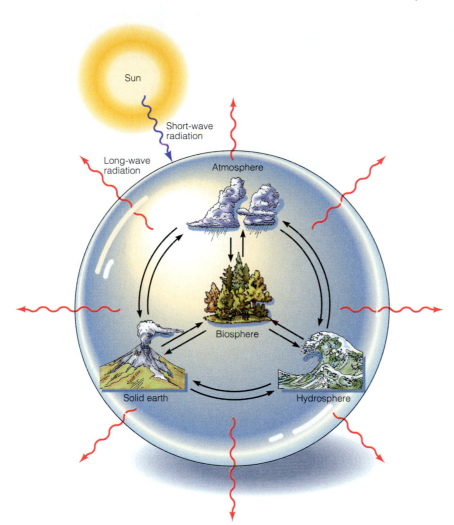

Sun

Short-wave
radiation

Long-wave
radiation

Atmosphere

Biosphere

Solid earth

Hydrosphere

◀ **F I G U R E 1.6**
**The Earth is essentially a closed
system. Energy reaches the Earth
from an external source and eventually returns to space as long
wavelength radiation. Smaller systems within the Earth, such as the
atmosphere, biosphere, hydrosphere, and lithosphere, are open
systems.**

DYNAMIC INTERACTIONS AMONG SYSTEMS

The causes and effects of disturbances in a complex closed system are very difficult to predict. Consider the anomalously warm ocean tide called El Niño, which occurs every few years off the west coast of South America. El Niños (discussed in greater detail in Chapter 9) are characterized by weakening of trade winds; suppression of upwelling cold ocean currents; worldwide abnormalities in weather and climatic patterns; and widespread incursions of biologic communities into areas where they do not normally occur. These features of El Niños are reasonably well known; what is *not* known is the triggering event. In other words, the interactions among processes in the atmosphere, hydrosphere, and biosphere are so complex, and these subsystems are so closely interrelated, that scientists can't pinpoint exactly what it is that initiates the whole El Niño process. It has even been suggested that changes originating in the lithosphere—in the form of localized heating of ocean water resulting from submarine volcanic activity—may create enough of an imbalance to trigger an El Niño.

From an environmental point of view, the significance

of interconnectedness is obvious: When human activities produce changes in one part of the Earth system, their effects—often unanticipated—will eventually be felt elsewhere. When sulfur dioxide is generated by a coal-fired power plant in Ohio, it can combine with moisture in the atmosphere and fall as acid rain in northern Ontario. When pesticides are used in the cotton fields of India, the chemicals can find their way to the waters of the Ganges River and thence to the sea, where some may end up in whale blubber. The chemicals also may be ingested by fish, which in turn may be caught and eaten. In this way, pesticides sometimes end up in the breast milk of mothers halfway around the world from the place where they were applied. Such processes can take a long time to happen, and that is why they have been all too easy to overlook in the past.

Cycling and Recycling

Since material is constantly being transferred from one of the Earth's spheres to another, you may wonder why those systems seem so stable. Why should the composition of the atmosphere be constant? Why doesn't the sea become saltier, or fresher? Why does rock 2 billion years old have

BOX 1.1

•

THE HUMAN PERSPECTIVE

INTERNATIONAL GEOSPHERE–BIOSPHERE PROGRAM

*T*oday a growing number of international, interdisciplinary research programs focus on developing and refining our understanding of Earth processes, the interrelationships between them, and how they are affected by human actions. One such research effort is the International Geosphere–Biosphere Program (IGBP), a far-reaching program operating under the aegis of the International Council of Scientific Unions (ICSU). The goals of the IGBP are to examine in detail some of the unknown links in global Earth cycles, to collect data and facilitate communications among researchers throughout the world, to promote the development and use of new research technologies, and to integrate the findings into a clearer and more directly applicable understanding of global processes and global change.

It's quite a tall order! But the task has been narrowed somewhat by focusing the program's international research efforts on a set of relatively specific questions, such as "How do ocean biogeochemical processes influence and respond to climatic change?" and "How do changes in land use affect the resources of the coastal zone, and will changes in sea level and climate alter coastal ecosystems?" Each of the core projects and participating research agencies focuses their efforts on one or more of these research priority questions. Information systems and data-management methodologies are also being developed to support the research efforts.

The IGBP has obtained funding from a variety of sources for the establishment of 10 regional centers for global change studies in developing countries. The regional centers are intended to facilitate collaboration and research on global change, emphasizing the processes of particular importance in each region. The IGBP is probably the most ambitious research program ever mounted on an international scale, and the list of its supporters is impressive: UNEP (the United Nations Environment Program), UNESCO (the United Nations Educational, Social, and Cultural Organization), the Third World Academy of Sciences, the Commission of the European Communities, the African Biosciences Network, the Organization of American States, and many other organizations and national governments. Most of the involvement to date has been through governmental, international, and scientific agencies. However, the IGBP is particularly anxious to promote the involvement of the private sector in the program because "In the next decades, with possibly great environmental changes, industry will need to be fully apprised of the best available scientific knowledge concerning environmental changes and the research that is planned to increase our understanding and predictive ability. Industry must know, on global as well as regional scales, the anticipated impacts of environmental change."

Source: Adapted from *On Common Ground* by Ranjit Kumar and Barbara Murck, John Wiley & Sons Canada Ltd., Toronto, 1992, p. 75. Quotation from: International Geosphere–Biosphere Program, "A Study of Global Change" (Initial Core Projects), IGBP Report 12 (June 1990) in Overview, pp. 1–3.

the same composition as rock only 2 million years old? The answers to these questions are the same: The Earth's natural processes follow cyclic paths. Materials flow from one system to another, but the systems themselves don't change much because the different parts of the flow paths balance each other: The amounts added equal the amounts removed. This cycling and recycling of materials and dynamic interaction among subsystems has been going on since the Earth first formed and it continues today.

The Cycling of Carbon

Carbon is a familiar example of a material that constantly cycles from one reservoir to another. Carbon can find a home in the biosphere (where it is the fundamental building block in virtually all molecules that make up living creatures); in the lithosphere (existing in rocks such as coal or limestone, which are made from the remains of living creatures); in the hydrosphere (where it is held in solution by a number of complex mechanisms); or in the atmosphere (as part of carbon dioxide gas, which helps keep the planet warm enough for life to continue). The complex set of interactions involving carbon can be depicted using a box model, as shown in Fig. 1.7.

As shown in the model, the atmosphere contains about 750 billion tons of carbon in the form of carbon dioxide. Photosynthesis by plants removes about 120 billion tons of

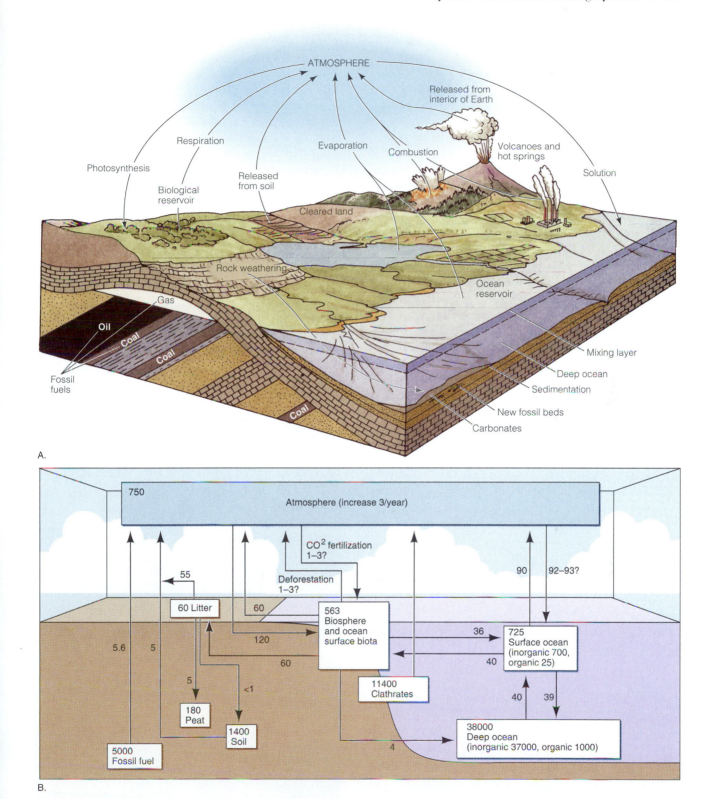

A.

B.

▲ **F I G U R E 1.7**
A. The global carbon cycle. B. Box model for carbon cycle. Each box represents a different reservoir; the numbers indicate the mass of carbon, in billions of tons, in that reservoir. Arrows and numbers between boxes represent exchanges (flows or fluxes) of carbon in billions of tons per year.

carbon per year from the atmosphere, but plant respiration plus the decay of dead plants and animals returns about the same amount. Living plants and animals (mostly plants) contain about 560 billion tons of carbon; plant detritus that has not yet decayed, together with organic matter buried in soils, amounts to 1400 billion tons. Approximately 11,000 billion tons of carbon are trapped in chemical compounds called *clathrates* [complexes of water and methane (CH_4)] found in sediments in the ocean floor and in some onshore regions of permafrost. The oceans contain 38,000 billion tons of carbon, most of it in the form of dissolved carbon dioxide.

By affecting such factors as the growth and death rates of plant and animals and the exchange of carbon dioxide between the oceans and the atmosphere, climatic changes cause the distribution of carbon among the reservoirs to vary. Thus, the atmospheric concentration of carbon dioxide will fluctuate naturally as climatic conditions cause more of it to be drawn into the oceans or the biosphere or, alternatively, as more is released from these sources back into the atmosphere. One of the striking features of past climatic variations is the remarkably close correlation between temperature and atmospheric concentrations of carbon dioxide (Fig. 1.8). What, then, are the potential effects on climate if human interventions cause imbalances in the global carbon cycle? We will consider some possible answers to this question in more detail in Chapter 18, but first let us consider some of the ways in which human activities are changing the natural carbon cycle.

With the advent of the Industrial Revolution 200 years ago, humans began burning massive and ever-increasing quantities of fossil fuels (oil, gas, and coal) for energy, thereby unlocking the vast amounts of carbon stored in these substances and releasing it to the atmosphere. At the same time, a rapidly expanding human population began to consume more and more of the world's forests as the demand for fuel, building materials, and—most of all—agricultural land grew. The burning of fossil fuels adds nearly 22 billion tons of carbon dioxide (about 6 billion tons of carbon) to the atmosphere every year. Deforestation is estimated to add a further 1.6 to 2.7 billion tons of carbon a year (Fig. 1.9). These amounts may seem small in comparison to the large fluxes of the natural carbon cycle. However,

▲ **F I G U R E 1.8**
Variation of temperature over Antarctica and of global atmospheric carbon dioxide concentration during the last 160,000 years. High temperatures are correlated with high CO_2 levels in the atmosphere.

▲ **F I G U R E 1.9**
Deforestation reduces the carbon reservoir of the forest. If the trees are burned or left to decay, their carbon content will be released into the atmosphere. The photograph shows the destruction of a luxuriant rain forest in the Amazon basin near Maraba, Brazil.

long-term imbalances in fluxes to and from the atmosphere resulting from the natural cycle generally amount to less than 1 billion tons a year. Compared to this small number, the fluxes of carbon resulting from human activities are, in fact, quite large. As a result, humankind is radically altering a delicately balanced natural system.

Cycles in the Earth System

As we have just seen in the discussion of the carbon cycle, it is useful to envisage interactions within the Earth system as a series of interrelated cycles. In the carbon example we discussed the movement of material between reservoirs. The movement of energy can be similarly treated. Both materials and energy can be stored in reservoirs, and the storage times can differ greatly. For example, carbon stored in plants may have a residence time of a few months or years, whereas carbon buried in the rock reservoir may have a residence time of millions of years. This means that a single cycle may include processes that operate on several different time scales.

A few basic cycles can serve to illustrate most of the Earth processes that are of importance in environmental geology. These include the *energy cycle,* the *hydrologic cycle,* the *rock cycle,* and *biogeochemical cycles* (of which the carbon cycle is an example). In the discussions that follow we will briefly consider each of these cycles. It is also possible to extend the concept of cycles to include human-controlled cycles that involve or affect natural processes; examples of such cycles will be introduced at appropriate places throughout this book.

THE ENERGY CYCLE

The **energy cycle** (Fig. 1.10) encompasses the great "engines"—the external and internal energy sources—that drive the Earth system and all its cycles. We can think of the Earth's energy cycle as a "budget": energy may be added to or subtracted from the budget and may be transferred from one storage place to another, but overall the additions and subtractions and transfers must balance each other. If a balance did not exist, the Earth would either heat up or cool down until a balance was reached.

Energy Inputs

The total amount of energy flowing into the Earth's energy budget is more than 174,000 terawatts (or $174,000 \times 10^{12}$ watts). This quantity completely dwarfs the 10 terawatts of energy that humans use per year. There are three main sources from which energy flows into the Earth system.

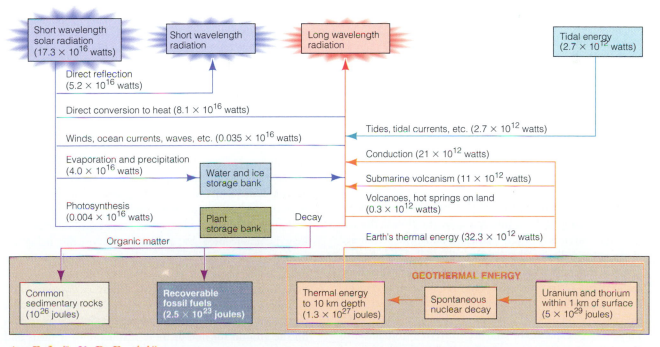

▲ **F I G U R E   1.10**
The energy cycle. There are three main sources of energy in the cycle: solar radiation, geothermal energy, and tidal energy. Energy is lost from the system through reflection and through degradation and reradiation.

Solar Radiation

Incoming short-wavelength solar radiation overwhelmingly dominates the flow of energy in the Earth's energy budget, accounting for about 99.986 percent of the total. An estimated 174,000 terawatts of solar radiation is intercepted by the Earth. Some of this vast influx powers the winds, rainfall, ocean currents, waves, and other processes in the hydrologic cycle. Some is used for photosynthesis and is temporarily stored in the biosphere in the form of plant and animal life. When plants die and are buried, some of the solar energy is stored in rocks; when we burn coal, oil, or natural gas, we release stored solar energy.

Geothermal Energy

The second most powerful source of energy, at 23 terawatts or 0.013 percent of the total, is **geothermal energy,** the Earth's internal heat energy. Geothermal energy eventually finds its way to the surface of the Earth, primarily via volcanic pathways. It drives the rock cycle (discussed below) and is therefore the source of the energy that uplifts mountains, causes earthquakes and volcanic eruptions, and generally shapes the face of the Earth.

Tidal Energy

The smallest source of energy for the Earth is the kinetic energy of the Earth's rotation. The Moon's gravitational pull lifts a tidal bulge in the ocean; as the Earth spins on its axis, this bulge remains essentially stationary. As the Earth rotates, the tidal bulge runs into the coastlines of continents and islands, causing high tides. The force of the tidal bulge "piling up" against land masses acts as a very slow brake, actually causing the Earth's rate of rotation to decrease slightly. The transfer of tidal energy accounts for approximately 3 terawatts, or 0.002 percent of the total energy budget.

Energy Loss

The Earth loses energy from the cycle in two main ways: reflection, and degradation and reradiation.

Reflection

About 40 percent of incoming solar radiation is simply reflected, unchanged, back into space, by the clouds, the sea, and other surfaces. For any planetary body, the percentage of incoming radiation that is reflected is called the *albedo*.

Each different material has a characteristic reflectivity. For example, ice is more reflectant than rocks or pavement; water is more highly reflectant than vegetation; and forested land reflects light differently than agricultural land. Thus, if large expanses of land are converted from forest to plowed land, or from forest to city, the actual reflectivity of the Earth's surface, and hence its albedo, may be altered.

Any change in albedo will of course have an effect on the Earth's energy budget.

Degradation and Reradiation

The portion of incoming solar energy that is not reflected back into space, along with tidal and geothermal energy, is absorbed by materials at the surface of the Earth, in particular the atmosphere and hydrosphere. This energy undergoes a series of irreversible degradations in which it is transferred from one reservoir to another and converted from one form to another. The energy that is absorbed, utilized, transferred, and degraded eventually ends up as heat, in which form it is reradiated back into space as long-wavelength (infrared) radiation. Weather patterns are a manifestation of energy transfer and degradation.

THE HYDROLOGIC CYCLE

For most people, the most familiar cycle is the **hydrologic cycle,** which describes the fluxes of water between the various reservoirs of the hydrosphere. We are familiar with these fluxes because we experience them as rain, snow, and running streams (Fig. 1.11). Like all the cycles in the Earth system, the hydrologic cycle is composed of pathways, the various processes by which water is cycled around in the outer part of the Earth, and reservoirs, or "storage tanks," where water may be held for varying lengths of time. The hydrologic cycle maintains a mass balance, which means that the total amount of water in the system is fixed and the cycle is in a state of dynamic equilibrium. There are fluctuations on a local scale—sometimes quite large fluctuations, such as those that cause floods in one area and droughts in another—but on a global scale these fluctuations balance each other out.

Pathways

The movement of water in the hydrologic cycle is powered by heat from the Sun, which causes *evaporation* of water from the ocean and land surfaces. The water vapor thus produced enters the atmosphere and moves with the flowing air. Some of the water vapor condenses and falls as *precipitation* (either rain or snow) on the land or ocean.

Rain falling on land may be evaporated directly or it may be intercepted by vegetation, eventually being returned to the atmosphere through their leaves by a process called *transpiration*. Or it may drain off into stream channels, becoming *surface runoff*. Or it may *infiltrate* the soil, eventually percolating down into the ground to become part of the vast reservoir of *groundwater*. Snow may remain on the ground for one or more seasons until it melts and the meltwater flows away. Snow that nourishes glaciers remains locked up much longer, perhaps for thousands of

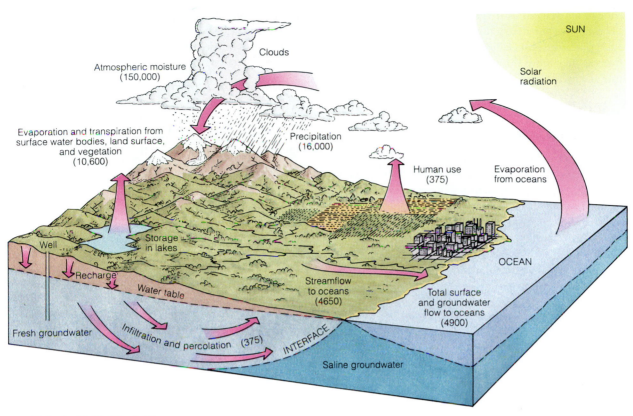

▲ F I G U R E 1.11
The hydrologic cycle.

years, but eventually it too melts or evaporates and returns to the oceans.

Reservoirs

The largest reservoir for water in the hydrologic cycle is the ocean, which contains more than 97.5 percent of all the water in the system. This means that most of the water in the cycle is saline, not fresh water—a fact that has important implications for humans because we are so dependent on fresh water as a resource for drinking, agriculture, and industrial uses. Surprisingly, the largest reservoir of fresh water is the permanently frozen polar ice sheets, which contain almost 74 percent of all fresh water. The ice sheets represent a long-term holding facility; water may be stored there for thousands of years before it is recycled. Of the remaining unfrozen fresh water, almost 98.5 percent resides in the next largest reservoir, groundwater. Only a very small fraction of the water passing through the hydrologic cycle resides in the atmosphere or in surface freshwater bodies such as streams and lakes.

There is a correlation between the size of a reservoir and the residence time of water in that reservoir: residence time

in the large-volume reservoirs, such as the oceans and the ice caps, is many thousands of years; in groundwater it is tens to hundreds of years, whereas in the small-volume reservoirs it is short—a few days in the atmosphere, a few weeks in streams and rivers.

BIOGEOCHEMICAL CYCLES

A **biogeochemical cycle** describes the movement of any chemical element or chemical compound among interrelated biologic and geologic systems. This means that biologic processes such as respiration, photosynthesis, and decomposition act alongside and in association with such nonbiologic processes as weathering, soil formation, and sedimentation in the cycling of chemical elements or compounds. It also means that living organisms can be important storage reservoirs for some elements. The carbon cycle is an important biogeochemical cycle; so are the nitrogen, sulfur, and phosphorus cycles, because each of these elements is critical for the maintenance of life.

It is difficult to produce a box model, even a highly simplified one, that accurately describes the biogeochemical behavior of an element as it cycles through the Earth system. These cycles potentially involve a wide variety of reservoirs and processes, and elements often change their chemical form as they move through the cycle. This complexity is illustrated by the nitrogen cycle, which we discuss briefly here.

The Nitrogen Cycle

Amino acids are essential components of all living organisms. They are given the name *amino* because they contain amine groups (NH_2), in which nitrogen is the key element. Nitrogen therefore is essential for all forms of life. The key to understanding the nitrogen cycle is understanding how nitrogen moves among the four major reservoirs of the Earth system—the atmosphere, biosphere, oceans, and soil and sediment. Figure 1.12 shows the reservoirs, the estimated number of grams of nitrogen in each reservoir, and the paths by which nitrogen moves among the reservoirs.

Nitrogen exists in three forms in nature. In the atmosphere it is present in the elemental form (N_2); reduced forms such as ammonia (NH_3) and oxidized forms such as nitrate (NO_3) also exist. Only in the reduced forms can nitrogen participate in biochemical reactions; N_2 cannot be used directly by organisms.

Nitrogen is removed from the atmosphere and/or made accessible to the biosphere in three ways:

1. Solution of N_2 in the ocean.
2. Oxidation of N_2 by lightening discharges to create NO_3, which is rained out of the atmosphere and into the soil and sea. Plants can reduce NO_3 to NH_3, thereby making nitrogen available to the biosphere.

3. Reduction of N_2 to NH_3 through the action of nitrogen-fixing bacteria in the soil or sea. The reduced nitrogen is quickly assimilated by the biosphere.

Once reduced, nitrogen tends to stay reduced, remain in the biosphere, and either be reused by other organisms or oxidized back to N_2 and returned to the atmosphere. The main route by which nitrogen returns to the atmosphere, however, is the reduction of nitrate. This route is kept open by bacteria that use the oxygen in nitrate during metabolism.

THE ROCK CYCLE

The hydrologic and biogeochemical cycles are driven by energy from the Sun. The most important cycle driven by geothermal energy is the **rock cycle.** Before we discuss this complex cycle it is necessary to consider some aspects of the internal structure of the Earth and the phenomenon of plate tectonics.

Heat Transfer in the Earth

Some of the Earth's internal heat makes its way slowly to the surface through the process of *conduction.* However, conduction (which basically works by passing thermal energy from one atom to the next) is a slow way to transfer heat. It is faster and more efficient for an entire packet of hot material to be transported, heat and all, from the hot part of the Earth's interior to the surface. This is essentially what happens when a fluid boils on a stovetop. If you watch a fluid such as pudding or spaghetti sauce as it boils, you will see that it turns over and over as packets of hot material rise from the bottom of the pot to the top. When it reaches the surface, the packet of hot fluid releases its heat and is swept back down to the bottom of the pot. The cycle of motion from bottom to top and back is called a *convection cell,* and this entire mode of heat transfer is called *convection.*

If you resume watching your pot of boiling pudding or sauce, you will see that a thin, hard film or skin forms on top of the fluid, where it is coolest. This film tends to ride around on the convecting fluid underneath. The same is true of the Earth: the lithosphere, or outer 100 km (approximately) of the Earth, is a cold, thin outer layer lying on top of hot, convecting material (Fig. 1.13). The thickness of the lithosphere relative to that of the Earth as a whole is about the same as that of the skin of an apple relative to the whole apple, or the glass sphere of a lightbulb relative to the whole bulb. Heat reaches the bottom of the lithosphere by convection; it passes through the lithosphere by conduction.

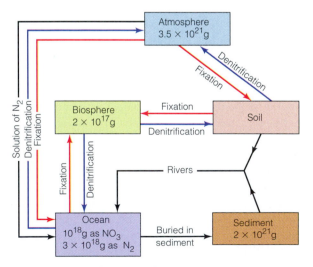

▲ **F I G U R E 1.12**
The nitrogen cycle.

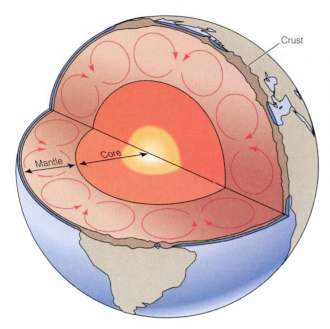

Crust

Mantle

Core

◀ **F I G U R E   1.13**
A sliced view of the Earth reveals a layered internal structure. The compositional layers, starting from the inside, are the core, the mantle, and the crust. Note that the crust is thicker beneath the continents than under the oceans. The outermost rocky layer, comprising the crust and the very top of the mantle, is called the lithosphere. The lithosphere "rides" around on the hot, convecting mantle beneath.

Think about this for a moment: If an exceedingly thin layer of cold, brittle material is riding and jostling around on top of a hot, mobile, convecting fluid, what can happen to the cold boundary layer? It can break, of course, and that is exactly what has happened to the rocky outer layer of the Earth. The lithosphere has broken into a number of jagged, rocky pieces called *plates*, which range from several hundred to several thousand kilometers in width (Fig. 1.14).

The lithospheric plates are riding around on an underlying layer of hot, ductile, easily deformed material called the *asthenosphere*, or "weak layer."

Some of the lithospheric plates are composed primarily of oceanic crustal material, whereas others are composed primarily of continental material. If it were possible to remove all the water from the ocean and view the dry Earth from a spaceship, we would see that the continents stand,

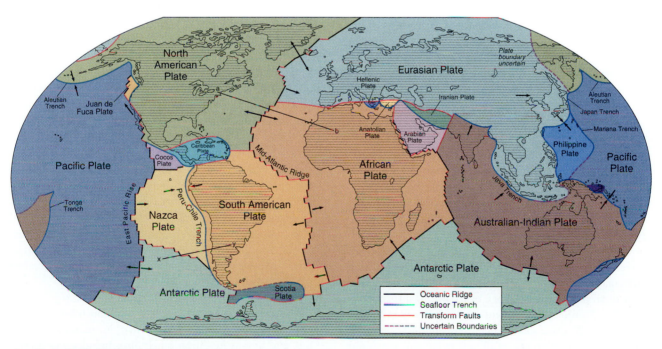

▲ **F I G U R E   1.14**
Six large plates and several smaller ones cover the Earth's surface and move steadily in the directions shown by the arrows.

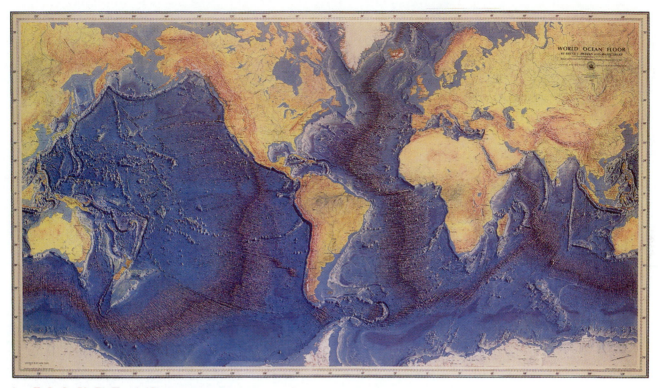

▲ **F I G U R E 1.15**
The topography of the continents and ocean floor.

on average, about 4.5 km above the floor of the ocean basins (Fig. 1.15). Continental crust is relatively light (density 2.7 g/cm³), whereas oceanic crust is relatively heavy (density close to 3.2 g/cm³). Because the lithosphere is floating on the weak asthenosphere, the plates capped by light continental crust stand high while those capped by heavy oceanic crust sit lower.

Plate Tectonics and the Earth's External Structure

Convection within the Earth is constantly moving the plates of lithosphere and slowly changing the Earth's surface. Such mountains as the Alps or Appalachians that seem changeless to us are only transient wrinkles when viewed from the perspective of geologic time. Mountain ranges grow when fragments of moving lithosphere collide and heave masses of twisted and deformed rock upward; then the ranges are slowly worn away, leaving only the eroded roots of an old mountain range to record the ancient collision (Fig. 1.16). The continents are still slowly moving at rates up to 10 cm a year, sometimes bumping into each other and creating a new mountain range and sometimes splitting apart so that a new ocean basin forms. The Himalaya is a range of geologically young mountains that began to form when the Indian subcontinent collided with Asia about 45 million years ago. The Red Sea is a young ocean that started forming about 30 million years ago when a split developed between the Arabian Peninsula and Africa as the two land masses began to move apart.

But it is not just the continents that move, it is the entire lithosphere. The continents, the ocean basins, and everything else on the surface of the Earth are moving along like passengers on large rafts; the rafts are huge plates of lithosphere that float on the underlying convecting material. As a result, all the major features on the Earth's surface, whether submerged beneath the sea or exposed on land, arise as either a direct or an indirect result of the motion of lithospheric plates.

Such motions involve complicated events, both seen and unseen, all of which are embraced by the term *tectonics,* derived from the Greek word, *tekton,* which means carpenter or builder. Tectonics is the study of the movement and deformation of the lithosphere. The special branch of tectonics that deals with the processes by which the lithospheric plates move and interact with one another is called **plate tectonics.**

Plate tectonics provides a unifying theory that can be used to explain hundreds of years of independent observations of the processes of rock formation, mountain building, and terrain modification. It also provides an effective framework for our discussion of geologic processes that affect people and form the environment in which we live.

Today the lithosphere is broken into six large plates and numerous small ones (Fig. 1.14), all moving at speeds ranging from 1 to 10 cm a year. As a plate moves, everything on it moves too. If the plate is capped partly by oceanic crust and partly by continental crust, then both the ocean floor and the continent move at the same speed and in the same direction. The term *continental drift* is sometimes used to describe continental movement, but it must be remembered that everything on a plate moves, not just the continents.

Plate Margins

Plates move as individual units, and interactions between plates occur along their edges. The most pronounced manifestations of these interactions are earthquakes and volcanoes, which occur primarily along plate margins. Through studies of these phenomena, particularly earthquakes, geologists have been able to decipher the shapes of the plates. Plate margins are of special interest in environmental geology because so many hazardous natural events tend to occur in these regions.

Plates have three kinds of margins (Fig. 1.17):

1. **Divergent margins,** which are also called *rifting* or *spreading centers* because they are fractures in the lithosphere where two plates move apart. Divergent margins are marked by submarine volcanism and frequent but weak earthquakes.

2. **Convergent margins,** where two plates move toward each other. Along convergent margins, one plate must either sink beneath the other, in which case we refer to

▲ F I G U R E 1.16

The scar of an ancient collision. Layers of rock, once horizontal, were twisted and contorted as a result of a collision between two plates. These eroded roots of an ancient mountain range north of Adelaide, South Australia, were recorded in a LANDSAT image in September 1983.

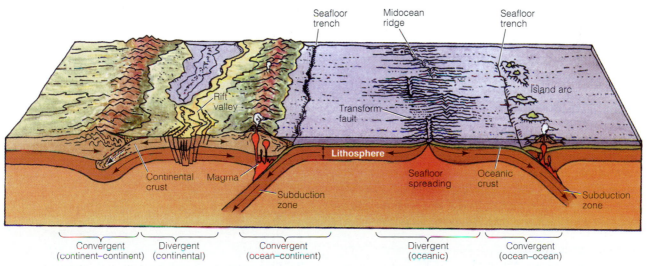

▲ F I G U R E 1.17

The various types of plate margins: divergent (spreading or rifting); convergent (subduction and collision zones); and transform (lateral motion). In an oceanic setting, divergent margins are marked by midocean ridges, like the Mid-Atlantic Ridge. In a continental setting, divergent margins are marked by rift valleys, like the East African rift. Ocean-ocean and ocean-continent convergent margins where subduction occurs are marked by deep trenches and lines of volcanoes. Continental collision zones, as in the Himalaya, are marked by mountain ranges. Transform margins vary in their topographic features; they are sometimes marked by a long linear valley.

the margin as a *subduction zone,* or the two plates must collide, in which case we refer to the margin as a *collision zone.* When oceanic crust is involved in the interaction (i.e., in an ocean–ocean or ocean–continent convergence), a subduction zone will occur. When only crustal material is involved, a collision zone forms. Convergent margins are often marked by explosive volcanism and powerful earthquakes.

3. **Transform fault margins** are fractures in the lithosphere where two plates slide past each other, grinding and abrading their edges as they do so. Transform fault margins are frequent sites of powerful earthquakes but are not associated with volcanism.

The Three Rock Families

An understanding of the rock cycle also requires familiarity with the major kinds of rocks. There are three large "families" of rocks, each defined by the processes that form them.

Igneous Rocks

The first major rock family consists of **igneous rocks** (named from the Latin *igneus,* meaning fire). Igneous rocks are formed by the cooling and consolidation of *magma,* or molten rock. Magma that cools and crystallizes underneath the ground becomes *plutonic rock* (after Pluto, the Greek god of the underworld). If the magma finds its way to the Earth's surface, erupting through volcanic conduits, we refer to the molten rock as *lava,* and when it solidifies it is called *volcanic rock.* Most igneous rock is formed along the spreading edges and convergent margins of plates.

Sedimentary Rocks

Rocks at the surface of the Earth are exposed to water, ice, and air, with which they interact both physically and chemically. The interactions cause rocks to break down into smaller particles, or *weather.* Some products of weathering are soluble in water and are carried away in solution by streams and rivers, but most are loose particles in the regolith. Loose particles that are transported by water, wind, or ice, and then deposited, are called **sediment.** Sediment eventually becomes **sedimentary rock,** a term that refers to any rock formed by chemical precipitation or by cementation of sediment. Sedimentary rocks constitute the second rock family and can be found anywhere on the Earth.

Metamorphic Rocks

The final major rock family is metamorphic rock (from the Greek *meta,* meaning change, and *morphe,* meaning form: hence, change of form). **Metamorphic rocks** are rocks whose original form has been changed as a result of high temperature, high pressure, or both. *Metamorphism*—the process that forms metamorphic rocks from sedimentary or igneous rocks—is analogous to the process that occurs when a potter fires a clay pot in an oven. The mineral grains in the clay undergo a series of chemical reactions as a result of the increased temperature; new compounds form, and the formerly soft clay becomes hard and rigid. Metamorphism occurs most noticeably along plate collision margins.

Rocks in the Crust

The Earth's crust is 95 percent igneous rock or metamorphic rock derived from igneous rock. However, as shown in Fig. 1.18, most of the rock that we actually see at the surface of the Earth is sedimentary. Sediments are products of weathering, and as a result they are draped as a thin veneer over the largely igneous crust. This distribution of rock types is one consequence of the rock cycle and the interactions between internal and external processes in the cycle.

The internal processes that form magma and in turn lead to the formation of igneous rock interact with external processes through weathering and erosion. When rock weathers, the particles form sediment. The sediment is transported and deposited. It may eventually become cemented, usually by substances carried in water moving through the ground, and in this way they are converted into new sedimentary rock. In places where such sedimentary rock forms, it can reach depths at which pressure and heat cause new compounds to form, with the result that the sedimentary rock becomes metamorphic rock. Sometimes metamorphic rock settles so deep that the high temperatures of the Earth's interior melt it and magma is formed. The new magma can then move upward through the crust, where it can cool and form another body of igneous rock. Eventually the new body of igneous rock can be uncovered

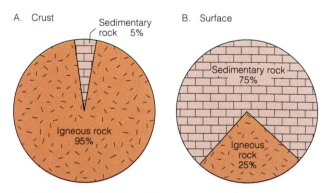

▲ **F I G U R E   1.18**
Relative amounts of sedimentary and igneous rock. (Metamorphic rocks are considered to be either sedimentary or igneous, depending on their origin.) A. The great bulk of the crust consists of igneous rock (95 percent) but sedimentary rock (5 percent) forms a thin covering at and near the surface. B. The extent of sedimentary rock cropping out at the surface is much larger than that of igneous rock, so 75 percent of all rock seen at the surface is sedimentary, and only 25 percent is igneous.

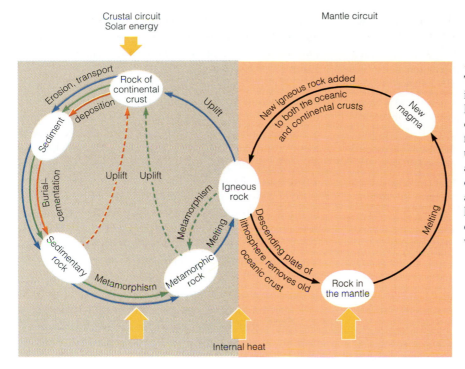

Crustal circuit
Solar energy

Mantle circuit

The rock cycle—an interplay of
internal and external processes.
Rock material in the continental
crust can follow any of the arrows
from one phase to another. At one
time or another it has followed
all of them. Within the mantle
circuit, magma rises from below
and forms new igneous rock in the
lithosphere. The old lithosphere
descends again to the mantle,
where it is eventually remixed.

and subjected to erosion, the eroded particles start once
more on their way to the sea, sediment is laid down, and
the cycle is repeated.

Rock Cycles and Circuits

The cycle involving igneous and sedimentary rock has occurred again and again throughout the Earth's long history.
It is not the only possible cycle, however; it is just one circuit among many that occur in the continental crust. As
the dashed lines in Fig. 1.19 show, other circuits involve
bodies of sedimentary rock that are neither metamor-
phosed nor melted before they are uplifted and eroded.
Whether the circuits are long or short, the continental crust
is being endlessly recycled as a result of erosion, on the one
hand, and plate tectonics, on the other. Because the mass of
the continental crust is large, the average time a rock takes
to complete the cycle is long. Estimates of the length of the
cycle vary, but the average age of all rock in the continental
crust seems to be about 650 million years.

The rock cycle of oceanic crust is faster than that of continental crust. When sinking lithosphere carries old oceanic
crust back down into the mantle, some of the crust melts
and rises to form volcanoes, and the rest is eventually
remixed into the mantle. Thus, the most ancient crust of
the ocean basins is only about 180 million years old, and
the average age of all oceanic crust is only 60 million years.

The magma that rises to form new oceanic crust forms
hot igneous rocks that react with seawater. In this reaction,
some constituents in the hot rock, such as calcium, are dis-

solved in the seawater and constituents already in the seawater, such as magnesium, are deposited in the igneous
rock. Because the magma that forms oceanic crust comes
from the mantle, the reactions between hot crust and seawater are one way in which the mantle plays a role in determining the composition of seawater. They are also an important example of the interaction between the rock and
hydrologic cycles.

UNIFORMITARIANISM AND EARTH CYCLES

Among the many important questions that have faced geologists who study Earth processes is one that concerns the
importance of small, slow changes, such as erosion caused
by a single rainstorm, as opposed to large-scale changes,
such as earthquakes and floods, that are infrequent but
cause dramatic changes in the landscape. During the seventeenth and eighteenth centuries, before geology became the
scientific discipline it is today, people believed that all the
Earth's features had been produced by a few great catastrophes. Those catastrophes were thought to be so huge that
they could not be explained by ordinary processes but must
have supernatural causes. This concept came to be called
catastrophism. The catastrophes were thought to be gigantic and sudden, and they were also thought to have occurred relatively recently and to fit a chronology of catastrophic events recorded in the Bible.

During the late eighteenth century the concept of catastrophism was reexamined, compared with geologic evidence, and found wanting. The person who assembled much of the evidence and proposed a countertheory was James Hutton (1726–1797). Hutton, a Scottish physician and gentleman farmer, was intrigued by what he saw in the environment around him. He observed the slow but steady effects of erosion: the transport of rock particles by running water and their ultimate deposition in the sea. He reasoned that mountains must slowly but surely be eroded away, that rocks must form from the debris of erosion, and that those rocks in turn must be slowly thrust up to form mountains. Hutton didn't know the source of the energy that caused mountains to be thrust up, but he argued that everything moved slowly along in a repetitive, continuous cycle.

Hutton's ideas evolved into what we now call the principle of **uniformitarianism,** which states that the same external and internal processes that we recognize in action today have been operating throughout the Earth's history. The principle of uniformitarianism provides a first and very significant step toward understanding the Earth's history. We can examine any rock, however old, and compare its characteristics with those of similar rocks that are forming today in a particular environment. We can then infer that the ancient rock very likely formed in the same sort of environment. For example, in many deserts today we can see gigantic sand dunes formed from sand grains transported by the wind. Because of the way they form, the dunes have a distinctive internal structure (Fig. 1.20A). Using the principle of uniformitarianism, we can safely infer that any rock composed of cemented grains of sand and having the same distinctive internal structure as modern dunes (Fig. 1.20B) is the remains of an ancient dune.

Geologists since Hutton's time have explained the Earth's features in a logical manner by using the principle of uniformitarianism. But in so doing they have made an outstanding discovery—the Earth is incredibly old. An enormously long time is needed to erode a mountain range, or for huge quantities of sand and mud to be transported by streams, deposited in the ocean, and cemented into new rocks, and for the new rocks to be deformed and uplifted to form a new mountain. Yet, slow though it is, the cycle of erosion, formation of new rock, uplift, and more erosion has been repeated many times during the Earth's long history.

Uniformitarianism and the Rates of Cycles

During the nineteenth century, geologists tried to estimate the duration of the rock cycle by estimating the thickness of all the sediments that have been laid down through geologic time. They assumed that the principle of uniformitarianism applied to the rates at which processes occur as well as to the processes themselves, and hence that rates of deposition of sediment have always been constant and equal to today's rates. Thus, they thought, it would be a simple calculation to estimate the time needed to produce all the sediments. The results, we now know, were greatly in error. One of the reasons for the error was the assumption of constancy of geologic rates.

The more we learn about the Earth's history and the more accurately we determine the timing of past events through *radiometric dating* (using rates of decay of naturally occurring radioactive atoms to determine the ages of rocks), the clearer it becomes that rock cycle rates have not always been the same. Some rates were once more rapid, others much slower. This means that the relative importance of different geologic processes has probably differed in the past. For example, just because glaciation is an important process today, we cannot assume that it has been equally important throughout geologic time. But we *can* assume that when glaciation did affect the Earth in geologically remote times, the processes and effects were the same as those we observe in glaciated regions today.

Uniformitarianism says that "the present is the key to the past"—that we can study present Earth processes in order to understand the processes that have shaped our environment in the past. Now we are discovering that the reverse is also true—the past holds important keys for understanding the present. For example, scientists are documenting changes in the chemical composition of the atmosphere that may signal major global climatic changes. But the Earth's climate system is highly complex, with cyclical variations that are as yet poorly understood. How can we be sure that we really understand the magnitude and significance of the changes we are witnessing? Studies of past climatic changes, in the scientific field of study known as *paleoclimatology,* are providing much-needed clues and a geologic baseline against which we can assess the significance of present changes.

Neo-Catastrophism

Uniformitarianism is a powerful principle, especially when it is constrained by radiometric dating and other geologic techniques, but should we abandon catastrophism as a totally incorrect hypothesis? Recent discoveries suggest not. For example, there is a growing body of evidence that at least once, and perhaps several times, the Earth has been struck by a meteorite large enough to wipe out many life forms. The impact of a large meteorite may have been responsible for the extinction of the dinosaurs 66 million years ago. (The effects of such an event are discussed more fully in Chapter 10.) Even more dramatic mass extinctions have occurred at other times in the past. The geologic record indicates that about 245 million years ago almost 90 percent of all living plants and creatures became extinct. We still have not discovered the cause of that occurrence, but we can be sure that it was a catastrophic disaster of some sort.

Such infrequent massive events fall somewhere between uniformitarianism and catastrophism. When we view the Earth's history as a series of such repeated but sporadic

A.

◀ F I G U R E 1.20
The internal structure of sand dunes, ancient and modern. A. A distinctive pattern of wind-deposited sand grains can be seen in a hole dug in this dune near Yuma, Arizona. B. The same distinctive pattern in rocks in Zion National Park, Utah, lets us infer that these rocks were once sand dunes too.

B.

events, there is no evidence to suggest that similar events will not occur in the future. Nor is there any evidence to suggest when another such event might occur. This new theory of catastrophism—based on the geologic record rather than on the Biblical record—is sometimes called *neo-catastrophism*. It recognizes uniformitarianism as the guiding principle in understanding Earth processes while acknowledging the role of infrequent catastrophic events in generating massive, far-reaching environmental changes on a very short time scale.

A fascinating but frightening possibility is that a catastrophe of a different kind may already be happening. It has been suggested that the collective activities of humans may be changing the Earth so rapidly, and so massively, that they may cause a catastrophe similar in magnitude to some of the major ones in the geologic record. This hypothesis—whether or not it proves true—emphasizes an important fact, and a major theme throughout the remainder of this book: Geology and the welfare of the human race are indissolubly linked.

SUMMARY

1. The concept of a system—a portion of the universe that can be isolated from the rest for the purpose of observing changes—can be a useful way to approach the study of complex problems. The first step in studying the Earth as a system is to identify the smaller systems that are its component parts: the atmosphere, hydrosphere, biosphere, and lithosphere.

2. Box models are often used to portray the rates at which material and/or energy enter or leave a system; the amount of matter or energy in the system; the various storage reservoirs for material and/or energy within the system; and the pathways by which matter and/or energy are transferred from one part of the system to another.

3. The Earth is a natural example of a closed system. This has two important implications for environmental geology: (1) The amount of matter in the system is fixed and finite. (2) When changes are made in one part of the system, the results will eventually affect other parts of the system.

4. Even though material is constantly being transferred from one of the Earth's systems to another, the systems themselves don't tend to change much because the different parts of the flow paths tend to balance one another. This is shown by the cycling of carbon, which is constantly transferred among the biosphere, hydrosphere, atmosphere, and lithosphere. The natural carbon cycle is balanced, but human activities since the Industrial Revolution have radically altered this state of balance.

5. The Earth cycles that are of most importance in environmental geology are the energy cycle, the hydrologic cycle, biogeochemical cycles, and the rock cycle.

6. The energy cycle encompasses the external and internal energy sources that drive the Earth system and all its cycles. The three sources of inputs into the energy cycle are solar radiation, geothermal energy, and tidal energy. The two sources of loss from the cycle are (1) reflection and (2) degradation and reradiation.

7. The hydrologic cycle describes the fluxes of water between the various reservoirs of the hydrosphere. The processes by which water is cycled around the Earth (such as evaporation and precipitation) are driven by energy from the Sun. The ocean is the largest reservoir in the hydrologic cycle. The polar ice caps are the largest reservoir for fresh water, and groundwater is the largest reservoir for unfrozen fresh water.

8. A biogeochemical cycle describes the movement of any chemical element or compound among interrelated biologic and geologic systems. In biogeochemical cycles, such as the nitrogen cycle, biologic processes (such as respiration, photosynthesis, and decomposition) act alongside and in association with nonbiologic processes (such as weathering, soil formation, and sedimentation).

9. Unlike the hydrologic cycle and biogeochemical cycles, which are driven by energy from the Sun, the rock cycle is driven by the Earth's internal energy—especially by the release of heat through the process of convection.

10. The Earth's lithosphere is a cold, brittle outer layer "floating" on hot, convecting material. As a result, the lithosphere has broken into large, jagged plates that range from several hundred to several thousand kilometers in width.

11. Plate tectonics refers to processes by which the Earth's lithospheric plates move around and interact with one another. Where plates interact, three types of plate margins can form: divergent margins, convergent margins (subduction zones and collision zones), and transform fault margins. Plate margins are of special interest in environmental geology because many hazardous natural events occur in these regions.

12. There are three major rock families: igneous rocks, which form by cooling and consolidation of molten rock (magma); sedimentary rocks, which form by chemical precipitation or by cementation of sediment; and metamorphic rocks, which form when rocks

change as a result of high temperature, high pressure, or both.

13. In the various circuits of the rock cycle, internal processes that form magma interact with external processes through erosion and sedimentation. Rocks do not always follow the same pathway through the rock cycle.

14. The principle of uniformitarianism states that the same external and internal processes that we recognize

in action today have been operating throughout the Earth's history. Thus, scientists can observe present Earth processes and draw conclusions about rocks that formed in similar environments long ago. Our understanding of past processes—climatic changes, for example—also can provide a baseline against which we can assess the magnitude and significance of current changes in the Earth system.

IMPORTANT TERMS TO REMEMBER

biogeochemical cycles (p. 29)
catastrophism (p. 35)
closed system (p. 21)
convergent margin (p. 33)
divergent margin (p. 33)
energy cycle (p. 27)
geothermal energy (p. 28)

hydrologic cycle (p. 28)
igneous rock (p. 34)
metamorphic rock (p. 34)
plate tectonics (p. 32)
reservoir (p. 22)
residence time (p. 22)
rock cycle (p. 30)

sediment (p. 34)
sedimentary rock (p. 34)
system (p. 20)
transform fault margin (p. 34)
uniformitarianism (p. 36)

QUESTIONS AND ACTIVITIES

1. The most important biogeochemical cycles are the carbon, nitrogen, phosphorus, sulfur, and oxygen cycles. (Some people consider the cycle of water—the hydrologic cycle—to be a biogeochemical cycle, too, because of the importance of water in supporting life.) Choose one of these cycles to investigate in detail. What are the main reservoirs and pathways in the natural cycle? Where do humans fit into the cycle, and what is the extent to which human actions have affected the natural balance? Try to draw a pre- and postindustrial box model of your biogeochemical cycle to highlight the impacts of human activities.

2. Our planet operates as a closed, dynamic system. In what ways do human actions demonstrate an understanding of this? Can you think of examples of human actions that seem to show ignorance of the implications of living in a closed system? How might we reorganize human activities and structures to better reflect this closed system and the finite nature of most Earth resources? These might be good topics for an essay or a class discussion.

3. Based on your own knowledge, see if you can construct a rough box model for a cycle that interfaces with natural Earth systems but is primarily controlled by human actions. Possibilities include the cycles of mineral resources or fossil fuels (from its extraction to processes, consumption, and waste disposal), or the

cycle of water use and wastewater treatment and disposal at a factory.

4. Figure 1.1 uses the example of a river flowing into a lake to illustrate the system concept. How many smaller systems can you think of that are component parts of the river–lake system? (The examples of a small volume of water or sediment are shown in the figure; try to think of as many others as you can.) Now, think big—what are some of the larger systems of which the river–lake system is a component part?

5. What kind of rock underlies your house (igneous, sedimentary, or metamorphic)? What about your school? Do you live near a plate margin? If so, what kind (divergent, convergent, or transform fault)? You can investigate the geology of your area through library research or by contacting the state or provincial geological survey.

6. Find out more about the International Geosphere–Biosphere Program. Are there any researchers at your college or university who participate in the program? You can receive the *IGBP Global Change Newsletter* by writing to the IGBP Secretariat, The Royal Swedish Academy of Sciences, Box 50005, S-104 05, Stockholm, Sweden. Can you find any other examples of scientific research programs that take an interdisciplinary approach to the study of the Earth system?

EARTH STRUCTURE AND MATERIALS

It isn't the size that counts so much as the way things are arranged.

• E. M. Forster, 1910

Sometimes the way a material is put together is its most important quality. This appears to be true for the mineral asbestos. "Asbestos" is not actually a true mineral name; rather, it is an industrial or commercial term applied to a group of minerals with different chemical compositions and crystal structures. A mineral is called *asbestos* (or "asbestiform") if it occurs in the form of fine, flexible, hairlike fibers. The fibrous form gives these minerals strength and flexibility as well as other qualities, such as incombustibility and low thermal conductivity. As a result, asbestos is a convenient material to use in any part of a building or machinery that requires insulation—the roofs and floors of building, brake linings, and many other applications.

Asbestos was widely used for such purposes in the period following the Second World War. Then in the 1960s evidence emerged that some people had suffered devastating health consequences as a result of exposure to asbestos. Incidents of lung cancer, asbestosis, and mesothelioma (a cancer of the lining of the chest or abdomen that is nearly always fatal) were undeniably linked to the inhalation of airborne fibers by miners and others who were exposed to asbestos, principally in occupational settings.

Removal of chrysotile asbestos from a public building. To avoid ingesting or inhaling fibers, the workman is completely covered and is breathing from a contained air supply.

The medical and epidemiologic evidence ignited an intense debate about the extent to which asbestos fibers in the environment should be regulated and exactly how they should be handled. Occupational hazard legislation now protects miners, firefighters, construction workers, and others who are routinely exposed to asbestos in the workplace. By 1973, regulations were passed limiting the spraying of asbestos insulation. In 1982, the United States passed legislation banning the use of asbestos in schools and mandating inspections of school premises (although the legislation does not specify any required corrective action).

One particular type of asbestos—crocidolite, which is infrequently encountered—has been shown to be more carcinogenic than others. Some studies have concluded that the fibers of the chrysotile form of asbestos, which accounts for about 95 percent of the asbestos currently in use, are as much as 10 times less hazardous than those of the crocidolite form. This finding has intensified the debate about how to deal with asbestos in the built environment. If the most commonly used form of asbestos is the safest, is the high cost of removal warranted? Should regulations (and funding) be focused on the most hazardous varieties? Or should all asbestos be removed, no matter what form it is? An additional problem is that the removal itself may contribute to the hazard. Asbestos causes health problems when it is inhaled; removing old, asbestos-bearing construction materials may generate dust with a high concentration of fibers.

The properties of asbestos minerals affect human lives on a daily basis. There are many unresolved questions—geologic, medical, political, and financial questions—about the hazards of asbestos and how they should be managed.

THE COMPOSITION OF THE EARTH

Environmental geologists study the physical structure and chemical composition of the Earth for three main reasons: (1) The structure of the Earth, particularly its internal structure, has a lot to do with shaping our landscape and causing events that may be hazardous for humans. (2) Humans are dependent on the materials of the solid Earth as resources. Because the Earth is a closed system, these resources are limited; we need to learn as much as possible about how they form, how to find them, and how to use them wisely. And (3) the materials of the Earth have distinct physical and chemical properties; we need to understand these materials because they can affect human lives in many different ways. In this chapter, therefore, we explore several aspects of the Earth's composition and structure.

What Is the Earth Made Of?

If you were given the task of figuring out what the Earth is made of, how would you go about it? The Earth is almost 13,000 km in diameter; even in the deepest hole ever drilled (to a depth of more than 12 km in Russia) we have direct access to only the top few kilometers of the Earth's outermost layer. Those few kilometers of material are so variable and inhomogeneous in composition, and some parts are so remote, that it is extraordinarily difficult to document the composition of even this outermost layer.

As it turns out, scientists have been able to compile information from a wide variety of sources—including not only sampling and direct observation but also experimentation and theoretical modeling—and have used that information to build up and refine our understanding of the composition and distribution of materials throughout the Earth. In the first part of this chapter, we examine the internal structure and the different compositional zones within the Earth, and some of the tools and approaches that Earth scientists have used to discover more about them. We then take a look at the Earth's basic building blocks, minerals and rocks, and learn something about the properties and characteristics of different Earth materials.

THE EARTH: INSIDE AND OUT

The Earth is a *differentiated* body. This means that material is not distributed evenly or homogeneously throughout the planet. *Differentiation* is the process whereby planetary objects develop concentric layers or zones that differ in their compositions. It is common to all of the terrestrial planets as well as a number of smaller bodies, such as the Moon. Planetary differentiation occurred early in the history of the solar system, when newly formed bodies like the Earth were very hot and partially or completely molten. The least dense of the melted materials, dominated by the elements silicon, aluminum, sodium, potassium, and oxygen, rose toward the surface. The rocks that form the outermost part of the Earth are still rich in these elements. The densest melted materials, principally molten iron and nickel, sank to the center of the planet. Between these two groups of materials is a vast mass of intermediate-density rocks that are rich in elements such as magnesium, calcium, and titanium. Escaping gases, mainly water vapor, carbon dioxide, methane, and ammonia, gave rise to the Earth's early atmosphere.

The processes of partial melting and differentiation turned the Earth into a layered planet (Fig. 2.1). Like other differentiated bodies, the Earth contains three major *compositional layers* (Fig. 2.2). At the center is the densest layer, the **core.** The core is a spherical mass, composed largely of metallic iron, with lesser amounts of nickel and traces of other elements. The thick shell of intermediate-density, rocky matter that surrounds the core is called the **mantle.** The mantle is less dense than the core. Above the mantle lies the thinnest and least dense layer, the **crust.**

The core and the mantle have nearly constant thicknesses. The crust is far from uniform, though, and may dif-

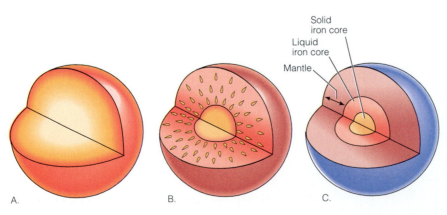

Solid iron core
Liquid iron core
Mantle

A.   B.   C.

◄ F I G U R E 2.1
Compositional differentiation of the early Earth. Early in Earth history, when the planet was partially molten, (A) the densest materials sank to the center and the lightest materials rose to the surface (B). This process, known as differentiation, made the Earth into a compositionally layered body (C).

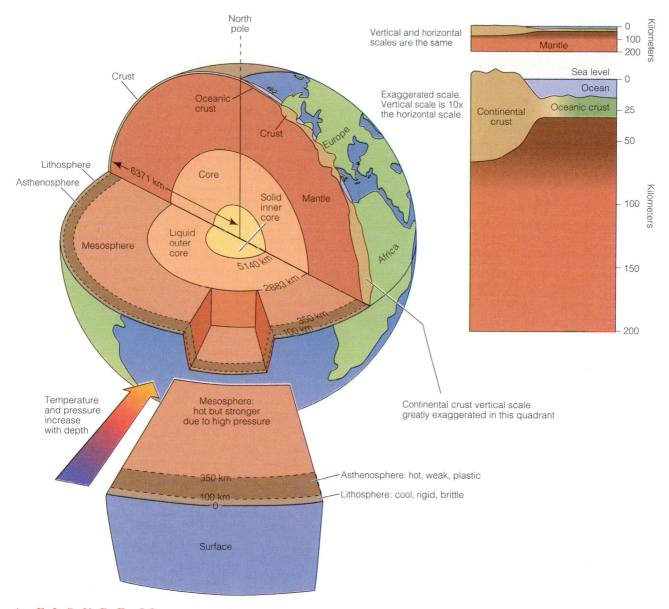

▲ FIGURE 2.2

The Earth's internal structure, including compositional layers and layers of differing rock properties. Note that boundaries between zones of differing rock strength—such as the lithosphere and the underlying, weaker asthenosphere—do not coincide with compositional boundaries.

fer in thickness from place to place by a factor of nine. Also, as discussed in Chapter 1, there are two different types of crust, each with its own average thickness, composition, and properties. *Oceanic crust* has an average thickness of about 8 km, a density of 3.2 g/cm^3, and a composition rich in calcium, magnesium, and iron. In contrast, *continental crust* (average density 2.7 g/cm^3) ranges in thickness from 30 to 70 km and has a composition richer in light elements, such as silicon, aluminum, sodium, and potassium.

Layers of Differing Physical Properties

It is important to understand the difference between compositional layers and other types of layering within the Earth. Physical properties such as the strength of rock vary with depth in the Earth. These changes are not due to compositional changes but are controlled mainly by the interplay between temperature and pressure. Strange as it may seem, the composition of a rock plays a much smaller role than temperature and pressure in determining its strength.

When a solid is heated, it loses strength. When it is compressed, it gains strength. The result of this interplay between temperature and pressure in the Earth is that the places where physical properties change do not coincide with the compositional boundaries between the crust, mantle, and core (Fig. 2.2).

The most profound change in physical properties is found deep within the Earth's core, where pressures are so great that iron is solid despite its high temperature. The solid center of the Earth is the *inner core*. Surrounding the inner core is a zone where temperature and pressure are so balanced that the iron is molten and exists as a liquid. This is the *outer core*. The difference between the inner and outer cores is not one of composition. Instead, the difference lies in the physical states of the two: one is a solid, the other a liquid.

Differences in temperature and pressure divide the rocky layers above the core into three distinct regions. In the lower part of the mantle, the rock is so highly compressed that it has considerable strength even though the temperature is very high. Thus, a solid region of high temperature but also relatively high strength extends from the core–mantle boundary (at 2883 km depth) to a depth of about 350 km and is called the *mesosphere* ("intermediate," or "middle, sphere") (Fig. 2.2). Above the mesosphere, from 350 to about 100 km below the Earth's surface, is the *asthenosphere* ("weak sphere"), where the balance between temperature and pressure is such that rocks have very little strength. Rock in the asthenosphere is weak and easily deformed, like butter or warm tar. Above the asthenosphere, and corresponding approximately to the outermost 100 km of the Earth, is a region where rocks are cooler, stronger, and more rigid than those in the plastic asthenosphere. This hard outer region, which includes the uppermost mantle and all of the crust, is the *lithosphere* ("rock sphere").

Keep in mind that even though the crust and mantle differ in composition, it is strength, not composition, that differentiates the lithosphere from the asthenosphere. Rocks in the lithosphere are strong and brittle; rocks in the asthenosphere are weaker and can be easily deformed. As we saw in Chapter 1, the lithosphere is broken into a number of plates. The lithospheric plates move about on the hot, convecting asthenosphere and interact with each other along plate boundaries. The rock cycle and plate tectonics are among the consequences of the differences in physical properties between the lithosphere and the asthenosphere.

Layers of Differing Chemical Composition

The composition of the Earth's crust is of particular interest to environmental geologists for a number of reasons. The crust is the only layer to which we have direct access. We live on it and are affected on a daily basis by the properties of crustal materials. We derive mineral and energy resources from the crust, and we draw water from it. Determining the composition of the crust has been challenging enough, but first let's take a quick look at how scientists have deduced the compositions of the layers to which we have no direct access—the mantle and core.

Composition of the Mantle and Core

Because we cannot see and sample either the core or the mantle, indirect measurements are used to find out about their composition and physical properties. One way to determine composition is to measure how the density of rock changes with increasing depth below the Earth's surface. We can do this by measuring the velocities with which earthquake waves pass through the Earth. The velocities of these waves are a function of the elastic properties and density of rocks. From such measurements, we discover that density increases with depth, but not smoothly. At some depths, abrupt increases in velocity indicate sudden increases in density. From this we infer that the solid Earth does not have a uniform composition but instead consists of distinct layers with different densities. Knowing these densities, we can estimate more precisely what the compositions of the different layers must be.

We can confirm the estimates of the composition of the mantle by comparing them with the composition of materials that have been brought to the surface from great depth by volcanoes. Molten rock that erupts as lava sometimes carries along fragments of rocks, called *xenoliths* (from the Greek for "foreign rock"), which have been torn away from the region deep in the Earth where the volcano originated. We find these xenoliths encased in volcanic rocks that have cooled and solidified at the Earth's surface (Fig. 2.3). However, the minerals in the xenoliths tell us that they once existed at extremely high temperatures and pressures. The composition of most xenoliths is consistent with estimates of mantle composition based on studies of earthquake waves. Taken together, all of the various sources of knowledge about the composition of the mantle suggest that it is composed of chemical compounds made primarily of the elements magnesium, silicon, iron, and oxygen.

Determining the composition of the core presents the greatest difficulty. The Earth's overall density is 5.52 g/cm³. The density of rocks at the surface is no more than 3.2 g/cm³. Even rock that is compressed at the bottom of the mesosphere is not as dense as 5.52 g/cm³. Scientists have therefore deduced that the Earth must have a very dense core. Some of the best evidence for the composition of the core comes from iron meteorites. Such meteorites are believed to be fragments from the core of a small differentiated planetary body that was shattered by a gigantic impact early in the history of the solar system. Scientists presume that this object must have had compositional layers similar to those of the Earth and the other terrestrial planets. Additional evidence comes from the Earth's magnetic field. The presence of a dipolar magnetic field requires the presence of a convecting, electrically conducting fluid in its interior; as in the other terrestrial planets, partially molten metallic

◄ F I G U R E 2.3
Kimberlite, a volcanic rock from the mantle which has carried up mineral fragments and rounded xenoliths of mantle rocks. One of the minerals in the sample is a large diamond (indicated by red arrow). Diamonds can only form under very high-temperature, high-pressure conditions in the mantle.

iron is the most likely material. Studies of the behavior of earthquake waves near the Earth's core confirm that the outer core is, indeed, in a liquid state.

Composition of the Crust

Slight compositional variations probably exist within the mantle and core, but it is difficult to determine very much about them. We can see and sample the crust, however. The samples reveal that the crust's overall composition and density are very different from those of the mantle, and that the boundary between them is distinct.

As we learned in our consideration of the rock cycle (Chapter 1), the composition of the crust is quite varied and the rocks that make up the crust are unevenly distributed. Those that are most common at the surface of the Earth, the sedimentary rocks, are not representative of the most common crustal rocks, the igneous rocks. Even among the igneous rocks, chemical compositions vary widely. If you were a geochemist, how might you go about estimating the overall composition of such a varied assortment of materials?

One early approach was to assume that the overall composition of the crust was essentially the same as the average composition of igneous rocks, given that igneous rocks were predominant in the crust (Fig. 1.18). But this did not produce a very refined estimate of crustal composition. Another approach, used in the 1920s by the Norwegian geochemist V. M. Goldschmidt, was to analyze the composition of fine glacial sediments deposited in lakes by vast ice sheets during the last glaciation. Goldschmidt reasoned that the moving glaciers had scoured fine rock powder from an extremely large area of igneous rock, blending the powder into a homogeneous mixture that was representative of the wide region of crustal material sampled by the glaciers.

Modern estimates of the composition of the crust usually take into account different tectonic environments, such as the deep ocean basins, young mountain chains, continental margins, and stable continental interiors. The relative proportions and average compositions of rock types in each of these regions are estimated, then weighted and added together according to the importance of each environment in the crust as a whole. Table 2.1 gives a modern estimate of the composition of the crust of the Earth.

T A B L E 2.1 • The Most Abundant Chemical Elements in the Continental Crust

Element	*Percentage by Weight*
Oxygen (O)	45.20
Silicon (Si)	27.20
Aluminum (Al)	8.00
Iron (Fe)	5.80
Calcium (Ca)	5.06
Magnesium (Mg)	2.77
Sodium (Na)	2.32
Potassium (K)	1.68
Titanium (Ti)	0.86
Hydrogen (H)	0.14
Manganese (Mn)	0.10
Phosphorus (P)	0.10
All other elements	0.77
Total	100.00

Compared to the Earth as a whole, the rocks that make up the crust of the Earth are rich in relatively light elements (Fig. 2.4). As shown in Table 2.1 and Figs. 2.4 and 2.5, the crust is overwhelmingly dominated by oxygen—over 45 percent by weight, over 60 percent by atomic proportions and over 90 percent by volume. The dominance of oxygen in the crust prompted Goldschmidt to comment that the crust of the Earth should be called the "oxysphere." Other relatively light elements, notably silicon and aluminum, make up the remainder of the material in the crust.

It may seem odd that so much oxygen—an element that we normally associate with the atmosphere—should be

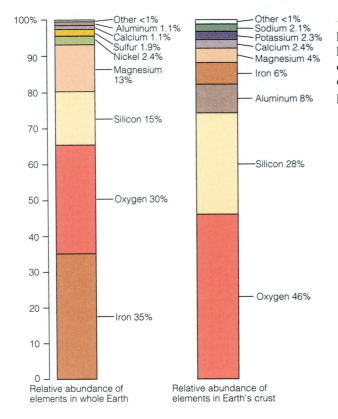

◀ **F I G U R E 2.4**

Relative abundance by weight of elements in the whole Earth and in the Earth's crust. Differentiation has created a light crust depleted in iron and magnesium and enriched in oxygen, silicon, aluminum, calcium, and potassium.

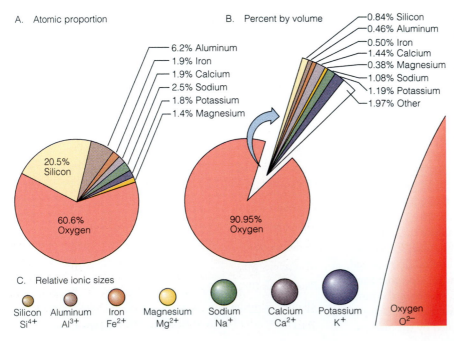

◀ **F I G U R E 2.5**

The abundances of elements in the Earth's crust. A. Abundances by atomic proportions. B. Percentages of elements by volume; oxygen makes up most of the volume because it has a large atomic radius. C. Relative ionic sizes of the most abundant chemical elements.

found in the solid crust of the Earth. But the oxygen in the crust differs significantly from that in the atmosphere in that crustal oxygen is very tightly linked or *bonded* to other elements in the form of naturally occurring chemical compounds called *minerals.*

MINERALS

The word **mineral** has a specific connotation in geology; it refers to any naturally formed, solid, inorganic chemical substance having a specific composition and a characteristic crystal structure. Minerals are chemical compounds, and as such they are made of chemical *elements*. Let's briefly review these before going on to discuss minerals and their properties in more detail.

Chemical Elements and Compounds

If you were a chemist and were asked to analyze a sample of mineral or rock, your report would list the kinds and amounts of chemical elements present in the sample. An **element** is the most fundamental substance into which matter can be separated by chemical means. For example, table salt (the chemical compound sodium chloride) is not an element because it can be separated into sodium and chlorine. But neither sodium nor chlorine can be further broken down chemically; thus, each is an element.

Every element is identified by a symbol, such as H for hydrogen and Si for silicon. Some symbols, such as that for hydrogen, come from the element's name in English. Other symbols come from other languages. For example, the symbol for iron is Fe, from the Latin *ferrum;* the symbol for copper is Cu, from the Latin *cuprum,* which in turn comes from the Greek *kyprios;* and the symbol for sodium is Na, from the Latin *natrium.* The 92 naturally occurring elements and their symbols are listed in Appendix B. The fundamental building blocks of the elements are discussed in further detail in Box 2.1.

The binding force between elements in a chemical compound is called *bonding.* The properties of compounds are quite different from those of their constituent elements. For example, the elements sodium (Na) and chlorine (Cl) are highly toxic, but the compound sodium chloride (NaCl, the mineral *halite* or table salt) is essential for human health. The character and strength of the bonds in a mineral help determine the specific physical and chemical properties of the mineral, as we will see shortly.

Definition of a Mineral

To be classified as a *mineral,* a substance generally must meet five requirements:

1. It must be *naturally formed.*
2. It must be a *solid.*
3. It must be *inorganic.*
4. It must have a *specific chemical composition.*
5. It must have a *characteristic crystal structure.*

Let's take a closer look at each of these requirements and its implications.

Naturally formed. The requirement that minerals be naturally formed excludes the vast numbers of substances produced in laboratories, such as, for example, synthetic ruby, steel, glass, or plastic. Technically, none of these substances is a mineral.

Solid. All liquids and gases—including naturally occurring ones such as oil and natural gas—are also excluded because minerals are solids. This requirement is based on the physical state of the material, not on its composition. Thus, for example, ice in a glacier is a mineral but water in the ocean and water vapor in the atmosphere are not, even though all three compounds have the same chemical formula (H_2O).

Inorganic. Materials that are derived from living organisms and contain organic compounds, such as leaves, are not minerals. This means that coal, for example, is not a mineral because it is derived from the remains of plant material and contains organic compounds. The teeth and bony parts of dead animals and the shells of sea creatures, which are preserved as fossils, present a slightly trickier case; in principle, such materials are not minerals because they are formed by organic processes. However, over the many millennia during which fossilization occurs, the organic remains are usually replaced by minerals, a process called *mineralization.* If you were to perform a chemical analysis of a dinosaur bone, for example, you would most likely find few or no organic compounds, only inorganic minerals, even though the fine internal structures of the bone may have been preserved by the mineralization process.

Specific chemical composition. The requirement that a mineral must have a specific chemical composition has several implications. Quartz, for example, has the chemical formula SiO_2. To a very limited extent, elements may be added to or substituted for the silicon and oxygen in quartz; however, if the chemical mixture strays too far from this simple formula, it will no longer be quartz. It won't look like quartz; its internal atomic structure will have changed; and it will no longer have the physical properties whereby we identify a mineral as "quartz."

However, the fact that the chemical compositions of minerals have specific limits doesn't necessarily mean that their chemical formulas are all simple ones like the formula for quartz. For example, the chemical formula of the mineral phlogopite, a common form of mica, is $K_2Mg_6Si_6Al_2O_{20}(OH)_4$, and many other minerals have even more complicated formulas. Nevertheless, the same thing is true for phlogopite as for quartz: if the chemical compound strays too far from this specific formula, the material will no longer have the same characteristics and will no longer be identifiable as phlogopite.

ATOMS AND IONS

A piece of a pure element, even a tiny piece no bigger than the head of a pin, consists of a vast number of identical particles called *atoms*. An **atom** is the smallest individual particle that retains all the properties of a given chemical element. Atoms are so tiny that they can be seen only with the most powerful microscopes ever invented, and even then the image is imperfect because individual atoms are only about 10^{-10} m in diameter. As you probably recall from high school chemistry, atoms are built up from *protons* (which have positive electrical charges), *neutrons* (which are electrically neutral), and *electrons* (which have negative electrical charges) (Fig. B1.1). The number of protons in an atom is what gives it its special physical characteristics, which, in turn, identify it as a specific element.

An atom that has an excess positive or negative electrical charge caused by the loss or addition of an electron is called an **ion**. When the charge is positive (meaning that the atom gives up electrons), the ion is called a *cation;* when the charge is negative (meaning that the atom adds electrons), it is an *anion.* The convenient way to indicate ionic charges is to record them as superscripts. For example, Li^+ is a cation (lithium) that has given up an electron, while F^- is an anion (fluorine) that has accepted an electron.

Chemical compounds form when one or more anions combine with one or more cations in a specific ratio. For example, lithium and fluoride combine to form the compound lithium fluoride, which is written LiF to indicate that for every Li atom there is a counterbalancing F ion. Similarly, two cations of H^+ combine with one anion of O^{2-} to make the compound H_2O. The smallest unit that retains all the properties of a compound is called a **molecule.**

◄ F I G U R E B1.1
Schematic diagram of an atom of carbon. At the center is the nucleus, containing protons and neutrons (six of each in the case of carbon). Six electrons orbit the nucleus.

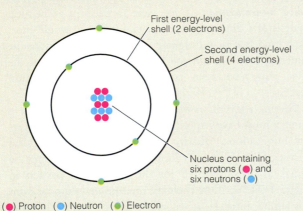

First energy-level
shell (2 electrons)

Second energy-level
shell (4 electrons)

Nucleus containing
six protons (●) and
six neutrons (●)

(●) Proton (●) Neutron (●) Electron

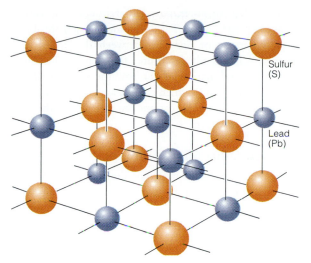

▲ **F I G U R E 2.6**
Arrangement of atoms in galena (PbS), the most common mineral containing lead. The orderly packing arrangement is repeated continuously throughout the crystal, and the atoms are so small that a cube of galena 1 cm on its edge contains 10^{22} atoms each of lead and sulfur.

The requirement of specific limits on the chemical compositions of minerals also serves to exclude materials, like glass, that vary in composition within a range that cannot be expressed by an exact chemical formula.

Characteristic crystal structure. Glass—even when it is naturally formed volcanic glass—is further excluded from being a mineral by the requirement that minerals must have characteristic crystal structures. The atoms in gases, liquids, and glasses are randomly jumbled, but the atoms in many solids are organized in regular, repetitious geometric patterns, as shown in Fig. 2.6. The geometric pattern that atoms assume in a solid is called the *crystal structure,* and solids that have a crystal structure are said to be **crystalline.** Solids that lack crystal structures are called *amorphous* solids (from the Greek for "without form") and are not minerals. All minerals are crystalline, and the crystal structure of a mineral is a unique property of that mineral. All specimens of a given mineral have an identical crystal structure. Extremely powerful, ultra-high-resolution microscopes enable scientists to look into the crystal structures of minerals and actually see the orderly arrangement of atoms in the mineral, which resembles that of eggs lined up in a carton (Fig. 2.7).

Common Minerals

Scientists have identified approximately 3000 minerals. Most occur in the Earth's crust, but a few have been identified in meteorites, and two new ones were discovered in the rocks brought back from the Moon by astronauts. The total number of minerals may seem large, but it is tiny when compared with the astronomically large number of ways in which a chemist can combine naturally occurring elements to form compounds. The reason for the disparity becomes apparent when we consider the relative abundance of the chemical elements. As Table 2.1 shows, only 12 elements—oxygen, silicon, aluminum, iron, calcium, magnesium, sodium, potassium, titanium, hydrogen, manganese, and phosphorus—occur in the crust in amounts greater than 0.1 percent. Together these 12 make up 99.23 percent of the crust's mass. The crust is constructed, therefore, of a limited number of minerals in which one or more of the 12 abundant elements is an essential ingredient. Minerals containing the scarce elements certainly do occur, but only in small amounts and only under special and restricted circumstances.

As you have already learned, two elements—oxygen and silicon—make up more than 70 percent of the crust by weight. Oxygen forms a simple anion, O^{2-} (see Box 2.1); compounds that contain the O^{2-} anion are called *oxides.* Oxygen and silicon together form an exceedingly strong grouping called a silicate anion $(SiO_4)^{4-}$. Minerals that contain the silicate anion are called *silicates,* or **silicate minerals.** These are the most abundant of all naturally occurring inorganic compounds; oxides are the second most abundant group. There are other mineral groups based on different anions—for example, carbonates $(CO_3)^{2-}$ and phosphates $(PO_4)^{3-}$—but they are much less common than the silicates and oxides.

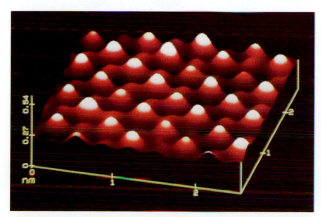

▲ **F I G U R E 2.7**
Atoms can be seen through special kinds of microscopes. Sulfur atoms (large) and lead atoms (small) at the surface of a galena (PbS) crystal are revealed with a scanning-tunneling microscope.

Silicate Minerals

The four oxygen atoms in a silicate anion are tightly bonded to the single silicon cation. As shown in Fig. 2.8A, the four oxygens sit at the corners of a polyhedral shape called a *tetrahedron,* and the small silicon cation sits in the space between the oxygens at the center of the tetrahedron. The structures and properties of silicate minerals are all determined by the way the $(SiO_4)^{4-}$ silicate tetrahedra are packed together in the crystal structure.

Two silica tetrahedra can link together by sharing an oxygen atom at one apex (Fig. 2.8B). In this way, the tetrahedra link to form double tetrahedra, rings, chains, sheets, or three-dimensional frameworks (Fig. 2.9). The linking process is called *polymerization.* Different common cations (such as Ca^{2+}, Al^{3+}, Mg^{2+}, Fe^{2+}, Na^+, and others) are able to fit into the spaces or *interstices* between the linked silica tetrahedra. What finally determines the identity of a silicate mineral is a combination of (1) how the silica tetrahedra are linked together; (2) which cations are present in the interstices; and (3) how the cations are distributed throughout the crystal structure.

Oxides and Other Mineral Groups

As just noted, the oxides represent the second most abundant group of minerals in the crust. Oxides are compounds based on the O^{2-} anion, which is bonded in various

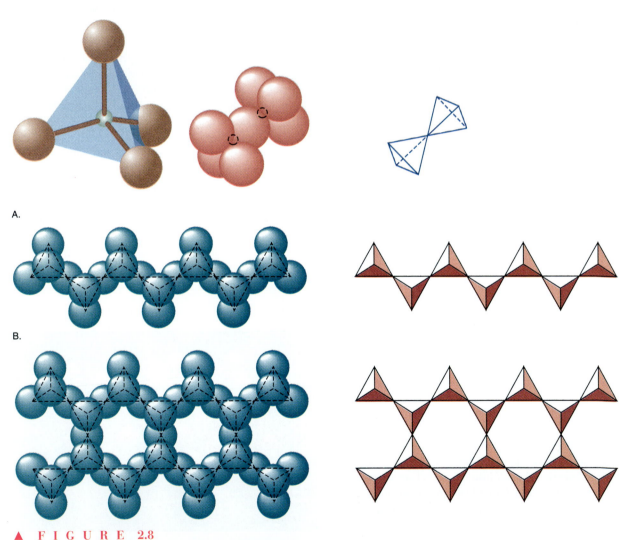

▲ **F I G U R E 2.8**
The silica tetrahedron and silicate mineral structures. A. The structure of the silica tetrahedron. The large oxygen anions are at the four corners, equidistant from a small silicon cation at the center. B. Silica tetrahedra can link together in pairs, chains, sheets, and other geometric structures. The properties of silicate minerals are determined, to a great extent, by the types of linkages between the silica tetrahedra.

Arrangement of silica tetrahedra		Formula of the complex ions	Typical mineral	
			Name	Composition
Isolated tetrahedra		$(SiO_4)^{4-}$	Olivine	$(Mg, Fe)_2SiO_4$
Isolated polymerized groups		$(Si_2O_7)^{6-}$	Epidote	$Ca_2Fe_2Al_2O[Si_2O_7][SiO_4](OH)$
		$(Si_6O_{18})^{12-}$	Beryl	$Be_3Al_2Si_6O_{18}$
Continuous chains		$(SiO_3)_n^{2-}$	Pyroxene	$CaMg(SiO_3)_2$ (Variety; diopside)
		$(Si_4O_{11})_n^{6-}$	Amphibole	$Ca_2Mg_5(Si_4O_{11})_2(OH)_2$ (Variety; tremolite)
Continuous sheets		$(Si_4O_{10})_n^{4-}$	Mica	$KAl_2(Si_3Al)O_{10}(OH)_2$ (Variety; muscovite)
Three-dimensional networks	Too complex to be shown by a simple two-dimensional drawing	(SiO_2)	Quartz	SiO_2

▲ **F I G U R E 2.9**
Summary of the ways silica tetrahedra link together to form the common silicate minerals. Linkages other than those shown are known but do not occur in common minerals.

arrangements with the common cations. Because iron is one of the most abundant elements in the crust, the iron oxides magnetite (Fe_3O_4) and hematite (Fe_2O_3) are the two most common oxide minerals. Magnetite takes its name from the Greek word *Magnetis,* meaning "stone of Magnesia," an ancient town in Asia Minor. Magnetis had the power to attract iron particles, and hence is the source of our words *magnet* and *magnetite.* The word hematite is derived from the red color of the mineral when powdered, the Greek word for red blood being *haima.* Magnetite and hematite are the main minerals from which the metal iron is derived. Other important oxides are uraninite (U_3O_8), the main source of uranium; cassiterite (SnO_2), from which tin is derived; and rutile (TiO_2), a widely used constituent of paints.

Another important mineral group is the *sulfides,* all of which contain the simple anion S^{2-} in combination with metal cations. Sulfide minerals are typically dense and have a metallic appearance. The two most common are pyrite (FeS_2) and pyrrhotite (FeS). Many of the sulfides are ore minerals (Fig. 2.10), which means that they are sought and processed for their valuable metal content. For example, most of the world's lead is won from galena (PbS), most of the zinc from sphalerite (ZnS), and most of the copper from chalcopyrite ($CuFeS_2$).

Like the silicates, oxides, and sulfides, each of the other mineral groups is based on combinations of cations with a particular anionic complex. The *carbonates* (Fig. 2.11)

contain the CO_3^{2-} anion, sulfates contain the SO_4^{2-} anion, and *phosphates* are based on the PO_4^{3-} anion. Each of these groups includes minerals that are important for one reason or another. For example, the phosphate mineral apatite $[Ca_5(PO_4)_3(F,OH)]$ is the substance from which bones and teeth are made. Gypsum ($CaSO_4 \cdot 2H_2O$), a sulfate, is the raw material from which plaster is made. And the carbonate mineral calcite ($CaCO_3$), from which organisms such as molluscs construct their shells, is the principal constituent of marble.

Finally, some materials occur in nature as *native elements,* that is, not in combination with other elements. Minerals that occur in this form include some metals, such as gold (Au) and silver (Ag), and some nonmetals, such as sulfur (S) and graphite (C).

Minerals as Indicators of the Environment of Their Formation

Finally, it is worth noting that minerals are not merely objects of beauty or economically valuable resources. Their makeup offers clues to the conditions under which they are formed. The study of minerals, therefore, can provide insight into the chemical and physical conditions in regions of the Earth that we cannot observe and measure directly.

Our understanding of the conditions in which minerals are formed has come largely from laboratory studies. For example, scientists have been able to determine the temper-

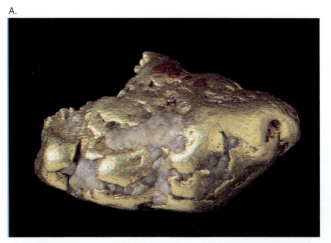

▲ **F I G U R E 2.10**
Examples of ore minerals. A. A nugget of metallic gold, from White Flats, Yukon River, Alaska. The nugget is 3 cm in length. B. A mixture of galena (gray) and chalcopyrite (yellow) from a richly mineralized vein in Peru. The specimen is 5 cm across. Galena (PbS) is the main ore mineral of lead, chalcopyrite (CuFeS₂) is the main ore mineral of copper. The two minerals must be separated into pure concentrates prior to treatment for recovery of the contained metals.

◄ F I G U R E 2.11

The two most common carbonate minerals, calcite ($CaCO_3$) on the left, dolomite ($CaMg(CO_3)_2$) on the right.

atures and pressures at which diamond forms instead of graphite. Because we know that diamonds are stable only at very high temperatures and pressures, and we understand how temperature and pressure increase at great depths within the Earth, we can state with certainty that rocks in which diamonds are found were formed in the mantle at least 150 km below the Earth's surface.

But studies of minerals are not limited to laboratory experiments. Field observations are also valuable. For example, the minerals that form in the regolith during weathering are controlled by the climate. Past climates can therefore be deciphered from the kinds of minerals preserved in sedimentary rocks. The composition of seawater in past ages can also be determined from the minerals formed when the seawater evaporated and deposited its salts.

ROCKS

In Chapter 1 we introduced the concept of the rock cycle and the three families of rocks: (1) *igneous rocks,* which have crystallized from molten rock, either deep underground or at the surface of the Earth; (2) *sedimentary rocks,* which have formed under low-temperature conditions near the Earth's surface, through chemical precipitation or cementation of sediment; and (3) *metamorphic rocks,* previously existing rocks that have been changed by the action of heat and pressure. But we did not present a precise definition of the term *rock.* Now that you have learned more about minerals—the basic components of rocks—we can be more specific.

A **rock** is a naturally formed, nonliving, firm, and coherent aggregate mass of solid matter that constitutes part of a planet. A pile of loose sand grains is not a rock because the grains are not locked together—that is, they are not coherent. A tree is not a rock, even though it is solid, because it is living. But coal, which is a compressed and coherent aggregate of twigs, leaves, and other bits of dead plant matter, is a rock.

To a nongeologist, the distinction between a rock and a mineral may not seem very clear, but it becomes clearer if one remembers that rocks are *aggregates.* This means that rocks are composed of lots of grains stuck together. The grains may be coarse or fine; they may all be the same type of mineral grain, or they may represent a variety of minerals; they may be tightly or loosely stuck together. All of these possible differences contribute to the description of different types of rock.

Describing Rocks

At first glance rocks seem confusingly varied. Some are either distinctly layered and have pronounced, flat surfaces covered with mica. Others are coarse and evenly grained and are not layered, yet they may contain the same kinds of minerals. Studying a large number of rock specimens soon makes it clear that no matter what kind of rock is being examined—sedimentary, metamorphic, or igneous—the differences between samples can be described in terms of two kinds of features: *mineral assemblage* and *texture.*

Mineral Assemblage

The first obvious feature of a rock is the kinds of minerals that are present. A few kinds of rock contain only one type of mineral, but most contain two or more types. The variety and the relative abundance of minerals present in rock, commonly called the *mineral assemblage* of the rock, are important pieces of information for interpreting how a rock was formed.

Texture

The second obvious feature is the rock's *texture*—the overall appearance produced by the size, shape, and arrangement of its constituent mineral grains. For example, the mineral grains may be flat and parallel to each other, giving the rock a platy or flaky texture like that of a deck of cards (Fig. 2.12A). In addition, the various minerals may be unevenly distributed and concentrated into specific layers, producing a texture that is both layered and platy. Alternatively, the

BOX 2.2
•
FOCUS ON...

PROPERTIES OF MINERALS

*T*he properties of minerals are determined by their composition and crystal structure. Once we know which properties are characteristic of particular minerals, we can use them to identify those minerals. It is not necessary, therefore, to analyze a mineral chemically or to determine its crystal structure to discover its identity. The properties most often used to identify minerals are obvious ones, such as color, external shape (the crystal form or *habit*), and hardness. Less obvious properties, such as *luster* (the quality and intensity of light reflected from the mineral), *cleavage* (the tendency of the mineral to break in preferred directions), and *density* (i.e., the "heaviness" of the mineral), are also used in identifying minerals.

The fibrous form of asbestos minerals results in properties that are both useful and hazardous for humans, as we saw in the opening paragraphs of this chapter. Here are a few more examples of the ways in which the composition and arrangement of atoms can dictate the physical properties of minerals. The mineral quartz (SiO_2) is normally colorless, with a glassy-looking luster (called *vitreous*) and elongate, six-sided crystals (Fig. B2.1A). When a tiny amount of manganese (as little as 0.01 percent) is introduced into the crystal structure of quartz, it takes on a deep purple color and we refer to it as amethyst (Fig. B2.1B). If the quartz occurs in minute grains and is associated with iron (Fe^{3+}), it looks completely different and we know it as chalcedony (Fig. B2.1C). Similarly, the mineral corundum (Al_2O_3) is normally white or grayish. When small amounts of Cr^{3+} are introduced, the corundum takes on a blood-red color and we call it ruby; if Fe and Ti are

A.

B.

C.

▲ **FIGURE B2.1**
The variable habits of quartz (SiO_2). A. Clear, six-sided, glassy-looking crystals of quartz. Individual crystals in this sample are about 5 cm in length. B. Beautiful purple-colored quartz crystals (variety *amethyst*). The color is caused by trace amounts of manganese. C. Colored bands of quartz comprised of grains so small they can only be seen with powerful microscopes. The colors are due to trace amounts of chemical elements such as iron and manganese. Banded, fine-grained quartz is called *chalcedony*.

present, the corundum becomes deep blue, and another prized gem, sapphire, is the result.

The hardness of a mineral and the way the mineral breaks are determined primarily by the way the atoms are linked together and the strength of the bonds. For example, diamond (Fig. B2.2A) has the same chemical composition as graphite (pure carbon), but in a diamond the carbon atoms are linked together by an extremely strong type of bond, a covalent bond. In graphite (Fig. B2.2B), the carbon atoms are tightly bonded together in flat sheets, but the Van der Waals bonding between the sheets is very weak. The result is a series of very strong, flexible sheets that separate easily. Graphite (used in pencil leads) feels slippery when it is rubbed between the fingers because the rubbing breaks the weak bonds and the sheets slide past each other. Other minerals with a sheetlike structure include talc, the mineral in "talcum" powder, micas, and a variety of clay minerals.

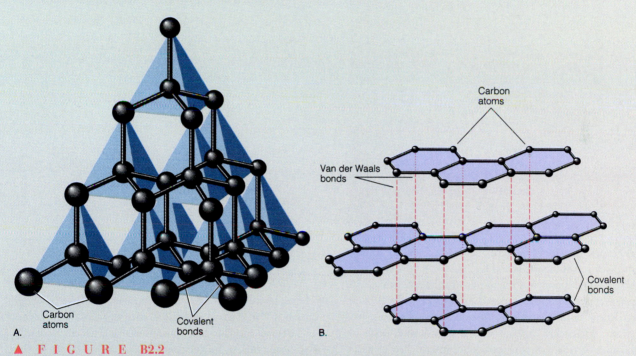

▲ F I G U R E B2.2

A. The three-dimensional geometric arrangement of carbon atoms in diamond. Note that each atom is surrounded by four others. B. The geometric arrangement of carbon atoms in graphite. Bonding within sheets is strong, but binding between sheets it is weak.

A.

B.

◀ **F I G U R E 2.12**

Examples of rock textures. A. A distinct layered texture produced by parallel, flattened mineral grains is characteristic of many metamorphic rocks. This sample, called a *gneiss,* consists of the minerals feldspar (pink), quartz (gray), and the mica (black). The sample is 6 cm in width. B. Oolitic limestone, a sedimentary rock consisting of small, rounded grains of calcite ($CaCO_3$) and broken shell fragments. All of the grains are about the same size (0.3 mm in diameter). The texture is homogeneous.

rock's texture may be homogeneous, made up of grains that are more or less the same size (Fig. 2.12B).

The mineral grains in some kinds of rock are firmly held together, whereas in other kinds the grains are easily broken apart. Rocks with a small proportion of empty space between the grains are igneous and metamorphic; both of these types of rock contain intricately interlocked mineral grains. During the formation of rocks of these types, the growing mineral grains crowd against each other, filling all spaces and forming an intricate three-dimensional jigsaw puzzle (Fig. 2.13A). A similar interlocking of grains holds together steel, ceramics, and bricks.

The forces that hold together the grains of sedimentary rocks are less obvious (Fig. 2.13B). Sediment is a loose aggregate of particles. It must be transformed into sedimentary rock in either of two ways. In one, water circulates slowly through the open spaces between sediment grains and deposits new materials that cement the grains together. In the other, sediment becomes deeply buried and the temperature rises, causing mineral grains to recrystallize; the growing grains interlock and form strong aggregates. The latter process is the same as that which occurs when ice crystals in a pile of snow recrystallize to form a compact mass of ice.

EARTH MATERIALS IN THE ENVIRONMENT

The characteristics of rocks and minerals that we have discussed in this chapter ultimately determine their properties and behavior in different settings in the crust of the Earth. For example, a rock composed primarily of olivine (Mg_2SiO_4), a mineral that is stable in rocks found in the mantle, will weather and eventually crumble into sediment much more readily than a rock composed of quartz, which is stable in rocks found in the crust. A rock in which the grains are loosely held together with lots of space between them is more likely to allow the passage of fluids than a rock with tightly interlocking grains. A rock that is strongly layered is more likely to split along the layering, which may lead to movement or, eventually, landsliding along the split.

The chemical and physical properties of a rock or a mineral also determine whether or not it is of interest as an economic resource. Does the rock contain gold? Is the gold concentrated enough, and in a form that can be extracted using current technologies? Is the rock packed loosely enough to permit oil to be stored in the spaces? Or water?

In addition to rocks and minerals, we discussed a variety of other Earth materials in the context of the rock cycle

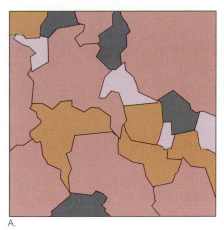 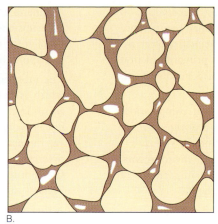

▲ **F I G U R E   2.13**
A. The close interlocking of grains in an igneous or metamorphic rock leads to a rock with few open spaces. Each color represents a different mineral species. B. Sediments and sedimentary rocks may have a significant amount of open space between grains. Some of the spaces may become filled through the processes of cementation and recrystallization. Sedimentary grains are shown in yellow, cement in brown.

(Chapter 1), including magma, lava, and sediment. Each of these materials has properties and characteristics with potential impacts on the environment or on other conditions that concern humans. For example, certain types of magma don't flow easily, so they tend to erupt explosively. Some sediments amplify seismic waves, leading to increased damage due to large earthquakes. Some sediments crack when they dry out, permitting radioactive gas to leak out. Some clay minerals chemically attract ions and therefore are useful for stopping leaks of toxic or nuclear waste. Other clay minerals chemically attract water, which may make them unstable as supporting foundations for buildings. And so on; each of these examples and many more will come up at appropriate places in later chapters. Instead of continuing with an inventory of the many ways in which the characteristics of Earth materials can affect our lives, let's go on to the next part of the book, in which we look at hazardous Earth processes in detail.

SUMMARY

1. Environmental geologists study the physical structure and chemical composition of the Earth because: (1) The internal structure of the Earth has a lot to do with shaping our landscape and causing events that may be hazardous for humans. (2) Humans depend on the materials of the solid Earth as resources. (3) The materials of the Earth have distinct physical and chemical properties that can affect human lives in many different ways.

2. The Earth is a differentiated body, consisting of three major compositional layers: the core, the mantle, and the crust.

3. In addition to being layered compositionally, the Earth has layers of differing rock properties. Most important are the layers of differing rock strength, which are determined primarily by the interplay between temperature and pressure at depth within the Earth. The Earth has a solid inner core; a liquid outer core; an intermediate zone called the mesosphere; a zone called the asthenosphere (from about 350 to 100 km below the Earth's surface), in which rocks have very little strength; and a stronger, more rigid outer layer, the lithosphere.

4. The core and mantle are not directly accessible, so their compositions have been determined by inference from indirect evidence such as the velocities of earthquake waves. The core is composed primarily of iron and nickel, whereas the mantle consists mainly of silicon, oxygen, magnesium, and iron.

5. The Earth's crust is highly variable in composition. Modern estimates of the composition of the crust are based on determinations of the abundances of different rock types in various tectonic environments, weighted with respect to the relative importance of each environment in the crust as a whole.

6. As a result of differentiation, in which the lightest elements rose to the top and the densest elements sank to the core, the Earth's crust is rich in light elements (such as silicon, aluminum, sodium, and potassium) compared to the Earth as a whole. The crust is particularly rich in oxygen (90 percent by volume); however, the oxygen in the crust differs significantly from that in the atmosphere in that it is tightly bonded to other elements.

7. A mineral is a naturally formed, solid, inorganic chemical substance having a specific composition and a characteristic crystal structure. Scientists have identified approximately 3000 minerals.

8. The most common mineral type is the silicate group, which is based on a grouping of silicon and oxygen atoms in the form of a tetrahedron. The silica tetrahedra link together in different ways to form the different minerals of the silicate group. The type of linkage determines the physical properties of the minerals. Other important mineral groups are the oxides, carbonates, sulfides, phosphates, and sulfates.

9. A rock is a naturally formed, nonliving, firm, and coherent aggregate mass of solid matter that constitutes part of a planet. The differences among rocks can be described in terms of their differing mineral assemblages and textures.

10. The chemical and physical characteristics of rocks and minerals determine their properties and, thus, their behavior in different settings in the crust of the Earth and their potential as economic resources.

IMPORTANT TERMS TO REMEMBER

atom (p. 48)
core (p. 42)
crust (p. 42)
crystalline (p. 49)

element (p. 47)
ion (p. 48)
mantle (p. 42)
mineral (p. 47)

molecule (p. 48)
rock (p. 53)
silicate minerals (p. 49)

QUESTIONS AND ACTIVITIES

1. Find out more about some of the minerals you encounter in your daily life. You could choose to investigate hazardous minerals, such as asbestos; or ore minerals, such as galena; or gems, such as emerald; or industrial minerals, such as diamond or garnet. Find out about the chemical and physical properties of the mineral that make it useful (or hazardous, or beautiful, or valuable).

2. Go on a walking tour of building stones in your town. Notice the types of rock that have been used to construct floors, countertops, and building facades. Office buildings, banks, and government buildings—old and new—are usually good places to start looking. Notice the different ways in which building stones respond to environmental forces, such as wind, water, or acid rain. Find out where the building stones used in your area have come from—are they from local quarries or imported from far away? Headstones in cemeteries can also provide interesting clues about rocks that have been used commercially in your area.

3. Do some research to find out more about how scientists use indirect techniques to study the interior of the Earth. Have some of these techniques also been applied to the study of other planetary interiors?

4. Grow your own crystals. Stir into a glass of warm water as much sugar as you can without allowing the sugar to collect on the bottom of the glass. Tie a string to a pencil and lay the pencil across the top of the glass, with the string hanging down into the water. Let the glass of water cool for a few days in a quiet place and then check the string for crystals. What is the shape of the crystals that formed on the string? Are these crystals actually minerals? Why or why not?

PART TWO

HAZARDOUS GEOLOGIC PROCESSES

"Civilization exists by geologic consent—subject to changes without notice."

• **Will Durant**

★ • E • S • S • A • Y • ★

ASSESSING GEOLOGIC HAZARDS AND RISKS

Geologic processes that we refer to as "hazardous" have existed throughout Earth history. Even the most destructive events are part of the normal functioning of this dynamic planet; to a great extent, they make the planet habitable. Earthquakes and volcanic eruptions, for example, are among the processes that have formed the continents, shaped the landscape, determined the positions of climatic zones, and allowed for the creation and stabilization of the atmosphere and oceans. The action of wind and water causes flooding, landslides, and windstorms but also replenishes soil and sustains life.

In other words, geologic processes affect our lives every day in ways that are both subtle and conspicuous, beneficial and harmful. Knowledge about those processes and the hazards associated with them should play an integral role in the planning of human activities. But before we address some of the challenges inherent in assessing geologic hazards and risks, we need to clarify the meanings of these and other terms.

TYPES OF HAZARDS

The terms *natural hazard* and *geologic hazard* denote a wide range of geologic circumstances, materials, processes, and events. **Geologic hazards** include earthquakes, volcanic eruptions, floods, landslides, and other processes and occurrences. They are included in the broader concept of **natural hazards,** which encompasses processes or events such

---

◀

Collapse of the Antelope Valley Freeway, California, as a result of the Northridge earthquake on January 18, 1994.

as locust infestations, wildfires, and tornadoes in addition to strictly geologic hazards.

Some natural hazards are catastrophic events—occurrences that strike quickly but with devastating consequences. For example, there is a calculable, though small, risk that a large comet or meteorite might strike the Earth, with potentially disastrous effects. Events that strike quickly and with little warning, such as earthquakes, flash floods, or sudden windstorms, are called **rapid onset hazards.** Other hazardous processes operate more slowly. Droughts, for instance, can last 10 years or more. The socioeconomic impacts of extended droughts are caused by the cumulative effects of season after season of below-average rainfall.

In general, natural processes are labeled "hazardous" only when they present a threat to human life, health, or interests, either directly or indirectly. In other words, we tend to take an anthropocentric approach to the study and management of natural and geologic hazards. This is human nature, of course; we are justifiably concerned with the protection of human life and property. But this approach has important implications because it can lead to an adversarial style of hazard management in which geologic processes are cast as the "enemy" and efforts are made to manipulate the environment into submission. A somewhat different approach, which is currently receiving a lot of attention, focuses on improving scientific understanding of natural processes and their triggering mechanisms in order to provide a foundation for better management.

A different category of hazard, sometimes referred to as **technological hazards,** is associated with everyday exposure to naturally occurring hazardous substances, such as radon, mercury, asbestos fibers, or coal dust, usually through some aspect of the use of these substances in our built environment. Still another type of hazard arises from pollution and degradation of the natural environment, which have led to problems such as acid rain, contamination of surface and underground water bodies, depletion of the ozone layer, and global climatic warming; we might refer to these as primarily **anthropogenic hazards,** or human-generated hazards.

Primary, Secondary, and Tertiary Effects

Geoscientists and other specialists often refer to hazardous events or processes as having primary, secondary, and tertiary effects. *Primary effects* result from the event itself: water damage resulting from a flood; wind damage caused by a cyclone; the collapse of a building as a result of ground motion during an earthquake. *Secondary effects* result from hazardous processes that are associated with, but not directly caused by, the main event. Examples include forest fires touched off by lava flows, house fires caused by gas lines breaking during an earthquake, and disruption of water and sewage services as a result of a flood. *Tertiary effects* and higher order effects are long term or even permanent. These might include the loss of wildlife habitat or permanent changes in a river channel as the result of flooding; regional or global climatic changes and resulting crop losses after a major volcanic eruption; or changes in topography or land elevation as a result of an earthquake. For example, the 1964 "Good Friday" earthquake near Anchorage, Alaska, caused permanent topographic changes along an extensive section of the Alaskan coastline.

VULNERABILITY AND SUSCEPTIBILITY

Exactly how vulnerable are we to the damaging effects of these and other natural hazards? During the past two decades as many as 3 million lives have been lost as a direct result of hazardous events, and at least 800 million people have suffered adverse effects such as loss of property or health. A United Nations committee has estimated that during the 1990s the Earth will experience tens of thousands of landslides and earthquakes; 1 million thunderstorms; 100,000 floods; and many thousands of tropical cyclones and hurricanes, tsunamis, droughts, and volcanic eruptions. According to researchers at the World Bank, natural disasters cause about US$40 billion each year in physical damage; windstorms, floods, and earthquakes alone cost about US$18.8 million *per day!*

The concept of **vulnerability** has a very specific meaning to hazard specialists. It encompasses not only the physical effects of a natural hazard but also the status of people and property in the area. A large number of factors can increase one's vulnerability to natural hazards, especially catastrophic events. Aside from the simple fact of living in a hazardous area, vulnerability depends on population density, scientific understanding of the area, public education and awareness of hazards, the existence of an early-warning

system and effective lines of communication, the availability and readiness of emergency personnel, construction styles and building codes, and cultural factors that influence public response to warnings.

Many of these factors help explain the fact that less developed countries are much more vulnerable to natural hazards than are industrialized countries. Whereas the actual dollar value of property damage from an event such as an earthquake or a flood may be higher in an industrialized country, the *relative* value of monetary losses is *much* greater, on average, in developing countries. There have even been occurrences in which single disasters have caused economic losses equal to a country's entire gross national product.

Poverty itself is a contributor to increased vulnerability. Inadequate housing and high population densities on sensitive lands contribute to much higher losses of life in natural disasters. Something as simple as an inoperative telephone line—an everyday reality in many countries—could cause an early-warning system to fail.

Human intervention in the functioning of natural systems can increase vulnerability in two ways: (1) through the development and habitation of lands that are sensitive or susceptible to hazards (e.g., floodplains or deltas), and (2) by increasing the severity or frequency of natural hazards (e.g., overintensive agriculture leading to increased erosion; mining of groundwater leading to subsidence; or global climatic change leading to increased intensity of tropical cyclones).

It is interesting to note that *either* poverty *or* affluence can cause these types of pressures on the environment. For example, extensive deforestation of lands bordering the Sahara Desert by poverty-stricken residents in desperate need of fuelwood has almost certainly accelerated the process of desertification, contributing to drought and famine in those areas. In contrast, in San Francisco economic pressures have led to the development and urbanization of large areas of land reclaimed from the Bay; these lands are particularly susceptible to failure through liquefaction (a quick-sand-like condition) in a major earthquake.

ASSESSING HAZARDS AND RISKS

In order to incorporate knowledge about natural processes into the planning of human activities, we have to assess the hazards and risks associated with them. Although the terms *hazard assessment* and *risk assessment* are often used interchangeably, they are not synonymous.

Hazard Assessment

Hazard assessment involves asking questions like, "How often can we expect a hazardous event to occur?" and "When such an event does occur, what effects is it likely to have?" Specifically, hazard assessment consists of the following activities:

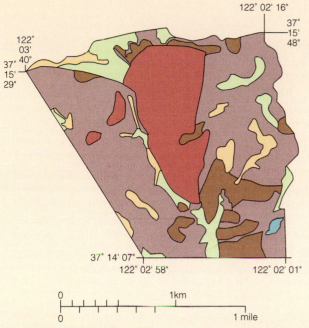

Relative stability	Map area	Geologic conditions	Recommended land use		
			Houses	Roads	
				Public	Private
Most stable		Flat to gentle slopes; subject to local shallow sliding, soil creep, and settlement	Yes	Yes	Yes
		Gentle to moderately steep slopes in older stabilized landslide debris; subject to settlement, soil creep, and shallow and deep landsliding	Yes	Yes	Yes
		Steep to very steep slopes; subject to mass-wasting by soil creep, slumping and rock fall	Yes	Yes	Yes
		Gentle to very steep slopes in unstable material subject to sliding, slumping and soil creep	No	No	No
		Moving shallow (>10 ft) landslide	No	No	No
Least stable		Moving, deep landslide, subject to rapid failure	No	No	No

▲ F I G U R E II.1

A landslide susceptibility map and recommended land use policies for the Congress Springs area near San Francisco. The results of hazard assessment are often portrayed in map format so they are understandable and useful for local planners and decision makers.

- Determining when and where hazardous events have occurred in the past.
- Determining the severity of the physical effects of past events of a given *magnitude* or size.
- Determining how frequently we can expect events that are strong enough to generate severe physical effects.
- Determining what a particular event would be like if it were to occur now, in terms of the type of effects it would have.
- Portraying all this information in a form that can be used by planners and decision makers.

The results of hazard assessment are (or should be) used by political leaders who have to make decisions concerning evacuation or contingency funding; by emergency personnel, who have to make decisions concerning levels of response and readiness; by planners and engineers, who have to make decisions concerning land use and zoning or building regulations; and by local scientific experts, who may be the first to sound an alert on the basis of such information. Often the results of hazard assessment are portrayed in the form of maps (as shown in Figs. II.1 and II.2) in order to make the information as accessible and understandable as possible.

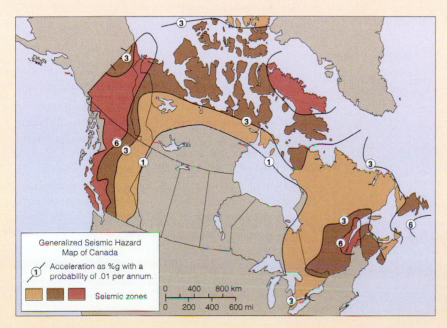

Generalized Seismic Hazard Map of Canada

① Acceleration as %g with a probability of .01 per annum

Seismic zones

◄ F I G U R E II.2

Earthquake probability map of Canada. The map clearly identifies those areas with the highest probability of exceeding a certain magnitude of ground motion during an earthquake.

Risk Assessment

Risk assessment differs from hazard assessment in some important respects. In the context of natural hazards, risk is a statement of the economic losses, injuries and deaths, and loss of functioning of urban support systems expected when a specific physical effect, such as ground shaking triggered by an earthquake, strikes a given area. Risk assessment starts by establishing the *probability* that a hazardous event of a particular magnitude will occur within a given period. It then goes on to take into account factors such as the following:

- The locations of buildings, facilities, and emergency systems in the community.
- Their potential exposure to the physical effects of the hazardous situation or event.
- The community's vulnerability—that is, potential loss of life, injury, or loss in value—when subjected to those physical effects (Fig. II.3).

Thus, risk assessment incorporates social and economic considerations in addition to the scientific factors involved in hazard assessment. Whereas hazard assessment focuses on characterizing the physical effects of a particular event, risk assessment typically focuses on the extent of the damage anticipated, control or mitigation of the damage, and actions that might reduce vulnerability. Risk assessment involves asking such questions as, "When a situation of this type exists or an event of this type occurs, what kinds of

ELEMENTS OF RISK

▼ F I G U R E II.3

The elements of risk. Risk assessment involves establishing the probability and characterizing the physical effects of natural hazards as well as incorporating societal and economic considerations such as location, exposure to effects, and vulnerability.

damages can the community expect?" and "Even though the likelihood of an event of this magnitude occurring is small, would the consequences be unacceptably severe?"

Risk is sometimes stated in terms of probabilities. For example, it has been estimated that smoking 1.4 cigarettes, drinking 0.5 liter of wine, having a single chest x-ray, or being exposed to earthquake hazards by living in Southern California for seven months all carry the same statistical risk: They all increase the chance of death by approximately 1 in a million. Alternatively, risk can be stated in terms of the cost (damages and injuries expressed in terms of dollar value lost) if a hazardous event were to occur. In either case, a risk assessment can help both decision makers and scientists compare and evaluate hazards, set priorities, and decide where to focus attention and resources.

PREDICTION AND WARNING

Predicting hazardous events and advising decision makers about their expected severity are other important aspects of dealing with geologic hazards. A **prediction** is a statement of probability based on scientific observation. Accurate prediction requires the continuous monitoring of geologic processes. Monitoring usually focuses on identifying anomalies that might be **precursors**—that is, small physical changes leading up to a catastrophic event. When observations of precursor phenomena have accumulated to the point at which they signal the imminent occurrence of a hazardous event, a prediction can be made. For example, before Mt. Pinatubo in the Philippines erupted violently in 1991, scientists observed a variety of precursor events, of which the most important were increases in the number and intensity of earthquakes in the area and changes in their locations, as well as changes in the quantity and composition of gases emitted by the volcano. Careful monitoring of trends in precursor phenomena allowed scientists to predict the time of eruption with great accuracy, saving many thousands of lives.

Forecasting

Sometimes the term **forecast** is used synonymously with *prediction;* in other contexts, it is used quite differently. For example, in the prediction of floods or hurricanes, forecasting generally refers to short-term prediction of the specific magnitude and time of occurrence of an event (days or hours ahead of time, rather than months or years). In the prediction of earthquakes, on the other hand, the term *forecast* is generally used to refer to a long-term, nonspecific statement of probability. Thus, before the Loma Prieta earthquake of October 17, 1989 (the "World Series earthquake"), the U.S. Geological Survey had issued a forecast indicating a 50 percent probability of a large earthquake occurring along the San Andreas fault in the region of Santa Cruz within 30 years. This was a relatively nonspecific long-range forecast based on general scientific understanding of seismicity and the geology of the area, rather than on the observation of specific precursor phenomena. (The threat of a major earthquake in the San Francisco–Santa Cruz area is still very real; in fact, after the Loma Prieta earthquake the USGS issued a report forecasting a 67 percent probability of a major earthquake occurring in the area within the next 30 years.)

Early Warning

The final step in preparing a community to deal with a hazardous event is the issuance of an **early warning.** A warning is quite different from a prediction or forecast and carries many implications. As one expert put it, a warning is "a public declaration that the normal routines of life should be altered for a period of time to deal with the danger posed by the imminent event." Warnings depend heavily on timeliness, effective communications and public information systems, and credible sources. If a warning is issued prematurely or irresponsibly and the event does not occur as predicted, the results can be disastrous. Like the boy who cried "Wolf!" the scientist may be unable to regain credibility, and the public may be slow to respond to a *real* threat.

RESPONSE AND THE ROLE OF GEOSCIENTISTS

Some natural hazards, such as meteorite impacts, are impossible to prevent and very difficult to predict within any useful time frame. Although we know that such an event might occur, there is virtually nothing we can do with current technologies to decrease the risk. However, each day we are faced with a wide range of natural hazards to which we *can* adapt by making certain choices and taking conscious actions to prepare ourselves or decrease our vulnerability. Many of these adaptive actions fall outside the realm of science; they involve economic, legal, political, or lifestyle choices. People who live with risk have adopted widely varying methods of responding to the risk, ranging from denial to acceptance to panic. What, then, should the role of the geoscientist be in shaping and informing the public's response to natural hazards?

There is a strong need on the part of the general public, especially decision makers, for knowledge concerning geologic hazards. Unfortunately, there are many gaps between the acquisition of scientific knowledge about hazardous processes and the effective transfer of this information to people who can use it in formulating plans and adopting policies for the reduction of hazards and risks (Fig. II.4). Scientists tend to have different priorities from those of

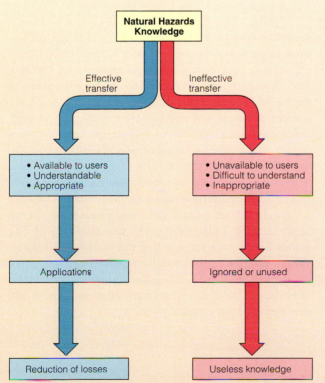

◄ F I G U R E II.4
Effective and ineffective transfer of knowledge about natural hazards.

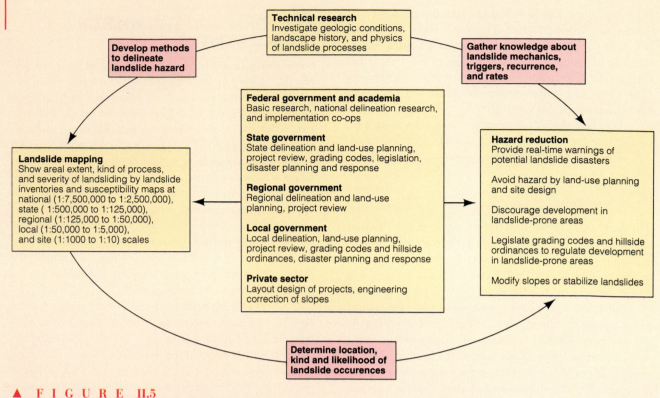

▲ FIGURE II.5

An integrated system designed to reduce landslide hazards. Geologic information can contribute to the establishment of an integrated system designed to reduce or mitigate the damage caused by natural hazards.

other members of the community and to use different modes of communication. They are concerned primarily with observing and understanding natural phenomena; in contrast, a government official is likely to be more concerned with the socioeconomic implications or with the feasibility and cost of hazard reduction strategies. Scientists typically communicate through technical papers, which are published in scientific journals and use specialized terminology; few government officials have the training or the time to keep up to date in the areas of study that are relevant to the management of natural hazards. These differing sets of priorities and communication styles can lead to gaps in communication and, thus, to the misdirection, misunderstanding, or misuse of important information.

Ideally, scientific understanding can contribute to the establishment of an integrated system in which geoscientists cooperate with government and private-sector decision makers to apply scientific and technical knowledge to the reduction of natural hazards. An example of such a system

is presented in Figure II.5. In this integrated approach to the management of landslide hazards, scientific and technical research contribute to the understanding of landslide mechanisms and the temporal and spatial delineation of the hazard. This understanding, in turn, leads to the development of hazard and risk reduction strategies and landslide hazard maps. All levels of government, academic institutions, and private-sector organizations would be involved in a balanced program of this type.

Geoscientists have a particular responsibility to contribute to such efforts. Because so many natural disasters are associated with geologic processes, geoscientists play an important role in furthering our understanding of hazardous Earth processes, assessing the hazards and risks involved, accurately predicting hazardous events, and assisting in prevention or the mitigation of impacts. All of these tasks depend on effective communication of scientific understanding about geologic processes.

IMPORTANT TERMS TO REMEMBER

anthropogenic hazard (p. 62)

early warning (p. 65)

forecast (p. 64)

geologic hazard (p. 61)

hazard assessment (p. 62)

natural hazard (p. 61)

precursors (p. 64)

prediction (p. 64)

rapid onset hazard (p. 61)

risk assessment (p. 64)

technological hazard (p. 61)

vulnerability (p. 62)

EARTHQUAKES

We rode on a sea of mountains and jungles, sinking in rubble and drowning in the foam of wood and rock. The Earth was boiling under our feet...making bells ring, the towers, spires, temples, palaces, houses, and even the humblest huts fall; it would not forgive either one for being high or the other for being low.

• **Survivor of an earthquake in Guatemala, 1773**

*I*n October 1989 climatologist Iben Browning predicted that a devastating earthquake would occur in the area of New Madrid, Missouri on December 2 or 3, 1990, plus or minus 2 days. His prediction was based on the idea that earthquakes may be triggered by peaks in the Earth's tides, a theory that has often been discussed in scientific journals but has never been substantiated. Scientists at the U.S. Geological Survey were quick to discredit Browning's prediction, stating that there was only a 1 in 60,000 chance that such an earthquake would occur in that location on that date.

In spite of the prediction's lack of scientific credibility, reporters across the United States jumped on the New Madrid story. Newspaper articles and radio talk shows about earthquakes proliferated. NBC aired a made-for-TV movie titled "The Big One." Government agencies in the central states, not wanting to be caught unaware, issued pamphlets about earthquake preparedness. Schools rehearsed emergency procedures, and most were scheduled to close on the day the quake was expected to occur. T-shirts and other quake memorabilia began to appear.

On January 17, 1995, an earthquake occurred in southern Honshu, the main island of Japan. The city most severely affected by the quake was Kobe. More than 5000 people were killed and a much larger number were injured when buildings and roadways collapsed. This high-speed elevated roadway through downtown Kobe buckled and collapsed without warning.

New Yorker correspondent Sue Hubbell, who owns a farm near New Madrid, reported with some irony on conditions in the area as "E day" approached:

When I got to my farm, I felt as though I had stumbled into the countdown for Armageddon. Everywhere, I heard stories of people buying gas cans, batteries, even emergency power generators that cost thousands of dollars, stocking their pickups with emergency food and driving them out to the middle of their fields. I heard of white roses blooming blue out of season. The Mississippi River was said to be bubbling sulfur. A kindly, grandmotherly woman told me of the Bag 'Em and Tag 'Em Project: an "outfit" (unspecified) was said to have offered plastic bags and Magic Markers to the school in New Madrid so that overbusy undertakers could enlist the help of children to go out, pick up the dead, and label them as best they could. When I reacted with disgust and disbelief, she said, "Well, it is just awful, and they turned down the plastic bags. Took the Magic Markers, though."

The predicted catastrophe did not materialize; even the most sophisticated techniques do not yet permit scientists to predict earthquakes with such precision. But the episode may have had some positive outcomes. New Madrid does in fact lie within an earthquake-prone region. In the winter of 1811–1812, the area was shaken by a series of three great quakes that caused tremendous damage and were felt over two thirds of the contiguous United States. Geologists believe that it is still possible for a major earthquake to occur in the region. As a result of Browning's prediction, residents and emergency personnel may be better prepared for such an event should it occur.

WHAT CAUSES EARTHQUAKES?

Try this experiment. Ask a friend to hit one end of a wooden plank or the top of a wooden table with a hammer while you press your hand on the other end. You will feel vibrations set up by the energy of the blow. The harder the blow, the stronger the vibrations. The reason you can feel those vibrations is that some of the energy imparted by the hammer is transferred to your hand by vibrations traveling through the solid wood. In the Earth, a bomb blast or a violent volcanic explosion will serve as an energy source. So, too, will the sudden slipping of rock masses along fault surfaces, which results in an earthquake.

When an earthquake occurs, it is as if someone has struck the Earth with a huge hammer. Energy that has been built up and stored in rocks, sometimes over a long period, is suddenly released. The more energy released, the stronger the quake. Just how the stored energy is built up, how it is released, and the impacts on people in the area are the subjects of this chapter.

Fractures and Faults

Most earthquakes are caused by the sudden movement of blocks of the Earth's crust. Such movement involves the fracture of brittle rocks and the movement of rock along the fractures. A fracture in a rock along which movement occurs is called a **fault.** Tectonic forces produce *stress,* or directional pressure, which causes large blocks of rock on either side of a fault to move past one another. But that movement is not a smooth, continuous slippage. Instead, stress builds up slowly until the friction between the two blocks is overcome; then the blocks slip abruptly again.

If stresses persist, the cycle of slow buildup followed by abrupt movement repeats itself many times. Although movement along a large fault may eventually total many kilometers, this distance is the sum of numerous small, sudden slips. Each of those slips may cause an earthquake and, if the movement occurs near the Earth's surface, disrupt and displace surface features (Fig. 3.1).

Figure 3.1 illustrates abrupt horizontal movement, but vertical movements along faults are also well documented. The largest abrupt vertical displacement ever observed occurred in 1899 at Yakutat Bay, Alaska. During a major earthquake, a stretch of the Alaskan shore was suddenly lifted as much as 15 m above sea level. The visible vertical displacement may be less than the total amount, because the fault is located offshore and the block of crust on the other side of it, lying beneath the sea, may have moved downward, thus adding to the total displacement of the stretch of beach.

Movement along a fault is usually, but not always, abrupt. Detailed measurements along the San Andreas Fault in California reveal places where gradual slipping occurs, sometimes by as much as 5 cm a year. Geologic studies reveal that movement has been occurring along the San Andreas Fault for at least 15 million years. The total extent of that movement is not known, but there is evidence suggesting that it amounts to more than 600 km.

It may be that no spot on the Earth is completely stationary. Measurements by surveyors over the past 100 years reveal large areas of the United States where the land is slowly sinking and other places where it is slowly rising. The causes of these vast, slow vertical movements are not well understood, but the movements prove that the Earth is not as rigid as it seems and that great internal forces are continuously deforming its crust.

◄ F I G U R E 3.1
An orange grove planted across the San Andreas Fault in southern California. Movement on the fault has displaced the rows of trees. The direction of plate motion is such that trees in the background have moved from left to right relative to the trees in the foreground.

Classification of Faults

Faults are classified according to (1) the inclination, or slope, of the fault surface (called its *dip*) and (2) the direction of relative movement along the fault. The common classes of faults are shown in Fig. 3.2. Movement along many faults is entirely vertical or entirely horizontal, but along some faults combined vertical and horizontal movement occurs.

Normal faults are caused primarily by *tension,* that is, by movement that tends to pull crustal blocks apart. The fault shown in Fig. 3.2B is a normal fault. Normal faulting tends to stretch and thin the crust. **Reverse faults,** in contrast, arise from *compression,* that is, movement that tends to push crustal blocks together. The relative movement on a reverse fault is such that one block of crustal material moves upward relative to the other. These relationships are illustrated in Fig. 3.2C. Reverse fault movement shortens and thickens the crust. A special class of reverse faults, called *thrust faults,* are low-angle (i.e., shallow, rather than steep) reverse faults (Fig. 3.2D). Such faults, which are common in great mountain chains, are noteworthy because along some of them the relative movement of crustal blocks may amount to many kilometers.

Strike-slip faults are those in which the principal movement is horizontal (Fig. 3.2E). Such faults arise from stresses that lead to horizontal, or *translational,* motions of the fault blocks; in other words, the two crustal blocks are sliding past one another. The San Andreas Fault is an example of a strike-slip fault. In Fig. 3.1, it is strike-slip movement on the San Andreas Fault that is offsetting the rows of orange trees.

Horizontal fault movement is designated as follows: To an observer standing on either block, the movement of the other block is *left-lateral* if it is to the left and *right-lateral* if it is to the right. The San Andreas Fault, for example, is a right-lateral strike-slip fault. If you are standing on the east side of the fault and looking across toward the ocean, the block on the other side (the Pacific Ocean plate) is moving in a northwesterly direction relative to where you are standing (i.e., toward your right), carrying a little bit of the California coast along with it. The relative motion in a right- or left-lateral fault is the same regardless of which block the observer is standing on. In other words, if you were to stand on the Pacific Ocean side of the San Andreas Fault and look across toward the continental United States, you would be looking at a block (the North American plate) that is moving toward the southeast relative to where you are standing (i.e., still moving toward your right).

Earthquake Mechanisms

Sudden movement along faults causes earthquakes, but some earthquakes are millions of times stronger than others. The reason for this is that in some cases stored energy is released by thousands of tiny slips and small earthquakes,

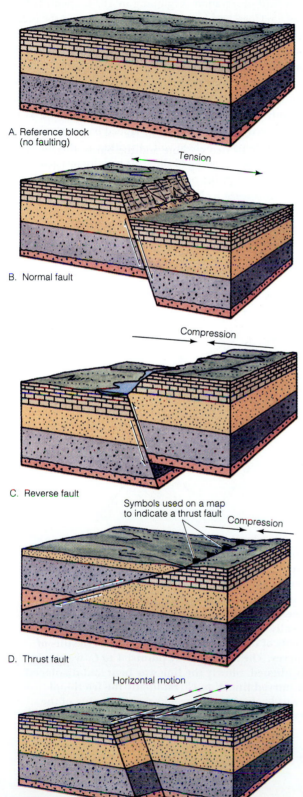

A. Reference block (no faulting)

Tension

B. Normal fault

Compression

C. Reverse fault

Symbols used on a map to indicate a thrust fault

Compression

D. Thrust fault

Horizontal motion

E. Slip-strike fault

◄ F I G U R E 3.2
The principal kinds of faults.

while in other cases it is released in a single immense quake. In this section we take a closer look at the mechanisms involved in earthquakes.

The Elastic Rebound Theory

The most widely accepted explanation of the origins of earthquakes is the **elastic rebound theory.** It is based on the mechanics of *elastic deformation* of rocks: reversible changes in the volume or shape of a rock (or other material) that is subjected to stress. When the stress is removed, the elastically deformed material returns to its original size and shape. You can demonstrate the storage of energy in an elastically deformed material with a heavy steel spring or a long metal ruler. When you compress the spring or bend the ruler across your knee, the material undergoes elastic deformation. When you suddenly release the spring or ruler, it bounces back to its original shape, releasing the built-up energy. Similarly, the elastic rebound theory suggests that energy can be stored in elastically deformed bodies of rock when they are subjected to stress along a fault. Eventually the stored energy is sufficient to overcome the

friction between the blocks. The energy is suddenly released in the form of an earthquake, and the elastically strained bodies of rock rebound to their original shapes.

The first evidence supporting the elastic rebound theory came from studies of the San Andreas Fault. During long-term field observations beginning in 1874, scientists from the U.S. Coast and Geodetic Survey determined the precise position of many points both adjacent to and distant from the fault (Fig. 3.3). As time passed, movement of the points revealed that the crust was slowly being bent. Near San Francisco, however, the fault was locked and did not slip. On April 18, 1906, the two sides of this locked fault shifted abruptly. The elastically stored energy was released as the fault moved and the bent crust snapped back, creating a violent earthquake. Subsequent repetition of the survey revealed that the bending of the crust had disappeared.

Most earthquakes occur in the brittle rock of the lithosphere. *Brittleness* is the tendency of a solid material to fracture when the deforming stress exceeds the material's *elastic limit,* that is, the limit beyond which deformation becomes permanent and the material will not be able to rebound to

A.

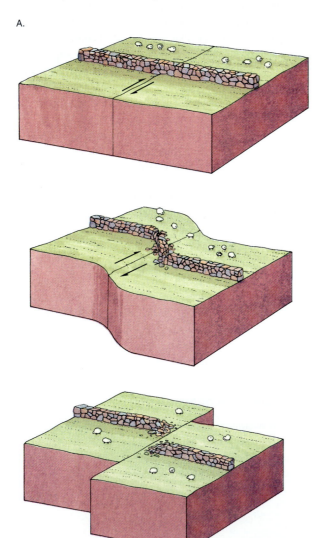

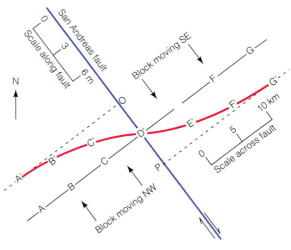

B.

◄ **F I G U R E 3.3**

An earthquake, caused by the sudden release of energy. A. Sketch based on detailed surveys near the San Andreas Fault, California, before and after the abrupt movement that caused the earthquake of 1906. A fence crosses the fault and is slowly bent as rock is elastically deformed. After the earthquake, the two segments of the fence are offset by 7 m. B. Results of the survey. The seven survey points, *A* to *G*, were originally aligned. Slowly the line was bent and displaced to the curved line *A'* to *G'*. Suddenly the frictional lock was broken and the rocks on either side of the fault rebounded. The surveyed points lay along the lines *A'O* and *PG'*.

A.

B.

▲ **F I G U R E 3.4**

Examples of rock deformation. A. Fracture of strata by brittle deformation. B. Bending of rock layers by ductile deformation.

its original shape or volume. At great depths below the Earth's surface, temperatures and pressures are so high that rocks will bend and fold but not break; this is called *ductile deformation* (Fig. 3.4). Under such conditions rocks can neither fracture nor store elastic energy. Instead, like putty, they undergo changes of shape that remain after the deforming forces have been removed. Earthquakes, therefore, are phenomena of the brittle, outer, cooler portion of the Earth.

HOW EARTHQUAKES ARE STUDIED

The study of earthquakes is known as **seismology,** from the ancient Greek word for earthquake, *seismos.* Scientists who study earthquakes are called *seismologists,* and the device used to record the shocks and vibrations caused by earthquakes is a **seismograph.**

Seismographs

An ideal way to record the vibrations and motions of the Earth would be to put a seismograph on a stable platform that is not affected by the vibrations. But a seismograph must stand on the Earth's vibrating surface, and it will therefore vibrate along with that surface. This means that there is no fixed frame of reference for making measurements. The problem is the same one that a sailor in a small boat faces when attempting to measure waves at sea. Because the boat moves up and down with each wave, there is no "platform" for measuring the height of the waves.

To overcome this problem, most seismographs make use of *inertia,* the resistance of a large stationary mass to sudden movement. If you suspend a heavy mass, such as a block of iron, from a light spring and suddenly lift the spring, you will notice that because of inertia the block remains almost stationary while the spring stretches (Fig. 3.5). This is the principle used in *inertial seismographs*

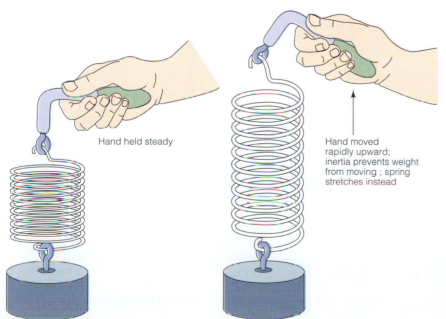

Hand held steady

Hand moved rapidly upward; inertia prevents weight from moving ; spring stretches instead

◀ **F I G U R E 3.5**
The principle of the inertial seismograph.

(Fig. 3.6). Vertical motion is measured by the device shown in Fig. 3.6A, in which a heavy mass is supported by a spring and the spring is connected to a support that in turn is connected to the ground. When the ground vibrates, the spring expands and contracts but the mass remains almost stationary. The distance between the ground and the mass can be used to sense the vertical displacement of the ground surface.

Horizontal displacement can be measured by suspending a heavy mass from a string to make a pendulum (Figure 3.6B). Because of its inertia, the mass does not keep up with the horizontal motion of the ground, and the difference between the movement of the pendulum and that of the ground serves as a measure of the ground motion. Inertial seismographs are commonly used in groups so that up–down, east–west, and north–south motions can be measured simultaneously.

Another device, a *strain seismograph,* employs two concrete piers about 35 m apart (Fig. 3.6C). When the Earth vibrates, the two piers move independently of each other

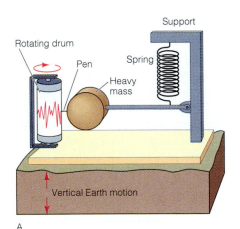

A.

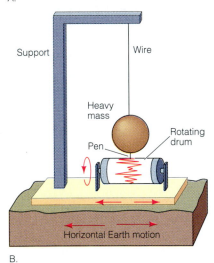

B.

C.

◀ F I G U R E 3.6
Seismographs measure vibrations generated by earthquakes. A. An inertial seismograph for measuring vertical motion. B. An inertial seismograph for measuring horizontal motion. C. A strain seismograph.

and the distance between them changes. Attached to one pier is a long, rigid silica-glass tube. The other pier carries a detector that can measure even the slightest movement in the end of the silica-glass tube. Strain seismographs are usually installed in mines, tunnels, and other places where temperature is relatively constant and wind disturbance minimal.

Modern seismographs are incredibly sensitive because any movement is amplified electronically. Vibrational movements as tiny as one hundred millionth (10^{-8}) of a centimeter can be detected. Indeed, many instruments are so sensitive that they can detect the ground depression caused by a moving automobile several blocks away.

Seismic Waves

The elastically stored energy released by an earthquake is transmitted to other parts of the Earth. As with any vibrating body, waves (vibrations) spread outward from the earthquake's point of origin. These waves, called **seismic waves,** spread in all directions, just as sound waves spread in all directions when a gun is fired. Seismic waves are elastic disturbances, so the rocks through which they pass return to their original shapes after the passage of the waves. The waves therefore must be measured and recorded while the rock is still vibrating. For this reason, continuously recording seismograph stations have been installed around the world. Wherever and whenever an earthquake occurs, the characteristic signatures and arrival times of each seismic wave are recorded by many seismographs; the records obtained in this way are called **seismograms.**

Seismic waves are of two main types. **Body waves** travel outward from the point of origin and have the capacity to travel through the Earth. **Surface waves,** on the other hand, are guided by and restricted to the Earth's surface. Body waves are analogous to light and sound waves, which travel outward in all directions from their points of origin. Surface waves are analogous to ocean waves because they are restricted to the vicinity of a free surface, such as the Earth's surface both where it meets the atmosphere and where it meets the ocean.

Body Waves

Rocks can be elastically deformed by a change in either volume or shape. One kind of body wave, **compressional waves,** deforms rocks through a change in volume. A compressional wave consists of alternating pulses of compression and expansion acting in the direction in which the wave is traveling (Fig. 3.7A). Sound waves are also compressional waves. When a sound wave passes through the air, it does so by alternating compression and expansion of the air. Compression and expansion produce changes in the volume and density of a medium. Compressional waves can pass through solids, liquids, or gases because all three can sustain changes in density. When a compressional wave passes through a medium, the compression pushes atoms closer together. Expansion, on the other hand, is an elastic response to compression that increases the distance between atoms. A solid subjected to compressional waves moves back and forth in the line of the wave's motion. Compressional waves have the greatest velocity of all seismic waves—6 km/s is a typical value for the uppermost

A. P wave

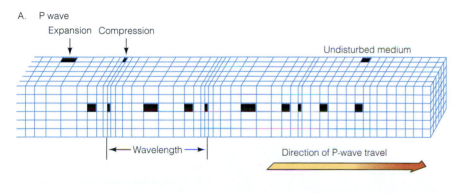

B. S wave

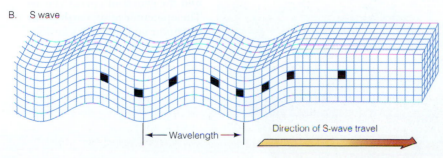

◀ **F I G U R E 3.7**
Seismic body waves of the P (compressional) and S (shear) types. A. P waves cause alternating compression and expansion in the rock the wave is passing through. An individual point in a rock will move back and forth parallel to the direction of P-wave propagation. As wave after wave passes through, a square will repeatedly expand to a rectangle, return to a square, contract to a rectangle, return to a square, and so on. B. S waves cause a shearing motion. An individual point in a rock will move up and down perpendicular to the direction of S-wave propagation. A square will repeatedly change to a parallelogram and back to a square.

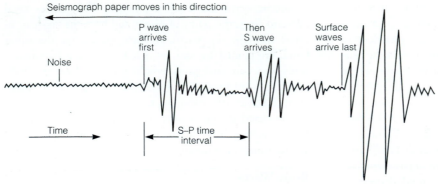

Seismograph paper moves in this direction

Noise

P wave arrives first

Then S wave arrives

Surface waves arrive last

Time

S–P time interval

▲ **F I G U R E 3.8**
Different travel times of P, S, and surface waves as shown on a typical seismogram. The P and S waves leave the point of origin at the same instant. The fast-moving P waves reach the seismograph first, and some time later the slower moving S waves arrive. The delay in arrival times is proportional to the distance traveled by the waves. The surface waves travel more slowly than either P or S waves.

portion of the crust—and they are the first waves to be recorded by a seismograph after an earthquake. They are therefore called **P** (for *primary*) **waves.**

Another kind of body wave, **shear waves,** deform materials through a change in shape. Because gases and liquids do not have the elasticity to rebound to their original shape, shear waves can be transmitted only by solids. Shear waves consist of an alternating series of sidewise movements, each particle in the deformed solid being displaced in a direction perpendicular to the direction of wave travel (Fig. 3.7B). A typical velocity for a shear wave in the upper crust is 3.5 km/s. Because shear waves are slower than P waves and reach a seismograph some time later, they are called **S** (for *secondary*) **waves.**

Surface Waves

Surface waves travel along or near the surface of the Earth like waves along the surface of a body of water. They travel more slowly than P and S waves, and they pass around the Earth rather than through it. Thus, surface waves are the

last to be detected by a seismograph. Figure 3.8 shows a typical seismogram, in which the P wave's arrival is seen first, followed by the arrival of the S wave and finally by that of the surface waves. Surface waves are important for planners and builders because they cause much of the ground shaking that causes damage to structures during large earthquakes.

Focus and Epicenter

The place where energy is *first* released to cause an earthquake is called the focus (plural *foci*). Because most earthquakes are caused by movement on a fault, the movement may eventually extend for several kilometers. The focus is where the movement starts. Earthquake foci lie below the surface, ranging from shallow (just below the surface) to deep (up to 700 km below the surface). Because of the variations in depth, it is more convenient to identify the site of an earthquake by its **epicenter,** the point on the Earth's surface lying vertically above the focus (Fig. 3.9). A good way

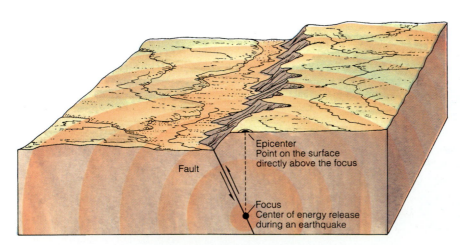

Epicenter
Point on the surface directly above the focus

Fault

Focus
Center of energy release during an earthquake

◀ **F I G U R E 3.9**
The focus of an earthquake is the site of first movement on a fault and the center of energy release. The epicenter of an earthquake is the point on the Earth's surface that lies vertically above the focus.

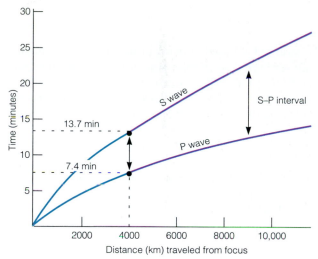

▲ F I G U R E 3.10

Average travel-time curves for P and S waves in the Earth, used to locate an epicenter. For example, when seismologists determine the S–P interval to be 13.7 min - 7.4 min = 6.3 min, they know that the epicenter is 4000 km away from their station.

to describe the location of an earthquake's focus is to state the location of its epicenter and its *focal depth,* that is, how far below the surface it lies.

Locating the Epicenter

If an earthquake's waves have been recorded by three or more seismographs, its epicenter can be determined through simple calculations. The first step is to determine how far the seismograph is from the epicenter. This is done by comparing the arrival times of the P and S waves (Fig. 3.10). The greater the difference between the arrival times, the greater the distance from the epicenter. After using a graph like the one shown in Fig. 3.10 to determine the distance from the seismograph to the epicenter, the seismologist draws a circle with a radius equal to the calculated distance, with the seismic station at the center of the circle. When similar information is plotted for three or more seismographs, the exact location of the epicenter can be determined by triangulation (Fig. 3.11).

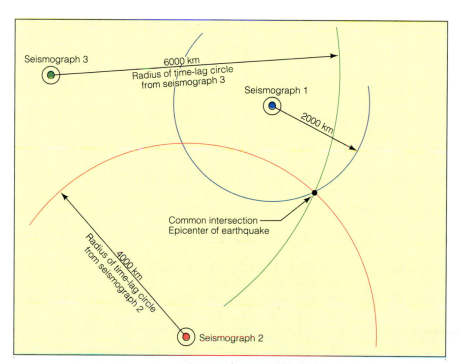

▲ F I G U R E 3.11

Locating an epicenter. The effects of an earthquake are felt at three stations. The time differences between the first arrival of the P and S waves depend on the distance of a station from the epicenter. The following distances are calculated by using the curves in Fig. 3.10.

	Time Difference	Calculated Distance
Seismograph 1	8.8 min – 4.7 min = 4.1 min	2000 km
Seismograph 2	13.7 min – 7.4 min = 6.3 min	4000 km
Seismograph 3	17.5 min – 9.8 min = 7.7 min	6000 km

On a map, a circle of appropriate radius is drawn around each station. The epicenter is located at the point where the three circles intersect.

Measuring Earthquakes

Very large earthquakes, such as the one that destroyed much of San Francisco in 1906, are relatively infrequent (Table 3.1). Such quakes occur more frequently in some areas than in others. In earthquake-prone regions such as San Francisco and the surrounding area, very large earthquakes occur, on average, about once a century. This means that it takes roughly 100 years to build up energy to the point at which the frictional locking of a fault is overcome. Small earthquakes may occur during this time as a result of local slippage, but even so, stored energy may be accumulating because parts of the fault remain locked.

Earthquake Magnitudes

Measurements of elastically deformed rocks before an earthquake, and of undeformed rocks after an earthquake, can provide an accurate indication of the amount of energy released by the quake. With that information in hand it is easy to compare two earthquakes and to say which one is the larger of the two. But the task of measuring deformation is time-consuming, and pre-earthquake measurements are rarely available.

A way to compare the magnitudes of earthquakes without resorting to pre-earthquake measurements was invented in 1935 by seismologist Charles Richter. The idea of the magnitude scale is that if two earthquakes occurred at the same place, and were recorded by the same seismograph, the bigger earthquake would produce a bigger *amplitude* wave record on the seismogram.

Richter measured the amplitudes of P and S waves, as recorded on a seismogram, for various earthquakes out to distances of 600 km. Because the wave amplitudes varied in size by factors of a million or more, Richter introduced a magnitude scale that is *logarithmic;* each increase in magnitude corresponds to a 10-fold increase in the amplitude of the wave signal. Thus, a magnitude 6 signal on Richter's scale has an amplitude 10 times larger than a magnitude 5, and 100 times larger than the signal of a magnitude 4 quake.

Magnitudes on what has come to be loosely called the Richter scale are corrected for depth and distance from the epicenter. This means that the magnitude calculated for a given earthquake is the same whether you are standing at the epicenter, where the effects of the earthquake would feel strongest, or 1000 km away, where the effects would feel less intense.

In order to really compare earthquakes accurately seismologists need to have a measure that takes account of the energy released. Unfortunately the straightforward Richter scale is not adequate for this purpose where very intense earthquakes are involved. While a wave amplitude is certainly a function of the energy released, it is not the sole function—the duration of shaking is also a function of the energy released. At magnitudes greater than about 5.5 more and more energy goes into a longer duration of shaking and the Richter scale is less and less reliable. Figure 3.12 shows how a magnitude is now calculated. The energy of a seismic wave is a function of both its amplitude (i.e. the height of the wave on a seismogram) and its frequency (the duration of a single oscillation), denoted T. Divide the maximum amplitude, X (measured in steps of 10^{-4} cm on a suitably adjusted seismograph) by T (measured in seconds). Then correct for the distance of the epicenter from the seismograph, Y (determined from the S-P interval). X/T is a measure of the maximum energy reaching the seismograph. The formula for an earthquake magnitude is:

T A B L E 3.1 • Earthquake Magnitudes and Frequencies of Occurrence, with Characteristic Damaging Effects

Richter Magnitude	Number per Year	Modified Mercalli Intensity Scale[a]	Characteristic Effects of Shocks in Populated Areas
<3.4	800,000	I	Recorded only by seismographs
3.5–4.2	30,000	II and III	Felt by some people who are indoors
4.3–4.8	4,800	IV	Felt by many people; windows rattle
4.9–5.4	1,400	V	Felt by everyone; dishes break, doors swing
5.5–6.1	500	VI and VII	Slight building damage; plaster cracks, bricks fall
6.2–6.9	100	VII and IX	Much building damage; chimneys fall; houses move on foundations
7.0–7.3	15	X	Serious damage, bridges twisted, walls fractured; many masonry buildings collapse
7.4–7.9	4	XI	Great damage; most buildings collapse
>8.0	One every 5–10 years	XII	Total damage, waves seen on ground surface, objects thrown in the air

[a]Mercalli numbers are determined by the amount of damage to structures and the degree to which ground motions are felt. These depend on the magnitude of the earthquake, the distance of the observer from the epicenter, and whether an observer is in or out of doors. Because magnitudes are not location-specific but Mercalli intensities are, a comparison between the two scales can only be made close to the epicenter and even there is only approximate.

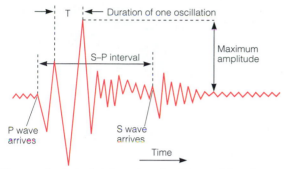

Measurements used for determining Richter magnitude (M) from a seismogram. Y is a correction factor that depends on the distance of the seismograph from the epicenter. It is calculated from the S–P interval.

$M = \log X/T + Y$

Each step on the Richter scale corresponds to a 10-fold increase in amplitude (X). However, the increase in energy released is proportional to X^2, that is, a hundredfold. When the oscillation frequency is considered it turns out that each step on the modern magnitude scale corresponds to only about a 30-fold increase in energy. Thus a magnitude 6 earthquake releases almost 100 times as much energy as a magnitude 4 quake (a 30-fold increase for each step in the scale, or $30 \times 30 = 900$). This means that a magnitude 8 earthquake, such as the New Madrid earthquakes of 1811–1812 mentioned at the beginning of the chapter, is not twice as large as a magnitude 4 quake but almost 1 million times as large ($30 \times 30 \times 30 \times 30 = 810,000$)! So even though it may happen infrequently, a single very large earthquake can release as much stored energy as many thousands of smaller quakes.

The Richter scale is open ended, which means that theoretically there is no upper limit on the possible size of an earthquake. The largest earthquakes recorded to date have Richter magnitudes of about 8.6. Many of the greatest earthquakes ever recorded occurred in subduction zones; they include the Chilean earthquake of 1960 (magnitude 8.5) and the Alaskan "Good Friday" earthquake of 1960 (magnitude 8.6). It is possible that earthquakes cannot be any larger because rocks cannot store more elastic energy. Before they are deformed further, they fracture and release the energy.

Through careful measurement of rocks along the San Andreas Fault, seismologists have found that about 100 J (joules) of energy can be accumulated in 1 m³ of elastically deformed rock. This is not very much—it is equivalent to only about 25 calories of heat energy—but when billions or trillions of cubic meters of rock are strained, the total amount of stored energy can be enormous. When the frictional lock is broken and an earthquake occurs, the elastic energy is released during a few terrible seconds. (Even near the epicenter of a very large earthquake, the duration of strong ground motion is rarely more than about 30 seconds.) The amount of elastically stored energy released during the 1989 Loma Prieta earthquake (the San Francisco

"World Series quake") was about 10^{15} J, and the 1906 San Francisco earthquake released at least 10^{17} J. For purposes of comparison, a hydrogen bomb blast also releases about 10^{17} J of energy.

The Modified Mercalli Intensity Scale

Another scale besides the Richter scale is sometimes used to measure the intensity of an earthquake. The **modified Mercalli intensity scale,** developed in the late 1800s and later modified by Father Giuseppi Mercalli and others, is based on the amount of vibration people feel during low-magnitude quakes and the extent of damage to buildings during high-magnitude quakes. The scale ranges from I (not felt except under favorable circumstances) to XII (waves seen on ground surface, practically all works of construction destroyed or greatly damaged).

Note that the Mercalli scale is not corrected for distance from the epicenter. This means that a single earthquake could have a Mercalli magnitude (i.e., a felt intensity) of IX or X near the epicenter, where the intensity is greatest, whereas 400 or 500 km away its intensity would be only I or II. The distance over which the effects are felt is partly a function of the size of the earthquake—that is, the intensity of the seismic waves—and partly a function of the efficiency with which the waves are transmitted by different materials. The correspondence among Mercalli intensity as measured at the epicenter, Richter magnitude, and the approximate frequency of earthquakes is shown in Table 3.1 (Fig. 3.13A).

The Mercalli scale has been particularly useful in the study of earthquakes that occurred before the development of modern measurement equipment. For example, the exact sizes of the series of earthquakes that hit New Madrid, Missouri, in 1811–1812 are unknown. However, historical eyewitness accounts indicate that all three quakes had Mercalli intensities of XI near the epicenter. Combining this with information about how far away the effects were felt, researchers have determined that the three shocks had Richter magnitudes of about 7.5, 7.3, and 7.8, respectively. Another great earthquake struck Charleston, South Carolina, on August 31, 1886, killing 60 people and de-

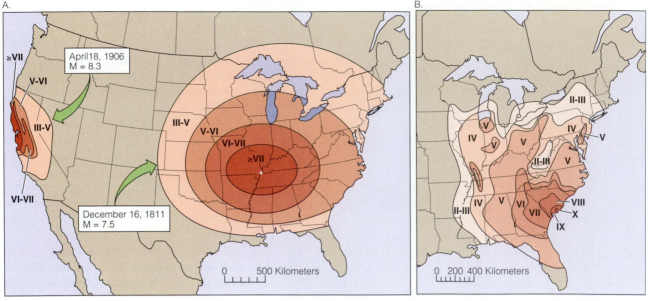

▲ F I G U R E 3.13

Intensities of great earthquakes on the modified Mercalli scale. A. Areas affected by the 1811–1812 New Madrid earthquakes (Richter magnitudes 7.5–7.8) and the 1906 San Francisco earthquake (Richter magnitude 8.3). The larger area around the New Madrid epicenter results from the greater efficiency of eastern crustal rocks in transmitting seismic waves. B. The earthquake that occurred on August 31, 1886, near Charleston, South Carolina, was felt throughout the eastern United States. The map shows intensities on the modified Mercalli scale. Estimates of Richter magnitude based on these observations range from 6.6 to 7.1.

stroying 90 percent of the city. The highest intensity of this quake on the Mercalli scale was X and was felt just northwest of Charleston (Fig. 3.13B). The effects of the earthquake were felt as far away as Minneapolis, Bermuda, and Cuba. The Richter magnitude estimated from these observations was in the range of 6.6–7.1.

Some Good News About Earthquakes

Before we turn our attention to the considerable hazards associated with earthquakes, it is worth noting that not everything about earthquakes is bad. Through measurements of earthquake vibrations, we can find out about parts of the Earth that are inaccessible to direct study. Used in this way, earthquake vibrations are like the X-rays a doctor uses to study the inside of a human body; they are the probes we use to sense and measure the world beneath our feet. The arrival times of seismic waves at seismographs around the world provide records of waves that have traveled along many different paths. From such records, it is possible to calculate how rock properties change with increasing depth below the Earth's surface. We can also identify the boundaries between layers with sharply different properties. For example, Danish seismologist Inge Lehmann used observations of the behavior of P waves passing through the center of the Earth to determine that the Earth has a solid inner core within its outer, liquid core. Her discovery was first published in a scientific paper entitled "P," which may qualify as the shortest article title ever published!

EARTHQUAKE HAZARDS AND RISKS

Each year there are many hundreds of thousands of earthquakes. Fortunately, only one or two are large enough, or close enough to major population centers, to cause loss of life. As discussed in the essay for Part II, population distribution, building codes, and emergency preparedness and response are significant factors in determining the risks associated with natural phenomena such as earthquakes. In this section we examine some of the risks and hazards associated with earthquakes and consider how scientific research can assist in addressing them.

Quantifying and Addressing the Risks

Seventeen earthquakes are known to have caused 50,000 or more deaths apiece (Table 3.2). The most disastrous one on record occurred in 1556 in Shaanxi Province, China. Many of the estimated 830,000 people who died in that quake lived in cave dwellings excavated from cliffs made of *loess* (fine, wind-deposited silt), which collapsed as a result of the quake. The worst earthquake disaster of the twentieth century also occurred in China. At 3:42 A.M. on July 28, 1976, while the 1 million inhabitants of T'ang Shan were asleep, a 7.8 magnitude quake leveled the city. Hardly a building was left standing, and the few that withstood the first quake were destroyed by a second one (*M* 7.1) that struck at 6:45 P.M. the same day. When the wreckage was cleared, 240,000 people were dead. The losses were so high because most of the buildings had unreinforced brick walls. When the ground started to shake, the walls collapsed, the roofs caved in, and the sleeping inhabitants were crushed.

Sometimes very large earthquakes occur in sparsely populated areas without causing significant damage. For example, in August 1993 a magnitude 8.1 quake occurred in the Pacific Ocean off the southeast coast of Guam. There were no deaths and only minor injuries, although some buildings sustained significant damage. The loss of life and property damage caused by the great Alaska earthquake (M 8.6) of 1964 were also much lower than they would have been had the quake struck in a more densely populated area.

Scientists have produced computer-simulated scenarios of what might happen if a great earthquake were to strike an area like Los Angeles or San Francisco. In one such model, a major earthquake (M >8.0) in a heavily populated area along the San Andreas Fault is projected to cause between 3000 and 13,000 deaths. The difference between the two extremes is simply the difference between day and night; the lower figure pertains to a scenario in which the quake occurs at 2:30 in the morning, whereas the higher one applies if the quake were to strike at 4:30 on a weekday afternoon. The reason for the difference is that wood-frame, single-family buildings are more earthquake resistant than the medium-rise, unreinforced masonry buildings that are common in many California towns. If Californians are at home in their beds when the "Big One" occurs, they will be less vulnerable overall than if they are at work or commuting to or from work.

Architecture and Building Codes

Certain areas, some of them heavily populated, are known to be earthquake prone, and building codes in those areas require structures to be as resistant as possible to earthquake damage. California, for example, imposes strict building codes and requires that special studies be conducted before any skyscraper, public building, or large construction project can be located near an active fault. Yet all too often earthquakes occur in areas in which buildings have not been constructed to withstand the effects of ground shaking. Examples include the T'ang Shan earthquake of 1976 and the magnitude 6.8 quake that struck Armenia in 1988. The Armenian earthquake occurred in a heavily populated area with poorly constructed buildings, killing an estimated 25,000 people. Another tragic example is the 1993 earthquake in Latur, India (M 6.4), in which more than 11,000 people were killed when their masonry homes collapsed into rubble. The horrendous casualties from these earthquakes belie the fact that their magnitudes were lower than those of both the 1989 Loma Prieta (San

T A B L E 3.2 • Earthquakes Occurring During the Past 800 Years That Have Caused 50,000 or More Deaths

Place	Year	Estimated Number of Deaths
Silicia, Turkey	1268	60,000
Chihli, China	1290	100,000
Naples, Italy	1456	60,000
Shaanxi, China	1556	830,000
Shemaka, USSR	1667	80,000
Naples, Italy	1693	93,000
Catalina, Italy	1693	60,000
Beijing, China	1731	100,000
Calcutta, India	1737	300,000
Lisbon, Portugal	1755	60,000
Calabria, Italy	1783	50,000
Messina, Italy	1908	160,000
Gansu, China	1920	180,000
Tokyo and Yokohama, Japan	1923	143,000
Gansu, China	1932	70,000
Quetta, Pakistan	1935	60,000
T'ang Shan, China	1976	240,000
Iran	1990	52,000

▲ **F I G U R E 3.14**
When a magnitude 7.0 earthquake struck Northridge, California, on January 20, 1994, an apartment house collapsed and crushed the automobiles parked beneath it.

BOX 3.1

•

THE HUMAN PERSPECTIVE

SEISMIC VERIFICATION OF NUCLEAR TESTING

*S*eismology is a global science by its very nature. Seismic data are collected and exchanged by scientists who collaborate with their international colleagues. Earthquakes generate seismic waves that are transmitted through the Earth and across its surface. The frequency, period, and amplitude of those waves create characteristic signatures that are received by an established global network of 125 monitoring stations known as the World-Wide Standard Seismological Network (WWSSN). The seismic waves of underground nuclear explosions detonated by countries testing nuclear devices are also received at the WWSSN stations. Scientists can interpret the seismic signatures of these waves to establish the location of a specific seismic event and determine whether it has a natural or nuclear source.

Since the period when only the United States and the former Soviet Union were nuclear opponents, decades of research have gone into developing the long-range verification system used by the WWSSN. The first completely underground test, RAINIER, was conducted in 1957 in Nevada. This test was also the first for which seismology was employed as a diagnostic tool. Following RAINIER, a global seismology program and network evolved between the academic institutions and government agencies of the United States and 30 other countries.

The Soviet Union and the United States had a bilateral seismic monitoring network throughout the Cold War period, with monitoring stations at predetermined distances from suspected underground nuclear testing sites. Seismic monitoring technology was designed and calibrated to detect seismic waves that traveled long distances through the Earth's deep interior. Now that more

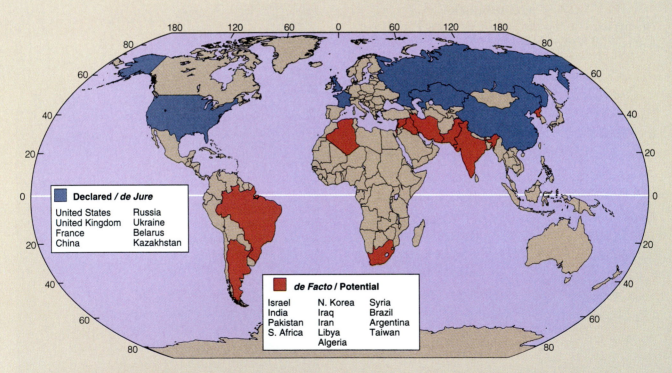

▲ F I G U R E B1.1
The new seismic monitoring task. The map shows countries of the world with declared (blue) and potential (orange) nuclear capabilities.

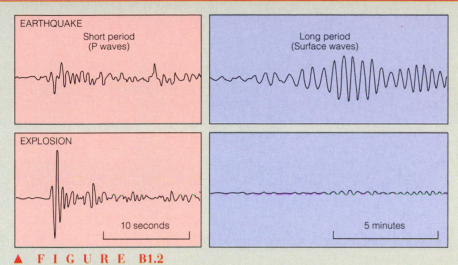

EARTHQUAKE

Short period
(P waves)

Long period
(Surface waves)

EXPLOSION

10 seconds

5 minutes

▲ F I G U R E B1.2

A comparison of seismic waves from explosions and earthquakes. Both events took place in October 1984, in the former Soviet Union. The most important difference between the signals from the two types of events is the much smaller surface waves produced by the explosion.

than 20 countries must be monitored for nuclear testing (Figure B1.1), the emphasis for the global monitoring network must be on detecting a variety of seismic signals from short-distance regional waves.

Earthquakes produce a more complicated set of seismic waves than do underground nuclear explosions. Earthquake waves emanate from a wide area over an interval of time, as plates move past one another deep within the Earth. Therefore, the seismic signal of earthquakes contains many long-period wavelengths. An underground nuclear explosion, by contrast, is focused within a relatively small area for a short time and produces more short-period wavelengths. For seismic events with large magnitudes (>4.5), the nuclear and earthquake signals are readily distinguishable (Figure B1.2). For events of smaller magnitude, a seismic signal of natural origin is more difficult to discriminate from one generated by an underground nuclear test.

Some countries have used these seismic principles to evade detection by the global monitoring states. For exam-

ple, the testing of small-magnitude tactical weapons creates high-frequency waves that are virtually impossible to detect over long distances. Nuclear devices have also been tested during natural earthquakes, which masks the seismic signal from the explosion. In other cases, countries have detonated a series of tests in order to mimic an earthquake signal, and sited tests in a large underground cavern in order to muffle the seismic signal. This last method, known as *decoupling,* blocks seismic waves emanating from the explosion because the stresses from the explosion on the rock will not exceed the elastic limit if the cavern is large enough.

The development of innovative seismic instrumentation and the ongoing cooperation among nations in the World-Wide Standard Seismological Network have led to a greater ability to discriminate between underground nuclear explosions and other seismic signals. The existence of this seismic network has critical importance today, both for the scientific exploration of the Earth's interior and for the deterrence of nuclear proliferation.

◀ **F I G U R E 3.15**
Collapse of the Nimitz Freeway, San Francisco, as a result of the Loma Prieta earthquake on October 21, 1989.

Francisco) earthquake (*M* 7.1), in which 62 people were killed, and the 1994 Northridge (Los Angeles) earthquake (*M* 6.9), in which 51 died (Fig. 3.14). The comparatively low death tolls in the California earthquakes can be attributed in part to structural engineering and geoscientific research into the effects of intense ground shaking on buildings.

Even where special building codes are in effect, however, unanticipated problems may arise, such as the collapse of the Nimitz Freeway and the Bay Bridge during the Loma Prieta earthquake (Fig. 3.15) and the collapse of Interstate Highway 5 during the Northridge quake. In 1985, an earthquake destroyed parts of Mexico City; many of the buildings that collapsed had been built to meet "California standards,"—the most exacting standards in effect at that time (Fig. 3.16). Further reminders were seen in the twisted rubble and collapsed buildings of Kobe, Japan, a city well known for earthquake preparedness, following a magnitude 7.2 earthquake in January 1995.

Structural engineers are continually refining their understanding of the factors that make tall buildings and other structures more resistant to collapse (Fig. 3.17). However, much of this understanding is based on theory and laboratory work; there are few real-life situations in which the response of a building to exceptionally intense ground motion can be tested and the standards verified.

Seismic Hazard and Risk Mapping

As discussed in the Part II essay, describing the nature and intensity of an expected event is only one part of risk and hazard assessment. Another important step is determining

◀ **F I G U R E 3.16**
When an earthquake struck Mexico City in September, 1985, this apartment house collapsed and sandwiched occupants between floors. The rescuer is searching the wreckage for survivors.

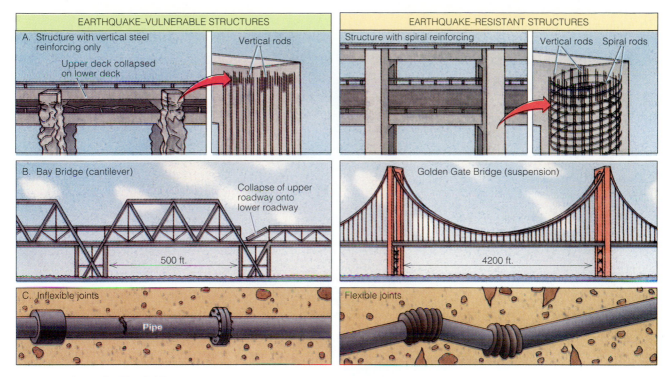

▲ F I G U R E 3.17

The way structures are designed and built can determine how they respond to ground shaking during an earthquake. A. The worst structural failure during the 1989 Loma Prieta earthquake was the collapse of the Nimitz Freeway, where a number of concrete columns gave way. The columns had been built in the 1950s with vertical steel rods inside, but they lacked the spiral reinforcing rods used in modern construction. B. During the Loma Prieta earthquake the Bay Bridge, with its inflexible cantilever-style design, collapsed, while the Golden Gate Bridge merely swayed. C. Flexible joints between gas and water pipes that give the pipes some flexibility during ground shaking can prevent gas leaks, water damage, and water shortages following an earthquake.

exactly where such an event is most likely to occur and comparing the expected event with human factors such as population distribution, building codes and zoning laws, and emergency preparedness in the area. This is often done with seismic hazard maps like those shown in Figs. 3.18 and 3.19.

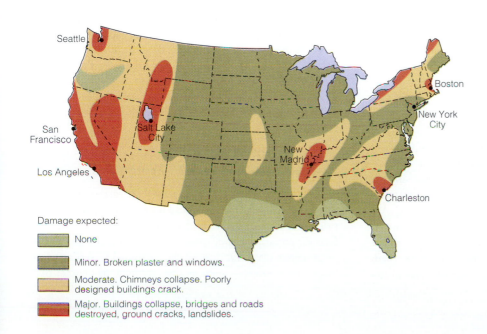

Damage expected:

None

Minor. Broken plaster and windows.

Moderate. Chimneys collapse. Poorly designed buildings crack.

Major. Buildings collapse, bridges and roads destroyed, ground cracks, landslides.

◄ F I G U R E 3.18

A seismic-risk map of the United States based on quake intensity. Zones refer to maximum earthquake intensity and therefore to maximum possible destruction. The map does not indicate the frequency of earthquakes. For example, frequency is high in southern California but low in eastern Massachusetts. Nevertheless, when earthquakes occur in eastern Massachusetts they can be as severe as the more frequent quakes in southern California.

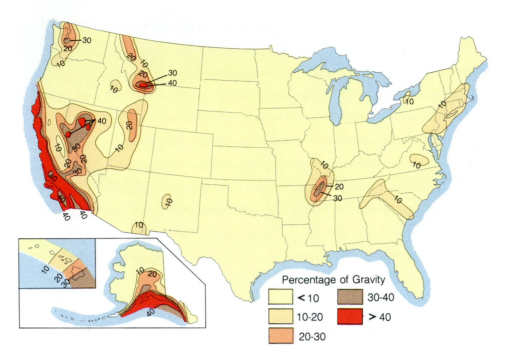

▲ **F I G U R E 3.19**

A seismic-risk map based on horizontal acceleration. Because acceleration during ground movement is the most important factor in the design of an earthquake-stable building, this type of risk map is preferred by builders and city planners. Numbers on contours refer to the maximum horizontal acceleration during an earthquake, expressed as a percentage of the acceleration due to gravity (980 cm/s^2). The probability that an acceleration of the amount indicated will occur in any given 50-year period is 1 chance in 10.

Active plate margins are a logical starting point for attempts to quantify the hazards and risks associated with earthquakes. For example, when earthquakes are mentioned, most people in the United States immediately think of California and the San Andreas Fault, which marks the active boundary separating the North American plate from the Pacific plate. However, some of the most intense earthquakes to jolt North America in the past 200 years were the ones centered near New Madrid, Missouri and Charleston, South Carolina, which are located far from active plate boundaries. Thus, the possibility that a major earthquake might occur in North America east of the Mississippi River is a topic of concern to seismologists and planners alike.

On the basis of known geologic structures (mainly faults) and the location and intensity of past earthquakes, the U.S. National Oceanographic and Atmospheric Administration (NOAA) prepared the seismic-risk map shown in Fig. 3.18. Although such a map is very informative, it is not particularly useful to people who plan and build roadways and public buildings or calculate insurance premiums.

Planners need to know just how strongly the ground is likely to shake and what the frequency of large earthquakes might be. To meet this need, the U.S. Geological Survey publishes seismic-risk maps like the one shown in Fig. 3.19. In this type of map the strength of a possible earthquake is compared to the acceleration from gravity, which is 980 cm/s^2 (1.0 G). Scientists have calculated that there is 1 chance in 10 that a given acceleration will be exceeded in a 50-year period. The maximum acceleration is likely to occur in California, reaching 80 percent of gravity (0.8 G). Damage begins at about 0.1 G.

Hazards Associated with Earthquakes

A variety of hazards are associated with earthquakes. The first two discussed here—ground motion and faulting—are primary effects; they cause damage directly. The other effects are secondary; they cause damage indirectly as a result of processes set in motion by the earthquake. Tertiary and longer term effects—such as displacement of people; disruption of services; loss of jobs; destruction of wildlife habitat; and permanent changes in groundwater levels, waterways, or coastlines—are also caused by large earthquakes.

Ground Motion

Ground motion results from the movement of seismic waves, especially surface waves, through surface rock layers and regolith. In the most intense earthquakes ($M > 8.0$) the surface of the ground can sometimes be observed moving in waves. Ground motion is the most significant primary cause of damage from earthquakes. It can damage and sometimes completely destroy buildings. As discussed earlier, proper design of buildings and other structures can do much to prevent these effects, but in a very strong earthquake even the best designed buildings may suffer some damage.

The wide variety of effects caused by earthquakes results partly from differences in the way Earth materials transmit seismic waves. For example, the New Madrid earthquakes of 1811–1812 were felt over a much wider area than the 1906 San Francisco earthquake of similar magnitude because seismic body waves are transmitted more efficiently by the crustal rocks in the central United States, as shown in Fig. 3.13A.

Sometimes local geologic factors can amplify the effects of ground motion. For example, the damage due to ground shaking during the Loma Prieta earthquake was of two types: that caused by normal ground shaking and that caused by enhanced (i.e., higher than expected) ground shaking. Enhanced ground motion is often attributable to the presence of soft sediments, like the landfill in the San Francisco Bay area. It has been calculated that enhanced shaking may have been responsible for as much as 70 percent of the losses incurred during the quake, whereas nor-

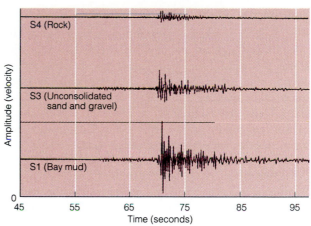

▲ **F I G U R E 3.20**

Seismograms recorded for a magnitude 4.1 aftershock to the Loma Prieta earthquake, on mud near the collapsed Nimitz Freeway (S1), on unconsolidated material (sand, gravel, etc.) near an uncollapsed portion of the freeway (S2), and on bedrock in the Oakland Hills (S3). All three seismograms are plotted on the same scale. The horizontal component of ground motion is amplified in the seismogram recorded at the mud site.

mal ground shaking caused only about 28 percent of total losses (Table 3.3). Enhanced ground shaking resulting from the presence of soft sediments may also be implicated in the collapse of the Nimitz Freeway during that event. As shown in Fig. 3.20, seismic signals recorded during a magnitude 4.1 aftershock show that ground motions were amplified by the soft bay muds, compared to other types of Earth materials in nearby sites.

T A B L E 3.3 • Losses Resulting from the Loma Prieta Earthquake, by Type of Hazard

Earthquake Hazard	Total Damages (millions)	Loss (% of Total)
Ground shaking		
Normal	$1635	28
Enhanced	$4170	70
Liquefaction	$97	1.5
Landslides	$30	0.5
Ground rupture	$4	0.0
Tsunami	$0	0.0
Total	$5936	100.00

From Holzer, Thomas L. *EOS*, vol. 75, no. 26, June 28, 1994, p. 300.

Faulting and Ground Rupture

Where a fault breaks the ground surface, buildings can be split, roads disrupted, and any feature that lies on or that crosses the fault broken apart. Large cracks and fissures can open in the ground (Fig. 3.21). Even the numerous small earthquakes resulting from slow, continuous movements along faults can contribute to structural damage. Such movement can be seen in Carrizo Plains, California, which sits astride the San Andreas Fault (Fig. 3.22).

Aftershocks

Earthquake crises are often exacerbated by **aftershocks,** the (usually) smaller earthquakes that occur shortly after a major quake. In the four months following the 1964 Alaska earthquake, for example, 1260 aftershocks were recorded. In some cases large earthquakes trigger aftershocks at locations distant from the original epicenter. The Landers (near Los Angeles) earthquake of 1992 (*M* 7.3) triggered secondary events at 14 locations, some of them as far as 1250 km away.

Fires

A secondary effect, but one that can pose a greater hazard than ground motion, is fire. Ground movement displaces stoves, breaks gas lines, and loosens electrical wires, thereby starting fires. Because ground motion also breaks water mains, there often is no water available to put out the fires.

In the earthquakes that struck San Francisco in 1906 and Tokyo and Yokohama in 1923, as much as 90 percent of the damage to buildings was caused by fire. In the 1906 San Francisco earthquake the fire destroyed 12 km², or 521 city blocks, in 3 days. For many years thereafter the quake was referred to as the "Great Fire." In 1989 fires caused by the Loma Prieta earthquake again ravaged the Marina district of downtown San Francisco (Fig. 3.23).

Landslides

In regions with steep slopes, earthquake vibrations may cause slipping of regolith, collapse of cliffs, and other rapid downslope movements of Earth material. This is particu-

▲ **F I G U R E   3.22**
Slow movement of the San Andreas Fault, Carrizo Plains, California, has detached the headwaters of a stream (upper right) from the downstream portion (lower left). The fault runs across the center of the photo. The land to the rear is moving right relative to the land in the foreground. Because motion on the fault is slow and continual, the two halves of the stream system remain in contact as water flows down the depression caused by the fault.

▲ **F I G U R E   3.21**
Fissures opened in Santa Cruz, California, as a result of the Loma Prieta earthquake, October 21, 1989.

◀ F I G U R E 3.23
Fire caused by gas lines that broke as a result of the Loma Prieta earthquake in 1989. This shot shows a fire in the Marina district of San Francisco.

larly true in Alaska, parts of southern California, China, and hilly places such as Iran and Turkey. Houses, roads, and other structures are destroyed by rapidly moving regolith. In 1970, for example, a devastating landslide in Yungay, Peru, in which at least 18,000 people were killed, was triggered by an earthquake of magnitude 7.75.

Liquefaction

The sudden shaking and disturbance of water-saturated sediment and regolith can turn seemingly solid ground into a liquidlike mass of quicksand. This process, called **liquefaction,** occurs when vibration causes sediment grains to lose contact with one another, allowing interstitial water to bubble through. Soil liquefaction and the resulting landslides were major causes of damage during the earthquake that destroyed much of Anchorage, Alaska, in 1964 (Fig. 3.24). In the same year liquefaction and uneven ground settling resulting from an earthquake caused apartment houses to sink and collapse in Niigata, Japan. Many of the buildings were not structurally damaged; they simply keeled over onto their sides. Apartment dwellers who were later permitted to enter the buildings retrieved their belongings by rolling wheelbarrows up the walls and lowering themselves through the windows.

◀ F I G U R E 3.24
Liquefaction of clay layers as a result of the Alaskan earthquake of March 27, 1964 destroyed these houses in Turnagain Heights.

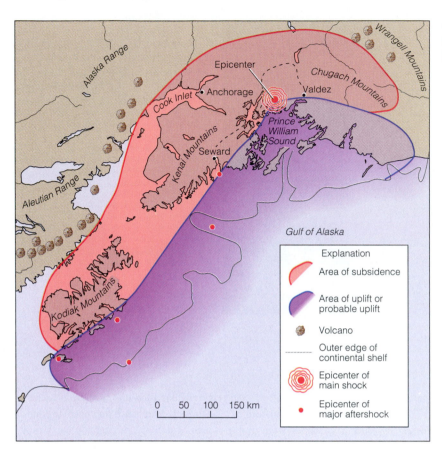

◄ F I G U R E 3.25
Changes in the crust after the Good Friday, 1964, earthquake in Alaska.

Changes in Ground Level

Sometimes the level of the ground changes over vast areas as a result of a very large earthquake (Fig. 3.25). As noted earlier, in the 1964 Alaska earthquake vertical displacements occurred along almost 1000 km of the coastline from Kodiak Island to Prince William Sound. The ground-level changes resulting from that earthquake included both uplift and subsidence. Subsidence, a lowering of ground level, was as much as 2 m, whereas vertical uplift reached 11 m in some places.

Tsunamis

Another secondary effect of earthquakes is **seismic sea waves,** also called *tsunamis.* Submarine earthquakes are the main cause of these waves, which are particularly destructive around the Pacific Ocean rim. A well-known example is the tsunami generated by a severe submarine earthquake near Unimak Island, Alaska, in 1946. The wave traveled across the Pacific Ocean at a velocity of 800 km/h, striking Hilo, Hawaii, about 4.5 hours later. Although the amplitude (height) of the wave in the open ocean was less than 1 m, it increased dramatically as the wave approached land. When it hit Hawaii, the wave had a crest 18 m higher than normal high tide. It demolished nearly 500 houses, damaged 1000 more, and killed 159 people. Another devastat-

ing tsunami was generated by an earthquake off the coast of Portugal in 1755. It caused the deaths of 60,000 people in Lisbon alone. The waves were observed a few hours later as far away as the West Indies. (Tsunamis are discussed in more detail in Chapter 5.)

Flooding

Flooding is a secondary or tertiary effect of earthquakes, usually resulting from ground subsidence (as in the 1964

▼ F I G U R E 3.26
Flooding in Portage, Alaska, due to ground subsidence, following the earthquake of March 27, 1964.

BOX 3.2
•
THE HUMAN PERSPECTIVE

NORTHRIDGE EARTHQUAKE: A PERSONAL ENCOUNTER

I used to think that earthquakes were fun and exciting. I used to tell my friends and students that I wanted to be in Los Angeles when the oft-predicted BIG ONE visits us. I no longer feel that way. I realize now that this attitude toward earthquakes stemmed from having been safely distant from the epicenters of past tremors. Earthquakes were fun, exciting, and made great lecture material for my class.

January 17, 1994, changed that perspective. At 4:31 A.M., I and several million of my closest neighbors were jarred from deep sleep. I knew instantly that the demonic shaking of my Valley condominium was earthquake induced, and I knew it was going to be bad. In total citywide darkness, my three-storey building shook and flexed and rolled with a ferocity so unimaginable to me that I was sure it would fall. A horrendously loud and ugly cacophony—the dissonant, overprinted sounds of furniture toppling, glass breaking, wood flexing, people screaming, and car alarms blaring—accompanied the intense lurching and vibrating of the building around me. With a flashlight, I ran down my swaying stairs to save a cabinet filled with prized art glass. That cabinet was my focus in the midst of a situation in which I had no control. The shaking and noises lessened, than ceased. With heart pounding, I stood in the almost quiet darkness beside the rescued cabinet and surveyed my littered living room. I remember my wonderment then—a feeling that continues three weeks later—that my home was still standing and that it was structurally undamaged. Just two blocks away and 10 miles from the epicenter, a major freeway interchange had suffered heavy damage and had nearly collapsed.

The Northridge earthquake. 25–30 seconds duration. 57 deaths. 9,300 injuries. 10,000 jobs lost. 5,700 mobile homes knocked off their foundation and another 240 burned when ruptured gas lines were ignited. 13,000 buildings destroyed or severely damaged. 22,000 dwelling units ordered vacated. 51,600 residences damaged. 55,000 persons homeless, at least temporarily. 304,000 people seeking financial aid. More than $10 billion in damage, and several million bruised states of mind.

The BIG ONE? Still wanna be here when it comes, Greg? All two or three minutes of it? Thanks, but I'll pass. Earthquakes just aren't fun anymore. And now I know, the big ones never were.

[Source: From *Geotimes,* May 1994. Contributed by Greg Davis. Editor's note from *Geotimes:* This piece was written three weeks after the Northridge earthquake, and some of the information cited has been officially updated since then.]

Alaska earthquake, Fig 3.26), the rupture of dams, or tsunamis. Reelfoot Lake, across from New Madrid on the Tennessee side of the Mississippi River, was created by flooding following ground subsidence during the series of 1811–1812 earthquakes.

PREDICTION AND CONTROL OF EARTHQUAKES

Charles Richter once said, "Only fools, charlatans, and liars predict earthquakes." Today, however, seismologists use sensitive instruments and sophisticated techniques to monitor seismically active zones. We still cannot predict the exact magnitude and time of occurrence of an earthquake, but our knowledge about seismic hazards and mechanisms and our awareness of the tectonic environments in which earthquakes occur have improved greatly since Richter's time. Long-term earthquake forecasting is based primarily on our understanding of the tectonic cycle and the geologic settings in which earthquakes occur. Short-term earthquake predictions, on the other hand, are usually based on observations of precursor phenomena, physical anomalies that may be early warning signs of earthquake activity.

The World Distribution of Earthquakes

The first step in earthquake prediction is to examine the world distribution of earthquakes and establish connections between the plate tectonic cycle and the mechanisms by which earthquakes are generated. Although no part of the Earth's surface is exempt from earthquakes, several well-defined **seismic belts** are subject to frequent shocks (Fig. 3.27). Of these, the most obvious is the *circum-Pacific belt,* for it is here that about 80 percent of all recorded earthquakes originate. The belt follows the mountain chains in the western Americas from Cape Horn to Alaska and crosses to Asia, where it extends southward down the coast through Japan, the Philippines, New Guinea, and Fiji and finally loops far southward to New Zealand. Next in prominence, giving rise to 15 percent of all earthquakes, is the Mediterranean-Himalayan belt, which extends from Gibraltar to Southeast Asia. Lesser belts follow the mid-ocean ridges.

A lot of the Earth's internal energy is released in these seismic belts. One might therefore expect other manifestations of internal energy also to appear in these regions, and indeed some of them do. Midocean ridges, deep-sea trenches, volcanoes, and many other features that outline the active margins of lithospheric plates either coincide with or closely parallel those margins. If you compare Fig. 3.27 with Fig. 1.14 you will see that seismic belts outline the plate boundaries.

Seismicity and Plate Tectonics

We learned in Chapter 1 that there are three kinds of plate boundaries: (1) *divergent boundaries,* or spreading centers, which coincide with rift valleys and midocean ridges; (2) *transform fault boundaries,* where two plates or portions of plates are sliding past each other; and (3) *convergent bound-*

aries, which coincide with oceanic trenches and/or continental collision zones. Each kind of boundary is characterized by earthquakes with distinctive fault motions and depth of foci. In addition, as we have seen, very large earthquakes occasionally occur in the interiors of plates, in the normally stable continental intraplate environment.

Divergent Boundaries

Along a divergent plate boundary or spreading center, two plates move apart from each other and the lithosphere is stretched by tensional stresses (Fig. 3.28A). The faults associated with tensional stresses are normal faults. Earthquakes at spreading centers tend to have low Richter magnitudes and foci that are invariably less than 100 km and usually less than 20 km deep. This indicates that the lithosphere is thin beneath a spreading center and that the ductile material underneath must come close to the surface.

Transform Fault Boundaries

Transform faults are huge, vertical strike-slip faults that cut down through the lithosphere. They mark the boundaries where two plates slide past each other (Fig. 3.28B). Earthquakes along transform faults always have shallow foci, no deeper than 100 km, and often have high Richter magnitudes. The locations of the earthquake foci suggest that when a transform fault cuts the continental crust, a system of parallel faults rather than a single fracture can develop. This seems to be true in the case of the San Andreas Fault, as shown in Fig. 3.29.

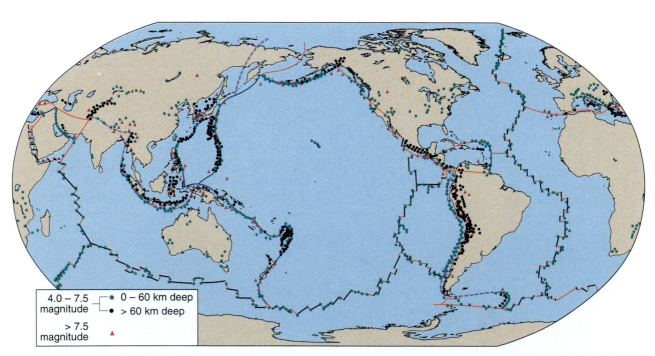

▲ F I G U R E 3.27
The Earth's seismicity outlines plate margins. This map shows earthquakes of magnitude 4.0 or greater from 1960 to 1989.

Convergent Boundaries

As discussed in Chapter 1, convergent plate boundaries are of two kinds: (1) *subduction boundaries,* where lithosphere capped by oceanic crust is subducted into the asthenos-phere and mesosphere; and (2) *collision boundaries,* where two continents collide. Each kind of convergent boundary has a characteristic pattern of seismic activity.

When oceanic lithosphere is subducted, it is subjected to complex stresses that can result in different kinds of earthquakes (Fig. 3.28C). Lithosphere bends downward as it is subducted, and the bending causes normal faults in the upper part of the plate. Earthquakes associated with such faults have very shallow foci and small Richter magnitudes. Subduction involves one plate sliding beneath the other, so the boundary between the two plates is a thrust fault. Above a depth of at least 100 km (the region where the two plates of lithosphere are in contact) earthquakes often have large Richter magnitudes and are invariably associated with thrust faulting. Below 100 km, where the subducted lithos-phere is sinking through the asthenosphere, earthquakes occur *in* the subducted slab. Some earthquakes indicate

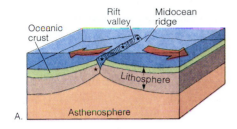

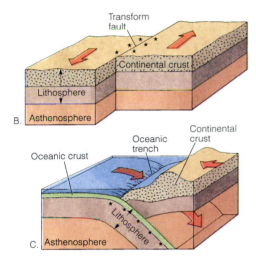

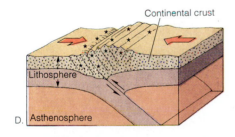

▲ **F I G U R E 3.28**
The relationship between seismicity and plate bound-aries. Stars indicate earthquake foci. A. Divergent boundary. Earthquakes have shallow foci, low Richter magnitudes, and normal fault motions. B. Transform fault boundaries. Earthquakes have foci down to 100 km, sometimes high Richter magnitudes, and strike-slip motions. C. Subduction boundary. Earthquakes are complex. Earthquakes with low Richter magni-tudes, shallow foci, and normal fault motions are ob-served seaward of the oceanic trench. Deeper earth-quakes can have high Richter magnitudes and thrust fault movement. D. Continental collision boundary. Earthquakes have foci down to 300 km, thrust fault motions, and sometimes high Richter magnitudes.

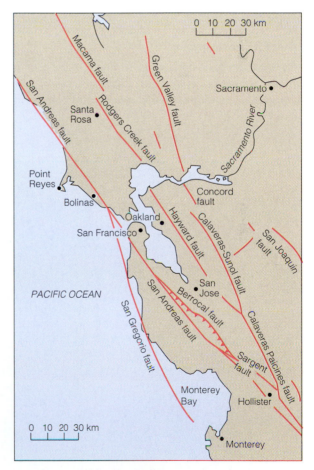

▲ **F I G U R E 3.29**
Map showing faults in the region of San Francisco, along which displacement has occurred within the last 10,000 years. The San Andreas Fault is not a sim-ple fault but a complex assemblage of fractures mark-ing the transform boundary between two plates.

tensional stresses (normal faults), whereas others indicate compressional stresses (reverse faults).

A few earthquakes originate at depths as great as 700 km. Exactly why and how such deep earthquakes occur in subducted slabs remains something of a puzzle. A very large (M 8.2), deep-seated (640 km) earthquake, the largest deep-focus earthquake ever measured, occurred in Bolivia in 1994. This event afforded scientists a rare opportunity to study the mechanisms associated with earthquakes that originate deep within the Earth, where high temperatures and pressures normally cause rocks to flow and deform ductilely under stress, rather than fracture.

Collision boundaries are the places where two continents collide. The Caucasus mountain chain between Russia and Asia Minor is a present-day collision boundary; the devastating 1988 earthquake in Armenia was a manifestation of this collision. A zone of collision tends to be several hundred kilometers wide, and within it rocks are intensely compressed and thrust-faulted, as shown in Fig. 3.28D. Within a collision zone, the crust is locally thickened. Earthquakes may have foci as deep as 300 km and high Richter magnitudes.

Intraplate Earthquakes

Whereas active plate boundaries are zones of frequent and intense earthquake activity, the stable interiors of continents tend to be seismically very quiet by comparison. But intraplate environments also can be the sites of large earthquakes. In some cases, such as that of the 1811–1812 New Madrid earthquakes, the seismic activity is believed to be related to ancient, deep-seated fault structures that are reactivated from time to time. In other cases, the earthquakes may be related to incipient continental rifting. In still others, such as the Latur earthquake of 1993, which occurred in the ancient, stable heart of the Indian subcontinent, the causes and mechanisms are simply not understood.

Approaches to Earthquake Prediction

As we have seen, some of the most dreadful natural disasters in history have been caused by earthquakes. It is hardly surprising, therefore, that a great deal of research around the world focuses on earthquake prediction. The hope is that through this research seismologists will be able to improve their forecasting ability to the point at which effective and accurate early warnings can be issued.

Long-Term Forecasting: Paleoseismology and Seismic Gaps

As noted earlier, the long-term prediction of earthquakes is based on knowledge of the tectonic cycle. In places where earthquakes are known to occur repeatedly, such as along plate boundaries, it is sometimes possible to detect a regular pattern in the recurrence interval of large quakes. To do so,

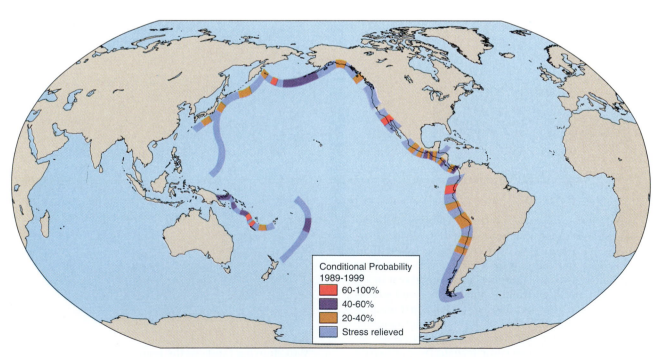

▲ **F I G U R E 3.30**
Seismic gaps in the circum-Pacific belt. In the areas indicated, earthquakes of magnitude 7.0 or greater are known to occur but have not done so in recent times. Strain is now building up in each seismic gap, raising the probability that a large quake will occur.

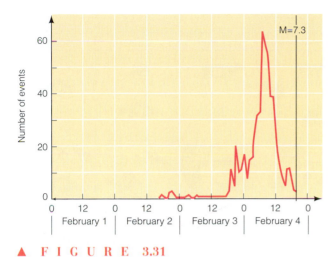

▲ F I G U R E 3.31

Frequency of foreshocks before the 1975 earthquake (*M* 7.3) at Haicheng, People's Republic of China. Foreshock activity peaked just before the main event.

however, seismologists require information about seismic activity going back much farther than historical records. This information is provided by experts in **paleoseismology,** the study of prehistoric earthquakes. The primary goal of paleoseismology is to search the sedimentary rock record for evidence of major earthquakes and, if possible, to discern the intervals between them. The evidence may include vertical displacement of sedimentary layers, as well as indicators of liquefaction or lateral offsetting of sedimentary features. If the pattern of recurrence suggests an interval of, say, a century between major quakes, it may be possible to predict where and when a large quake may happen next. Certainly it is possible to monitor such areas closely when a big quake is thought to be imminent.

Studies of recurrence patterns have identified a number of **seismic gaps** around the Pacific rim (Figure 3.30). These are places along a fault where earthquakes have not occurred for a long time even though tectonic stresses are still active and elastic energy is steadily building up. Seismic gaps receive a lot of attention from seismologists because they are considered to be the places most likely to experience large earthquakes. A major seismic gap has been identified along the San Andreas Fault in the vicinity of Parkfield, California, about halfway between Los Angeles and San Francisco. Through paleoseismological study and close examination of historical records of earthquake activity, seismologists have determined that large earthquakes tend to occur along this stretch of the fault at intervals of about 20 to 27 years. The last large earthquake near Parkfield occurred in 1966, suggesting that the area may be overdue for another major quake. In 1985 seismologists at the U.S. Geological Survey predicted that a magnitude 6 earthquake would strike Parkfield in 1988 ± 5 years. The predicted time has come and gone with no large quakes along the

Parkfield portion of the fault. It may be that some of the stored energy was dissipated by large earthquakes in nearby areas, such as Landers and Loma Prieta. In the meantime, however, seismologists are keeping a close eye on Parkfield. If they cannot accurately predict the quake, at least they hope to be ready and waiting, with instruments in place, when it finally does occur.

Short-Term Prediction: Physical Precursors

Short-term prediction of earthquakes is based on observations of precursor phenomena: anomalous physical occurrences that may serve as early warning signs of earthquake activity. Most research on short-term earthquake prediction involves monitoring changes in the properties of elastically strained rocks—properties such as rock magnetism and electrical conductivity. Even simple observations, such as the level of water in wells or the amount of radon (an inert gas) in well water, might indicate changes in the properties of underlying rocks. Some researchers have also reported the detection of unusual radio waves near the epicenter of the Loma Prieta earthquake just before the quake.

Tilting or bulging of the ground and slow rises and falls in elevation are among the most reliable indications that strain is building up. Most significant are the small cracks and fractures that develop in severely strained rock. These can cause swarms of tiny earthquakes—**foreshocks**—that may be a clue that a big quake is coming. One of the most famous successful earthquake predictions, made by Chinese scientists in 1975, was based on slow tilting of the land surface, fluctuations in the magnetic field, and the numerous small foreshocks that preceded a magnitude 7.3 quake that struck the town of Haicheng (Fig. 3.31). Half the city was destroyed, but because authorities had evacuated more than a million people before the quake, only a few hundred were killed.

Because the People's Republic of China has suffered so many terrible earthquakes, Chinese scientists have used every technique they can think of in their efforts to predict quakes. They have even observed animal behavior, and on one occasion animal behavior apparently did foretell a quake. On July 18, 1969, zookeepers at the People's Park in Tianjin observed highly unusual behavior among animals in the park. Normally quiet pandas screamed, swans refused to go near water, yaks did not eat, and snakes would not go into their holes. The keepers reported their observations to the earthquake prediction office, and at about noon on the same day a magnitude 7.4 earthquake struck the region. Throughout the world there have been many reports, both documented and informal, of strange animal behavior before earthquakes (Fig. 3.32), and Japanese researchers have conducted extensive laboratory experiments to investigate the connections between animal behavior and earthquakes. However, the Tianjin earthquake remains the only well-documented case in which the behavior led to a successful prediction.

	\multicolumn{7}{c}{Time before earthquake}
	1–2 minutes	10–30 minutes	1–4 hours	6–12 hours	1 day	Few days	Few weeks
Epicenter area							
20–50 km							
70–100 km							
150–200 km							
>250 km							

▲ **F I G U R E   3.32**

Spatial and temporal distribution of reported incidents of anomalous animal behavior before the main shocks of 36 major earthquakes in Europe, Asia, North America, and South America. The symbols indicate unusual behavior by catfish, eels, other fish, frogs, snakes, turtles, sea birds, chickens, other birds, dogs, cats, deer, horses, cows, rats, and mice; the circled numbers indicate the number of reported incidents.

The long-term prediction of earthquakes (i.e., predictions of quakes in particular areas within the next few years or decades) has met with reasonable success. Seismologists know where the most hazardous areas are; they can calculate the probability that a large earthquake will occur in a particular area within a given period, and they have a theory of earthquake generation that successfully unites their predictions and observations in a plate tectonic context. The short-term prediction of earthquakes (i.e., pinpointing the actual day or time of the event and issuing an early warning) has been rather less successful. The difficulties with short-term prediction stem partly from the fact that the mechanisms and processes associated with earthquakes are hidden under the ground, where they are not amenable to study and monitoring. But a more significant problem is that earthquakes are highly inconsistent in terms of their precursor phenomena. For example, after several notable successes in predicting earthquakes on the basis of foreshocks, radon anomalies, ground tilting, and anomalous animal behavior, Chinese scientists suffered a major setback when the T'ang Shan quake of 1976 struck with no discernible warning signs.

One of the most promising earthquake monitoring techniques involves the use of satellites to measure very small changes in the surface of the ground. Through the Global Positioning System (GPS), a network of ground-based receivers tracks distance ranging signals from satellites orbiting high above the Earth's surface. If there is a change in the distance from the ground station to the satellite, displacement or deformation of the ground surface has occurred. In an even more sophisticated application, satellite radar images taken immediately before and after the 1992 Landers, California, earthquake were superimposed on one another, yielding an "interferogram"—essentially a contour map of changes in the ground surface between the times when the two radar images were acquired (Fig. 3.33). This approach is particularly exciting because of its very high resolution and because it can be done entirely by satellite, without requiring the use of ground-based instrumentation. Such detailed measurements of changes in the ground surface allow seismologists to refine their understanding of the crustal deformation resulting from earthquakes and may one day allow them to issue early warnings based on observations of changes in the rate or extent of crustal deformation.

Controlling Earthquakes

If earthquakes are caused by the sudden, catastrophic release of huge amounts of stored elastic energy, might it not

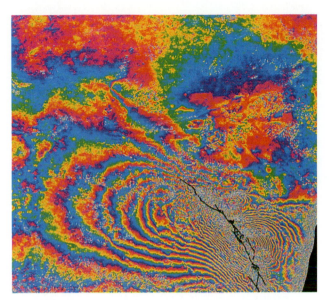

▲ F I G U R E 3.33
An "interferogram" of the 1992 Landers, California, earthquake. This image was created by D. Massonnet and colleagues by superimposing a satellite radar image of the region made immediately before the quake on an image made just after the quake. The result is essentially a contour map revealing the extent of changes in the position of ground reflectors between the time of acquisition of the two images. The fault itself can be seen as a jagged line running diagonally across the right-hand side of the image. Ground deformation was most severe in the areas nearest the fault, as evidenced by the close spacing of the interference contours in those regions. Each curved color fringe represents a vertical movement of 2.5 cm. The area shown is approximately 90 km by 90 km.

be possible to control earthquakes by finding a way of releasing that energy more gradually? Scientific investigations into the possibility of control have centered on earthquakes that were unwittingly caused by human actions. The first such quakes were reported after the construction of the Hoover Dam in 1935. About 600 local earthquakes were reported in the 10 years following construction of the dam, during which time Lake Mead was being created. Most of the tremors were small, but one had a Richter magnitude of

about 5 and two others had magnitudes of about 4. The cause of these quakes was thought to be the increased load of water in Lake Mead, which lubricated and reactivated some ancient underlying faults. Many other examples of *induced seismicity* associated with the filling of large reservoirs have since been documented, with magnitudes ranging as high as 6.5. It has been suggested that the filling of a nearby reservoir may have played a role in the Latur, India, earthquake, although this theory has yet to be substantiated.

In the late 1960s swarms of unexplained small earthquakes occurred in the area of Denver, Colorado, which normally is not seismically active. A correlation was soon made between the occurrence of the quakes and the pressure in a hazardous waste disposal well at the Rocky Mountain Arsenal, into which the United States Army was periodically injecting fluid wastes (a fully legal procedure). The epicenters of the earthquakes were located near the disposal well, and their focal depths ranged from 4 to 6 km, just below the depth of the 3.8 m well. In this case, too, the cause of the earthquakes was thought to be the reduction of frictional pressure and the resulting reactivation of old faults as a result of increased fluid pressure. Similar occurrences were later observed at the Rangely oil field in northwestern Colorado and near the Perry nuclear power plant in Cleveland, Ohio.

It has also been observed that earthquakes can be triggered by underground nuclear explosions. In the late 1960s the U.S. Geological Survey conducted seismologic monitoring of underground testing of explosives at the Nevada Test Site. These studies revealed the occurrence of thousands of small aftershocks (generally $M < 5$; the test itself yielded a M of 6.3) with focal depths as great as 7 km and epicenters as far as 13 km from ground zero. Most of the aftershocks occurred within the first week after the explosion, but a general increase in seismic activity was still evident 3 weeks later. The aftershocks appeared to have occurred along a system of fractures that was not evident at the surface.

These events and observations have not led to a mechanism for controlling or avoiding major earthquakes. What they do show is that it is possible to generate earthquakes by injecting fluids or setting off explosions underground. This, in turn, suggests that someday it may be possible to develop a way of artificially inducing slip along a locked section of a major fault in order to avoid the dangerous accumulation and sudden release of elastic energy.

SUMMARY

1. Earthquakes are caused by abrupt movements of the Earth's crust that release stored energy.

2. Fractures in rocks along which movement occurs are called faults. Normal faults are caused by tensional stresses that tend to pull the crust apart, whereas reverse faults arise from compressional stresses that squeeze, shorten, and thicken the crust. Strike-slip faults are caused by shear stresses; they are vertical fractures that have horizontal motion.

3. According to the elastic rebound theory, friction between two blocks on either side of a fault may inhibit smooth, continuous slippage along the fault. When this happens, the rocks in the blocks deform, changing their shape and/or volume and storing elastic energy in the process. When enough energy has built up to overcome the frictional lock, abrupt slippage occurs along the fault, energy is released suddenly in the form of an earthquake, and the rocks return to their original shapes.

4. The study of earthquakes is called seismology. Scientists who study them are seismologists, and earthquake vibrations are measured and recorded by seismographs.

5. Energy released at an earthquake's focus radiates outward as body waves, which are of two kinds—P waves (primary waves, which are compressional) and S waves (secondary waves, which are shear waves). Earthquake energy also causes the surface of the Earth to vibrate because of surface waves.

6. The focus and epicenter of an earthquake can be located by measuring the differences in travel time between P and S waves.

7. Earthquakes are measured on the Richter magnitude scale, which is based on the amount of energy released during an earthquake as determined from seismic wave amplitudes recorded on seismograms. Earthquake effects can also be quantified using the modified Mercalli intensity scale, which is based on reports of felt intensity of vibrations and damage to structures.

8. Earthquakes can be devastating in terms of loss of life and damage to structures. Factors such as population density, building codes and construction styles, and emergency preparedness are important in determining earthquake risk. Engineers are continually refining their knowledge about how structures respond to intense ground motion.

9. Seismic hazard mapping is based on known geologic structures, especially faults, as well as on records of previous earthquakes and calculations of the probability that ground motion will occur in a given location.

10. Primary hazards associated with earthquakes include ground motion, faulting, and ground rupture. The original shock may also trigger aftershocks. Secondary and tertiary hazards associated with earthquakes include fires, landslides, liquefaction, changes in ground level, tsunamis, and flooding.

11. Ninety-five percent of all earthquakes originate in the circum-Pacific seismic belt (80%) and the Mediterranean seismic belt (15%). The remaining 5 percent are widely distributed along the midocean ridges and elsewhere.

12. Zones of seismic activity define active plate boundaries. Each type of plate boundary is characterized by earthquakes that are distinctive in terms of fault motions and depth of foci. Very large earthquakes occasionally occur in the normally stable continental intraplate environment.

13. The association of earthquakes with the tectonic cycle and with specific geologic environments is the first step in the long-term prediction of earthquakes. Paleoseismology is used to determine the average interval between earthquakes in a given region. The most powerful tool in long-term prediction is the identification of seismic gaps—that is, regions along a fault where no earthquakes are occurring even though elastic energy is accumulating.

14. Short-term prediction of earthquakes is based on observations of precursor anomalies such as sudden changes in rock magnetic or electrical properties, levels of water or amounts of radon in wells, foreshocks, tilting or bulging of the ground surface, and even anomalous animal behavior. Short-term prediction is difficult because earthquake foci are deep underground and therefore are difficult to study and monitor, and because earthquakes are inconsistent in terms of precursor phenomena.

15. Someday it may be possible to control earthquakes by finding a technique through which built-up energy can be released slowly and safely. Possibilities include increasing fluid pressure along a fault and setting off controlled explosions.

IMPORTANT TERMS TO REMEMBER

aftershock (p. 88)
body wave (p. 75)
compressional wave (p. 75)
elastic rebound theory (p. 72)
epicenter (p. 76)
fault (p. 70)
focus (p. 76)
foreshock (p. 95)
liquefaction (p. 89)

modified Mercalli intensity scale (p. 79)
normal fault (p. 71)
P wave (p. 76)
paleoseismology (p. 95)
reverse fault (p. 71)
Richter magnitude scale (p. 78)
S wave (p. 76)
seismic belt (p. 92)
seismic gap (p. 95)

seismic wave (p. 75)
seismic sea wave (p. 90)
seismogram (p. 75)
seismograph (p. 73)
seismology (p. 73)
shear wave (p. 76)
strike-slip fault (p. 71)
surface wave (p. 75)

QUESTIONS AND ACTIVITIES

1. Do you live in a seismically active area? Are you sure? What is the nature of the geologic setting in which you live? Start your investigation by examining the seismic hazard maps in Fig. II.2, 3.18, and 3.19. You can find out more about the geology of your area by contacting the state, provincial, or national geologic survey or by visiting the library at your college or university and looking through the geology periodicals.

2. If there is a chance—even a remote one—of an earthquake occurring in your area, you might want to learn the answers to these questions: Do building codes and zoning regulations reflect the seismic hazard? Are possible secondary hazards, such as landslides, taken into account? Have emergency personnel rehearsed any response procedures? Would you know what to do if an earthquake occurred in your town? What steps might you take to prepare for such an event? Where would you go for emergency assistance?

3. Why are some earthquakes so devastating, whereas other, equally strong ones, are not? Write a brief account comparing the Latur, India earthquake of 1993 and the Northridge, California earthquake of 1994. What were the most significant differences between these two earthquakes in terms of the geologic setting and the impacts on people in the area?

4. Why do you think seismologists are so cautious about making predictions? Are their reasons valid, in your opinion? Write a fictional account of a predicted earthquake that never came to pass. What effects might a false prediction have? If you were a seismologist and thought you had evidence that an earthquake was imminent, would you issue a prediction? Consider this question from the points of view of science and society (including the public as well as leaders and decision makers) and from a personal perspective.

VOLCANIC ERUPTIONS

A blast of burning sand pours out in whirling clouds.
Conspiring in their power, the rushing vapours
Carry up mountain blocks, black ash, and dazzling fire.

• Lucilius Junior (A.D. 50)

*I*n Indonesia during the summer of 1883 an apparently dormant volcano, Krakatau, began emitting steam and ash. On Sunday, August 26, the volcano's activity increased, and on the next day Krakatau literally blew up and disappeared. As a telegram of the time tersely reported, "Where once Mount Krakatau stood, the sea now plays." The explosion was heard on an island in the Indian Ocean, 4600 km away. Gigantic waves 40 m high spread out from the site of the explosion and crashed into the shores of Java, where more than 36,000 people lost their lives.

The effects of Krakatau were felt around the world. About 20 km^3 of volcanic debris was ejected during the eruption, with some particles blasted as high as 50 km into the stratosphere. Within 13 days the stratospheric dust had encircled the globe, and for months afterward there were strangely colored sunsets—sometimes green or blue and other times scarlet or flaming orange. One November sunset over New York City looked so much like the glow from a massive fire that numerous calls were made to fire departments. The suspended dust made the atmosphere so opaque to the Sun's rays that the average temperature around the Earth dropped at least 0.5°C during 1884. Within 5 years all the dust had fallen to the ground and the climate returned to normal conditions.

Every now and then a truly devastating eruption reminds us of the enormous magnitude of volcanic forces. The 1991 eruption of Mount Pinatubo in the Philippines stands out as one such event. Although the volcano had slumbered for more than 400 years, there were ominous signs of renewed activity in April 1991. By the end of the month the activity was increasing in intensity, and Filipino geologists requested the assistance of colleagues from the U.S. Geological Survey. From April to June the joint team of geologists monitored Pinatubo's activity and worked frantically to delineate the hazard and issue warnings to area residents. In the end, the top of the volcano was blasted off in an immense explosion of gas, ash, and rock fragments. The monitoring team had successfully predicted the date of the eruption. At least 58,000 people were evacuated and tens of thousands of lives saved.

VOLCANOES AND MAGMA

Magma is molten rock, sometimes containing suspended mineral grains and dissolved gases, that forms when temperatures rise sufficiently high for melting to occur in the Earth's crust or mantle. When magma reaches the Earth's surface, it does so through a **volcano,** a vent from which magma, solid rock debris, and gases are erupted. The term *volcano* comes from Vulcan, the Roman god of fire. For most people, the thought of a volcano conjures up visions of incandescent **lava**—magma that reaches the Earth's surface and pours out over the landscape.

Although lava often takes the form of hot, flowing streams, magma can also be erupted in other forms, such as clouds of tiny red-hot fragments, or large clots of molten material blasted high into the air. In fact, volcanoes and the magma that erupts from them are much more varied than is commonly realized. This chapter begins by giving some insight into the properties of magma and how they influ-

Mt. Pinatubo, Philippines, during the eruption cycle of 1991. A vast cloud of hot, dust-laden gas has rolled down the flanks of the volcano and is spreading rapidly across the surrounding plains. The car and driver escaped harm, but the houses, trees, and fields were smothered with a blanket of hot, volcanic dust.

ence the style and explosiveness of volcanic activity. We then proceed to a discussion of volcanic eruptions and landforms; the regions where different types of volcanoes occur; the hazards and benefits associated with volcanoes; and finally how volcanologists monitor volcanoes and predict eruptions.

Characteristics of Magma

The characteristics and properties of magma have a significant influence on the style and, in particular, the explosiveness of volcanic eruptions. By observing and studying eruptions of lava, we can draw three important conclusions concerning magma:

1. Magma is characterized by a *range of compositions* in which silica (SiO_2) is predominant. Most magma contains some dissolved gases, as well as liquid (molten rock) and crystals.

2. Magma is characterized by *high temperatures.*

3. Magma has the properties of a liquid, including the *ability to flow.* This is true even though some magma is almost as stiff as window glass.

In the rest of this section we will examine each of these observations in greater detail.

Composition

The composition of magma is controlled by the most abundant elements in the Earth—Si, Al, Fe, Ca, Mg, Na, K, H, and O. (As we learned in Chapter 2, O^{2-} is the most abundant anion. Therefore it is usual to express the compositional variations of magmas in terms of oxides, such as SiO_2, Al_2O_3, CaO, and H_2O.) The most abundant component of magma is SiO_2.

The most common types of magma are *basaltic, andesitic,* and *rhyolitic* magmas. The first type contains about 50 percent SiO_2, the second about 60 percent, and the third about 70 percent. Three types of volcanic (extrusive) rocks—**basalt, andesite,** and **rhyolite** (Table 4.1)—are derived from these three types of magma.

For each of the common volcanic rocks there is a corresponding intrusive or plutonic rock. As discussed in Chap-

ter 1, plutonic rocks form from magmas that never make it to the surface of the Earth, instead cooling and crystallizing more slowly underground. The plutonic rock types corresponding to basalt, andesite, and rhyolite are *gabbro, diorite,* and *granite,* respectively.

The three common types of magmas are not formed in equal abundance. Approximately 80 percent of all magma erupted by volcanoes is basaltic, whereas andesitic and rhyolitic magmas each account for about 10 percent of the total. Hawaiian volcanoes such as Kilauea and Mauna Loa are basaltic; Mount St. Helens, Mount Pinatubo, and Krakatau are andesitic volcanoes; and the now dormant volcanoes at Yellowstone National Park are rhyolitic. As we will see later in the chapter, the origin and distribution of these different kinds of volcanoes are largely a function of plate tectonics.

Dissolved Gases

Small amounts of gas (0.2 to 3 percent by weight) are dissolved in all magmas. Even though they are present in very low abundance, these gases can strongly influence the properties of the magma, which in turn influence the style and explosiveness of the eruption.

As you can probably imagine, volcanic gases are particularly difficult to study. Aside from the dangers associated with eruptions and high temperatures, volcanic gases can be toxic or caustic. It is also exceedingly difficult to obtain a sample of volcanic gas without its becoming contaminated by the surrounding air (Fig. 4.1). According to popular accounts, two Russian volcanologists collected gas samples while riding down a river of lava on a "raft" of solidified volcanic rock during an eruption of Bilinkai, a volcano on the peninsula of Kamchatka. Other investigators have been less fortunate. In A.D. 79 the Roman statesman Pliny the Elder died, possibly from gas poisoning or asphyxiation, while trying to collect samples of volcanic gas during an eruption of Mount Vesuvius. Despite the obstacles, however, volcanologists have managed to gather large amounts of information about volcanic gas emissions.

The principal volcanic gas is water vapor, which, together with carbon dioxide, accounts for more than 98 percent of all gases emitted from volcanoes. Other volcanic

T A B L E 4.1 • Volcanic Rock Name, Plutonic Rock Name, Silica Content, Viscosity, and Melting Temperature for the Three Main Types Of Magma

Volcanic Rock Name and Lava Type	Corresponding Plutonic Rock Name	Silica Content (approx.)	Relative Viscosity	Melting Temperature (experimental) (°C)
Basalt	Gabbro	≈50%	Low	≈1400
Andesite	Diorite	≈60%	Intermediate	≈1100
Rhyolite	Granite	≈70%	High	≈ 800

Geologist in a fire- and heat-resistant garment moving in to collect samples of volcanic gas during an eruption of Mauna Loa volcano, Hawaii. The highest temperature recorded, 1,145°C, was measured at the time the gas samples were taken on the bright orange region in the center of the photo.

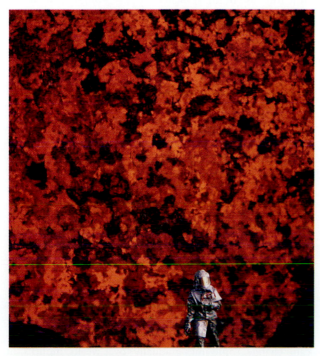

▲ **F I G U R E 4.2**
Fountaining starts an eruption of Krafla, a basaltic volcano in Iceland. Use of a telephoto lens foreshortens the field of view. The geologist in a protective suit is making measurements several hundred meters away from the fountain.

gases include nitrogen, chlorine, sulfur, and argon, which are rarely present in amounts exceeding 1 percent. Throughout geologic history, volcanic gas emissions have been the primary mechanism whereby the Earth has released the volatile material from its interior. The chemical evolution of the atmosphere and the origin of the oceans are intimately linked to this outgassing process.

Temperature

It is difficult to measure the extrusion temperature of magma, but it can sometimes be done during volcanic eruptions. For obvious reasons, such measurements are usually made from a distance using optical devices (Fig. 4.2). Magma temperatures during eruptions of volcanoes such as Kilauea in Hawaii and Mount Vesuvius in Italy have been recorded as ranging from 800 to 1200°C. Experiments on synthetic magmas in the laboratory suggest that under some conditions magma temperatures might be as high as 1400°C (Table 4.1).

Viscosity

The degree to which a substance is resistant to flow is its **viscosity;** the more viscous a substance, the less fluid it is. The viscosity of a magma depends on its temperature and composition, especially its silica and dissolved gas content. Dramatic film footage of lava flowing rapidly down the sides of a volcano prove that some magmas are very fluid

▲ F I G U R E 4.3
This stream of low-viscosity basaltic lava moving smoothly away from an eruptive vent demonstrates how fluid and free flowing lava can be. The temperature of the lava is about 1100°C. The eruption occurred in Hawaii in 1983.

(Fig. 4.3). Low-silica basaltic lava moving down a steep slope on Mauna Loa in Hawaii has been clocked at 64 km/h, indicating a very low viscosity. On the other hand, rhyolitic magma containing 70 percent or more SiO_2 flows so slowly that its movement can hardly be detected. The properties of such magma are more akin to those of solids than to those of liquids.

The SiO_4^{4-} anions that form the foundation for most common minerals are also present in magmas. Just as they do in minerals, these anions link together or *polymerize* by sharing oxygens. Unlike the anions in silicate minerals, however, those in magma form irregularly shaped groupings of chains, sheets, and networks. As the polymerized groups become larger, the magma becomes more viscous—that is, more resistant to flow—and behaves increasingly like a solid. The higher the silica content, the larger the polymerized groups. For this reason, high-silica rhyolitic magma is always more viscous than low-silica basaltic magma, whereas andesitic magma has a viscosity somewhere between those of rhyolitic and basaltic magma (Table 4.1).

Temperature and gas content also affect the viscosity of magmas. The higher the temperature and, in general, the higher the gas content, the lower the viscosity and the more readily the magma flows. A very hot magma erupted from a volcano may flow readily, but it soon begins to cool, becomes more viscous, and eventually comes to a halt. In Fig.

▲ F I G U R E 4.4
The flow of lava is controlled by viscosity. Two different flows are visible here. They have the same basaltic composition. The lower flow, on which the geologist is standing, is a pahoehoe flow formed from a low-viscosity lava such as that shown in Fig. 4.3. The upper flow (the one being sampled by the geologist), which is very viscous and slow moving, is an aa flow that erupted from Kilauea in 1989. The pahoehoe flow erupted in 1959.

4.4 the smooth, ropy-surfaced lava, called *pahoehoe* (a Hawaiian word), formed from a hot, gas-charged, very fluid lava. The rubbly, rough-looking lava, by contrast, formed from a cooler, gas-poor lava with a high viscosity. Scientists call this rough lava *aa* (another Hawaiian word). Interestingly, these two types of lava are so distinctive that there are two equivalent words in the Icelandic language: *helluhraun* (for pahoehoe-type lavas) and *apalhraun* (for aa-type lavas).

VOLCANIC ERUPTIONS

It is impossible to produce a strict classification of volcanic eruptions. During most eruptive episodes the types of activity and the nature of the materials ejected from the volcano change, sometimes gradually (over weeks, months, or years) and sometimes from one day to the next or even from hour to hour. Nevertheless, it is possible to categorize volcanic eruptions on the basis of the volcano's eruptive style, the types of materials erupted, and the type of edifice or landform built up by the volcano. A general classification scheme is shown in Table 4.2. The table divides vol-

T A B L E 4.2 • Types of Eruptions

Eruptive Style	Characteristics	Type of Landform and Products	Examples
Principle Kinds of Central Eruptions			
Hawaiian	Discharge of fluid basaltic magma from vents within summit caldera and fissures on flanks Minor pyroclastics Fountaining Nonexplosive	Shield volcano Lava flows Calderas Lava tubes Spatter cones	Mauna Loa Kilauea
Strombolian	Mild explosive eruptions of incandescent bombs, ash, and lapilli Extrusion of lava from vents on flanks Fumarolic activity	Spatter cones Cinder cones	Stromboli Parícutin
Vulcanian	Viscous, silica-rich, gas-rich magmas Dark eruption clouds composed of ash and lapilli fragments Pyroclastic flows	Pumiceous cones surrounded by wide sheets of fine pyroclastics and flows of glassy lava	Vulcano Bárcena
Peléan	Viscous, silica-rich, gas-rich magmas Violent, destructive eruptions Glowing avalanches	Steep-sided domes Short, thick flows Hot flows of blocks and ash	Mont Pelée Santiaguito Domes
Plinian (or Krakatoan)	Exceptionally powerful, continuous gas blast Gas-rich, siliceous magma Voluminous pyroclastics, glowing avalanches Summit caldera Can be interlayered with flows	Composite volcanoes (stratovolcanoes)	Vesuvius Krakatau Mount Mazama Pinatubo
Surtseyan (or phreatomagmatic)	Rising magma comes into contact with groundwater or seawater Violent explosive eruptions	Ash cones Cinder cones	Surtsey Taal
Steam-blast (or phreatic)	Immensely powerful blasts No fresh magma is discharged, only fragments of solid rock mixed with steam Result from groundwater coming into contact with hot rocks at depth	Domes Collapse caldera Glowing avalanches Hot lahars	Arenal Bezymianny Lassen Peak Vesuvius Mount St. Helens
Other Kinds of Eruptions			
Fissures related to central vents	Caused by local or regional stresses on cones or shield volcanoes Nature of eruption and composition of magma can vary	Radial fractures and lines of vents, often related to central vents	Laki Hekla
Fissures unrelated to central vents	Never witnessed in historic times Pyroclastic flows; pumice mixed with lapilli and ash	Linear or arcuate fissures Form vast sheets of welded siliceous tuffs (ignimbrites)	Valley of Ten Thousand Smokes Yellowstone Plateau
Plateau (flood) basalts	Most voluminous of all volcanic eruptions Fluid basaltic lavas All erupted during an interval of 2 to 3 million years about 15 million years ago	Extensive sheets of basalt Swarms of parallel fissures and multiple coalescing vents Fields of low, coalescing shield volcanoes	Columbia Plateau Deccan Plateau
Oceanic volcanism	Related to "hot spots" or sea-floor spreading along mid-ocean rifts Fluid basaltic lavas Essentially no pyroclastic debris	Pillow lavas Glassy flows Extensive fissure systems related to seafloor spreading	Mid-Atlantic Ridge East Pacific Rise Hawaiian–Emperor Seamount chain

[a] The central eruptions increase in explosiveness from top to bottom in the table. Note that a single volcano could combine several of these explosive styles, varying either over the long term or from day to day or hour to hour.

canic eruptions into those related to a central volcanic vent and those related to long, linear or arcuate fractures in the Earth's crust, called **fissures.**

Factors Influencing Eruptive Styles

Like most other liquids, magma is less dense than the solid rock from which it is formed. Once formed, therefore, the lower density magma exerts upward pressure on the enclosing higher density rock and slowly forces its way up. There is, of course, pressure on a rising mass of magma created by the weight of the overlying rock. This pressure is proportional to depth. As a magma rises, the pressure of the overlying rock decreases.

Pressure controls the amount of gas a magma can dissolve—more gas at high pressures, less at low pressures. Gas dissolved in a rising magma acts like gas dissolved in soda water. When a bottle of soda is opened, the pressure inside the bottle drops, gas comes out of solution, and bubbles form. Gas dissolved in an upward-moving magma also comes out of solution and forms bubbles. What happens to the bubbles is determined by the viscosity of the liquid. Thus, dissolved gases and viscosity are the two factors that are most influential in determining the eruptive style and explosiveness of a volcano.

Nonexplosive Eruptions

Understandably, people tend to regard any volcanic eruption as a hazardous event and to view active volcanoes as dangerous places that should be avoided. However, geologists have discovered that some volcanoes are comparatively safe and relatively easy to study. Nonexplosive eruptions, such as those we can witness in Hawaii, are relatively safe compared to violent, explosive events like the 1980 eruption of Mount St. Helens in Washington, the 1982 eruption of El Chichón in Mexico, and the 1991 eruption of Mount Pinatubo in the Philippines, each of which caused substantial destruction and loss of life. The differences between nonexplosive and explosive eruptions are largely a function of magma viscosity and dissolved gas content. Nonexplosive eruptions are associated with low-viscosity magmas and low dissolved-gas contents.

Even nonexplosive eruptions may appear violent during their initial stages. Gas bubbles in a low-viscosity basaltic magma rise rapidly upward like the gas bubbles in a glass of soda water. If a basaltic magma rises rapidly, resulting in a rapid decrease in pressure, gas can bubble out of solution so fast that spectacular fountaining occurs (Fig. 4.2).

When fountaining dies down, hot, fluid lava emerging from the vent flows rapidly downslope (Fig. 4.3). Because heat is lost quickly at the top of a flow, the surface forms a crust beneath which the liquid lava continues to flow downslope along well-defined channels. These enclosed lava tubes inhibit the upward loss of heat and enable a low-viscosity lava to move along just below the surface for great distances. As the lava cools and continues to lose dissolved gases, its viscosity increases and the character of the flow changes. The very fluid lava initially forms thin pahoehoe flows, but with increasing viscosity the rate of movement slows and the stickier lava may be transformed into a clinkery aa flow that moves very slowly. Thus, during a single, nonexplosive, Hawaiian-type eruption both pahoehoe and aa flows may be formed from the same batch of magma (Fig. 4.4).

As a basaltic lava cools and its viscosity rises, it becomes increasingly difficult for gas bubbles to escape. When the lava finally solidifies into rock, the last-formed bubbles become trapped, forming bubble holes called *vesicles.* The term *vesicular* is used to describe the texture produced by vesicles in an igneous rock (Fig. 4.5). In many vesicular basalts the vesicles are later filled with minerals deposited by heated groundwater.

Explosive Eruptions

In andesitic or rhyolitic magmas, gas bubbles can rise only very slowly because they are held back by the viscosity of the fluid. As these magmas move upward toward the Earth's surface, the decrease in pressure eventually allows the dissolved gas to expand and escape explosively. Bubbles that form quickly in a huge mass of sticky rhyolitic magma can shatter into a froth of innumerable tiny glass-walled bubbles, producing a rock called *pumice.* Some pumice has such a low density that it will float in water. Beaches on mid-Pacific islands are often littered with pumice that has floated in on currents from distant volcanic eruptions.

When little or no dissolved gas is present, a magma will be erupted as a lava flow regardless of its composition. If dissolved gas is present, however, it must escape somehow. The higher the viscosity, the more difficult it is for the gas to form bubbles and the greater the likelihood the escaping gas bubbles' causing an explosive eruption. We can extend

▲ **F I G U R E 4.5**
Vesicular basalt. The holes in the rock were left by small bubbles of gas in the magma. The specimen is 4.5 cm across.

T A B L E 4.3 • Names for Tephra and Pyroclastic Rock

Average Particle Diameter (mm)	Tephra (unconsolidated material)	Pyroclastic Rock (consolidated material)
>64	Bombs	Agglomerate
2–64	Lapilli	Lapilli tuff
<2	Ash	Ash tuff

the analogy of gas dissolved in soda by imagining that the soda bottle is full of a very thick, viscous liquid—such as wet cement—with gas held in solution under pressure. If we open the bottle and cause the pressure to drop very rapidly, gas bubbles will quickly form and escape—but with much more vigorous spattering of the liquid than would occur if the bottle had contained soda.

Tephra and Pyroclastic Rocks

A fragment of rock ejected during a volcanic eruption is called a **pyroclast** (from the Greek words *pyro*, meaning heat or fire, and *klastos*, meaning broken; hence, hot, broken fragments). Rocks formed from pyroclasts are **pyroclastic rocks.** Geologists also commonly refer to pyroclasts as **tephra,** a Greek word for ash. Tephra is employed as a collective term for all airborne pyroclasts, including fragments of newly solidified magma as well as fragments of older broken rock. It includes individual pyroclasts that fall directly to the ground and those that are transported over the ground as part of a hot, moving flow. Abundant tephra and pyroclastic materials are characteristic of violent, explosive eruptions.

The terms used to describe tephra of different size— *bombs, lapilli,* and *ash*—are listed in Table 4.3 and illustrated in Fig. 4.6. The term *volcanic ash,* referring to the fine pyroclastic material thrown out by volcanoes, is somewhat misleading. Strictly speaking, ash is the solid matter

A.

B.

C.

▲ **F I G U R E 4.6**
Tephra. A. Large spindle-shaped bombs up to 50 cm long cover the surface of a tephra cone on Haleakala volcano in Maui. B. Intermediate-sized tephra called lapilli cover the Kau Desert, also located in Hawaii. The coin is about 1 cm in diameter. C. Volcanic ash, the smallest tephra, blankets a farm in Oregon after the eruption of Mount St. Helens in 1980.

that remains after something flammable, such as wood, has burned. However, fine tephra looks so much like true ash that it has become customary to use the word for this material also.

Volcanologists are fond of saying that tephra is igneous on the way up but sedimentary on the way down, meaning that the pyroclasts are ejected from the volcano in a molten state and solidify as igneous rocks in midair, but are deposited on the ground in the form of sedimentary fragments. As a result, pyroclastic rocks are a transitional form between igneous and sedimentary rocks. The names of pyroclastic rocks, like those of sedimentary rocks, are keyed to the size of the mineral grains of which they are composed. Pyroclastic rocks are called *agglomerates* when the tephra particles are large (i.e., bomb sized), and *tuffs* when they are smaller, either lapilli or ash sized (Table 4.3). Unconsolidated tephra can be converted into pyroclastic rock in two ways. The first, and most common, way is through the addition of a cementing agent such as quartz or calcite introduced by groundwater. The second way is through the welding of hot, glassy ash particles. When ash is very hot and plastic, the individual particles can fuse together to form a glassy pyroclastic rock.

Eruption Columns and Tephra Falls

The largest and most violent explosive eruptions are associated with silica-rich magmas that are high in dissolved gases. As the rising magma approaches the surface, rapid decompression causes the gases to expand and produces the violent upward thrust of a dense mixture of hot gas and tephra. This hot, turbulent mixture rises rapidly in the cooler air above the vent to form an **eruption column** (sometimes called a *Plinian column*) that may reach as high

as 45 km in the atmosphere. The photo at the beginning of the chapter shows the base of a huge eruption column rising from Mount Pinatubo in the Philippines.

Eruption clouds are typically dark, tempestuous, ominous-looking masses. Their internal dynamics are complex: the heat of the cloud and the force of the eruption create a tendency to rise, but the density of the ash produces a tendency toward collapse. At a height at which the density of the material in the column equals that of the surrounding atmosphere, the column will spread out laterally to form a mushroom-shaped cloud of the type seen in pictures of nuclear bomb explosions. As the cloud begins to drift with the upper atmospheric winds, particles of debris fall out and eventually accumulate on the ground as tephra deposits (Fig. 4.7). During exceptional explosive eruptions, tephra can be spread over distances of 1500 km or more.

Pyroclastic Flows

A hot, highly mobile flow of tephra that rushes down the flank of a volcano during a major eruption is called a **pyroclastic flow.** Often the material in the flow is dense, gas-charged, and so hot that it incandesces, in which case it may be referred to as a **glowing avalanche** (or *glowing ash cloud* or *nuée ardente*). These are among the most devastating and lethal forms of volcanic eruptions. The historical record of pyroclastic flows reveals that they can travel 100 km or more from source vents and reach velocities of more than 700 km/h. Pyroclastic flows can be caused by the gravitational or explosive collapse of a mass of hot lava near the top of a volcano, which produces a dense, downrushing mass of blocks, lapilli, ash, and hot gases. Geologists call the resulting poorly sorted deposit an *ignimbrite*. Pyroclastic flows can also be generated by the partial or continuous

A.

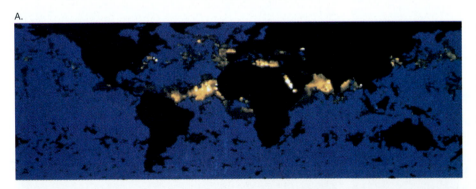

B.

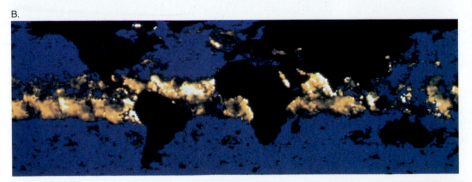

◄ F I G U R E 4.7
The ash-cloud from the June, 1991 eruption of Mt. Pinatubo, Philippines, spreading around the world. A. Image prior to the eruption. Optical density, a measure of particulate matter suspended in the atmosphere, is determined by satellite. High density areas (bright yellow) are regions of high population density or places where dust storms or some other local disturbance is producing a high level of suspended aerosols. B. After the eruption, between July 4 and 10, 1991, a bright band encircles the globe due to the spread of particulate matter erupted high into the stratosphere by Mt. Pinatubo.

collapse of an eruption column. During the 1980 eruption of Mount St. Helens, for example, a number of hot (850°C) pyroclastic flows, probably caused by column collapse, traveled up to 8 km down the north side of the mountain and covered an area of about 15 km².

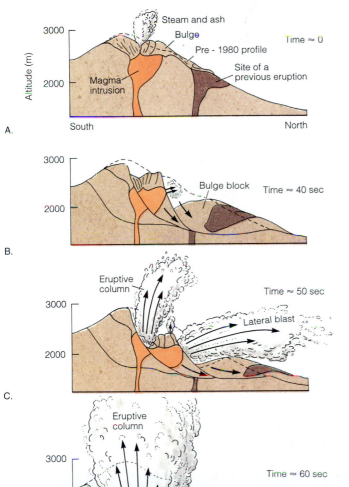

A.

B.

C.

D.

▲ F I G U R E 4.8

The sequence of events leading to the eruption of Mount St. Helens in May 1980. A. Earthquakes and then puffs of steam indicate that magma is rising; a small crater forms, and the north face of the mountain bulges alarmingly. B. On the morning of May 18, an earthquake shakes the mountain and the bulge breaks loose and slides downward. This reduces the pressure on the magma and initiates the lateral blast. C. The violence of the eruption causes a second block to slide downward, exposing more of the magma and initiating an eruption column. D. The eruption increases in intensity. The eruption column carries volcanic ash as high as 19 km into the atmosphere.

Lateral Blasts

The 1980 eruption of Mount St. Helens displayed many of the features of a typical large explosive eruption. Nevertheless, the magnitude of the event caught geologists by surprise. The events leading to this eruption are shown diagrammatically in Fig. 4.8.

As magma moved upward under the volcano, the north flank of the mountain began to bulge upward and outward. Finally the slope became unstable, broke loose, and slid toward the valley as a gigantic landslide of rock and glacier ice. The landslide exposed the mass of hot magma in the core of the volcano. With the lid of rock removed, dissolved gases underwent such rapid decompression that a mighty blast resulted, blowing a mixture of pulverized rock and hot gases sideways as well as upward (Fig. 4.9). The sideways eruption, referred to as a **lateral blast,** initially traveled at the speed of sound. It roared across the landscape, killing every living thing in the blast zone. Within the devastated area, which extends as much as 30 km from the crater and covers some 600 km², trees in the formerly dense forest were blasted to the ground and covered with hot debris.

▲ F I G U R E 4.9

Plinian ash column from the May 1980 eruption of Mount St. Helens. At least 63 people died as a result of this eruption.

VOLCANIC LANDFORMS

During its lifetime a volcano builds up a volcanic landform or edifice whose shape is determined by the composition and characteristics of the volcanic materials and the types of eruptions that have occurred. These are summarized in Table 4.2. In this section we discuss each of the major types of volcanoes and the landforms it creates.

Shield Volcanoes

Perhaps the easiest kind of volcano to visualize is one built up of successive flows of fluid lava. Such lavas are capable of flowing great distances down gentle slopes, forming thin sheets of nearly uniform thickness. Eventually the pile of lava builds up a **shield volcano,** a broad formation, resembling a shield lying horizontally, with an average slope of only a few degrees (Fig. 4.10). Shield volcanoes are characteristically formed by the eruption of basaltic lava; the proportions of ash and other tephra are small. Hawaii, Tahiti, Samoa, the Galapagos, and many other oceanic islands are actually the upper portions of large shield volcanoes. The Hawaiian volcano Mauna Kea rises only about 4.2 km above sea level, but its true height when measured from the sea floor exceeds that of Mount Everest, making it the tallest mountain on Earth.

The slope of a shield volcano is slight near the summit because the magma is hot and very fluid; it will readily run down even a very slight slope. The farther the lava flows down the volcano's flanks, the cooler and more viscous it becomes and the steeper the slope must be in order for it to flow. Slopes on young, growing shield volcanoes, such as Kilauea in Hawaii, typically range from less than 5° near the summit to 10° on the flanks.

Stratovolcanoes

Stratovolcanoes, also called *composite volcanoes,* emit both tephra and viscous lava and build up steep, conical mounds of interlayered lava and pyroclastic deposits. They are most often composed of andesitic material. The volume of tephra in a stratovolcano may equal or exceed the volume of the lava.

The slopes of stratovolcanoes, which may be thousands of meters high, are steep. Near the summit the slope is typically about 30°, whereas toward the base the slope flattens to about 6° to 10°. The steep slope near the summit of a stratovolcano is due in part to the short, viscous lava flows that are erupted, and in part to tephra. The lava flows are a major factor in distinguishing between *tephra cones* (described below) and stratovolcanoes. As a stratovolcano develops, lava flows act as a cap to slow down erosion of the loose tephra, and as a result the volcano becomes much larger than a typical tephra cone.

Virtually all major continental volcanoes are stratovolcanoes. In general, however, stratovolcanoes are much smaller than the great oceanic shield volcanoes. For example, the total volume of Mauna Loa, the Earth's largest shield vol-

▼ F I G U R E 4.10
Mauna Kea, a 4200-m-high shield volcano in Hawaii, as seen from Mauna Loa. Note the gentle slopes formed by highly fluid lava. The view is almost directly north. A pahoehoe flow is in the foreground on the northeast flank of the volcano.

▲ F I G U R E 4.11
Mount Fuji, Japan, a snow-clad giant towering over the surrounding countryside, displays the classic profile of a stratovolcano.

cano, is more than 300 times that of Mount Fuji in Japan, one of the most voluminous of all stratovolcanoes.

The beautiful, steep-sided cones of stratovolcanoes are among the most picturesque sights on Earth (Fig. 4.11). The snow-capped peak of Mount Fuji has inspired poets and writers for centuries. Mount Rainier and Mount Baker in Washington and Mount Hood in Oregon are majestic examples of such volcanoes in North America.

Other Features of Volcanoes

Tephra Cones

Rhyolitic and andesitic volcanoes tend to eject large quantities of tephra. As the debris showers down, a **tephra cone** builds up around the vent (Fig. 4.12). The slope of the cone is determined by the size of the tephra. Fine ash will stand at a slope angle of 30° to 35°, whereas lapilli generally stand at an angle of about 25°. The gradual decrease in the

A.

B.

► F I G U R E 4.12
Tephra cones. A. Two small tephra cones forming as a result of an eruption in Kivu, Zaire. Arcs of lights are caused by the eruption of red-hot lapilli and bombs. B. Tephra cone in Arizona built from lapilli-sized tephra. Note the small basaltic lava flow coming from the base of the cone.

◄ **F I G U R E 4.13**
Volcanic gas streams from a vent follow-ing a 1965 fissure eruption in Hawaii. Sulfur condensed from the acrid gases kills vegetation and covers the ground with a yellow coating of elemental sulfur.

volume of fallout material at greater distances from the vent leads to gentler slopes near the base of the cone.

Geysers, Fumaroles, and Thermal Springs

Gases that bubble up from magma far below the surface may emerge either from a central volcanic vent or from small satellite vents on the flanks of the volcano. A vent or fracture from which volcanic gases are emitted is referred to as a **fumarole.** The gases tend to be mostly water vapor, but other gases, such as evil-smelling sulfurous gas, may be present as well. Such gases can alter and discolor the rocks with which they come into contact (Fig. 4.13).

When active volcanism finally ceases, the rock in an old magma chamber may remain hot for hundreds of thousands of years. Descending groundwater that comes into contact with the hot rock is heated and tends to rise to the surface along rock fractures, forming a **thermal spring.** Thermal springs at volcanic sites in Italy, Iceland, Japan, and New Zealand, as well as many other locations, have become famous health spas.

A thermal spring equipped with a natural system of plumbing and heating that causes intermittent eruptions of water and steam is a **geyser.** The name comes from the Icelandic word *geysir,* meaning to gush, for Iceland is the home

◄ **F I G U R E 4.14**
The great Geysir, Iceland, from which all geysers take their name.

◄ **F I G U R E 4.15**
Crater Lake, Oregon, occupies a caldera 8 km in diameter that crowns the summit of a once lofty stratovolcano called Mount Mazama. Wizard Island, a small tephra cone and related lava flows, was formed after the collapse that created the caldera.

of many geysers (Fig. 4.14). Most of the world's geysers outside Iceland are located in New Zealand and in Yellowstone National Park in the United States.

Craters and Calderas

Near the summit of most volcanoes is a **crater,** a funnel-shaped depression opening upward, from which gases, tephra, and lava are ejected. However, many volcanoes, both shield and stratovolcanoes, have a much larger depression known as a **caldera,** a roughly circular, steep-walled basin that may be several kilometers or more in diameter. Calderas are formed as a result of the partial emptying of a magma chamber. Rapid ejection of magma during a large

lava or tephra eruption can leave the magma chamber empty or partly empty. The unsupported roof of the chamber then collapses under its own weight, like a snow-laden roof on a shaky barn, dropping downward on a ring of steep vertical fractures.

Crater Lake in Oregon occupies a circular caldera 8 km in diameter that was formed after a great tephra eruption about 6600 years ago by a volcano posthumously called Mount Mazama (Fig. 4.15). The tephra deposits from that eruption can still be seen at many places in Crater Lake National Park and over a vast area of the northwestern United States and adjacent parts of Canada (Fig. 4.16). After the outpouring of about 75 km^3 of tephra, what remained of

◄ **F I G U R E 4.16**
The Pinnacles, Crater Lake National Park. Striking erosional forms developed in the thick tephra blanket left by the catastrophic eruption of Mount Mazama 6600 years ago.

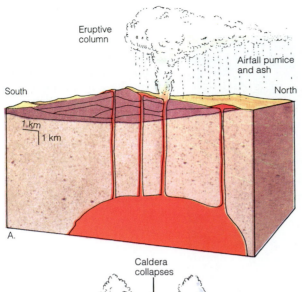

◀ F I G U R E 4.17
The sequence of events that formed Crater Lake following the eruption of Mount Mazama. A. An eruptive column of tephra rises from the flank of Mount Mazama. B. The eruption reaches a climax. Dense clouds of ash fill the air and hot pyroclastic flows sweep down the mountainside. It was at this stage that the deposits shown in Fig. 4.16 were formed. C. The top of Mount Mazama collapses into the partly empty magma chamber, forming a caldera 10 km in diameter. D. During a final phase of eruption, Wizard Island is formed. The water-filled caldera is Crater Lake, shown in Fig. 4.15.

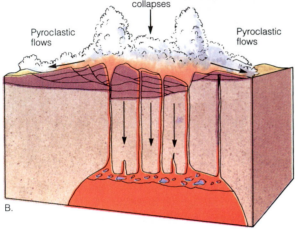

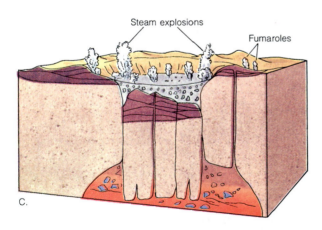

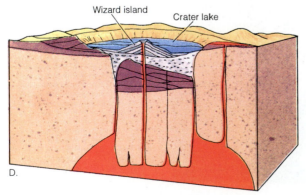

the roof of the magma chamber collapsed into the partly empty chamber (Fig. 4.17).

Resurgent Domes

A volcano does not necessarily become inactive after the formation of a caldera. Magma starts reentering the chamber and in the process causes uplifting of the collapsed floor of a caldera to form a structural dome. Such a feature is called a **resurgent dome.** Subsequently, small tephra cones and lava flows build up in the interior of the caldera. Wizard Island in Crater Lake is a cone that was formed in this way.

When lava is extruded after a major volcanic eruption, it tends to have very little dissolved gas left in it. If the lava is sticky and viscous, it squeezes out to form a **lava dome.** A growing lava dome more than 200 m high can now be seen in the center of the crater of Mount St. Helens (Fig. 4. 18).

Fissure Eruptions

Some lava reaches the surface via elongate fissures through the crust. These events are called **fissure eruptions** (Fig. 4.19). Such eruptions, which can be very dramatic, are characteristically associated with basaltic magma. Successive lavas that emerge as fissure eruptions on land tend to spread widely and may create flat lava plains called *basalt plateaus.* In 1783 a fissure eruption at Laki in Iceland occurred along a fracture 32 km long. Lava flowed 64 km outward from one side of the fracture and nearly 48 km outward from the other side. Altogether the flow covered an area of 588 km^2. The volume of the lava extruded was 12 km^3, the largest lava flow in historic times and also one of the most deadly. The flow destroyed homes and food supplies, killed livestock, and covered fields. In the ensuing famine 9336 people died. There is good evidence to prove that even larger eruptions occurred in prehistoric times. The Roza flow, a great sheet of basaltic lava in eastern Washington State, can be traced over an area of 22,000 km^2 and has been shown to have a volume of 650 km^3.

◄ F I G U R E 4.18
A resurgent lava dome in the crater of Mount St. Helens, Washington, in May 1982. The plume rising above the dome is composed of steam.

VOLCANOES AND PLATE TECTONICS

We come now to some of the most interesting questions concerning magmas and volcanoes: How and where do magmas form, why do the different types of volcanoes occur in distinct tectonic environments, and why are there three major kinds of magmas—basaltic, andesitic, and rhy-

olitic? Many clues can be obtained from the distribution of different kinds of volcanoes. As noted earlier, there is a close relationship between plate tectonics and the locations of the different types of volcanoes. Understanding volcanism in the context of plate tectonics is important not only for improving our scientific knowledge of these processes but also for our ability to assess the hazards associated with volcanoes and to predict eruptions. A summary of present thinking about the distribution of the kinds of volcanoes in

◄ F I G U R E 4.19
Aerial view of a fissure eruption, Mauna Loa, Hawaii, in 1984. Basaltic lava is erupting from a series of parallel fissures. Note the tephra cone (upper left) formed during an earlier eruption.

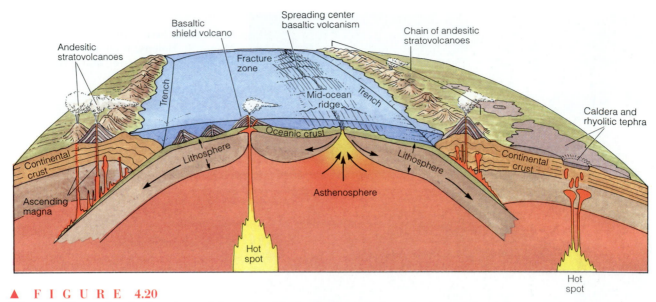

▲ **FIGURE 4.20**
Diagram illustrating the locations of the major kinds of volcanoes in a plate-tectonic setting.

the context of plate tectonics is presented in Fig. 4.20 and in the following discussion.

Global Distribution of Volcanoes

Volcanoes that erupt rhyolitic magma are found primarily on the continental crust, suggesting that the processes that form rhyolitic magma must occur in continental crust but not in oceanic crust. Nor, presumably, do the processes that form rhyolitic magma occur in the mantle, because if they did, the magma would rise to the surface regardless of the kind of crust above it. Rhyolitic magmas are thought to form through partial melting of continental crust. Because the minerals in continental rocks are rich in water and carbon dioxide, as well as silica, the magmas that result from their melting also tend to have a high gas and silica content. Rhyolitic volcanoes therefore tend to erupt explosively. An example is the prehistoric eruption of Mount Mazama discussed earlier.

Volcanoes that erupt andesitic magma are found on both oceanic and continental crust. This suggests that andesitic magma forms in the mantle and rises, regardless of the nature of the overlying crust. An additional piece of information comes from the geographic distribution of andesitic volcanoes (Fig. 4.21). A ring of volcanoes surrounds the Pacific Ocean, forming the so-called *Ring of Fire*. The Ring of Fire, also called the *Andesite Line* by geologists, is home to many of the world's most active and explosive volcanoes, including Mount Pinatubo (Philippines), Mount Unzen and Mount Fuji (Japan), Krakatau and Tambora (Indonesia), and Mount Spurr (Alaska). The Ring of Fire is exactly parallel to the Pacific-rim plate subduction margins shown in Fig. 1.14. Andesitic magma is probably formed as

a result of the melting of old oceanic crust that has been subducted back into the mantle.

Volcanoes that erupt basaltic magma also occur on both oceanic and continental crust. Therefore, the mantle must be the source of basaltic as well as andesitic magma. The geographic distribution of basaltic volcanoes does not seem to be related to specific features of the crust, suggesting that basaltic magma must be formed by melting of the mantle itself, regardless of the kind of crust above it.

Two observations concerning basaltic volcanoes suggest something further about the origin of basaltic magma, however. First, everywhere along the midocean ridges volcanoes erupt basaltic magma. Midocean ridges coincide with plate spreading margins, so we must consider the possibility that plate motion and the generation of spreading-margin magma might somehow be connected. The second observation concerns large basaltic volcanoes that are not located along midocean ridges. An example is found in the volcanoes of Hawaii, which are located on oceanic crust in the middle of the Pacific Plate, far from any plate edges. The Hawaiian volcanoes that are active today—Mauna Loa, Kilauea, and Loihi (a submarine volcano)—are the youngest members of a chain of mostly extinct volcanoes. To the northwest, the volcanic rocks in the Hawaiian chain are progressively older (Fig. 4.22). The Hawaiian volcanic chain is believed to have formed as the Pacific Plate moved slowly northwest across a midocean "hot spot" above which frequent and voluminous eruptions built a succession of volcanoes. The hot spot is thought to have been building basaltic volcanoes on the moving plate for at least 70 million years. The exact causes of hot spot magmas are not yet known, but both spreading-margin magmas and hot-spot magmas are believed to be caused by convection in the mantle.

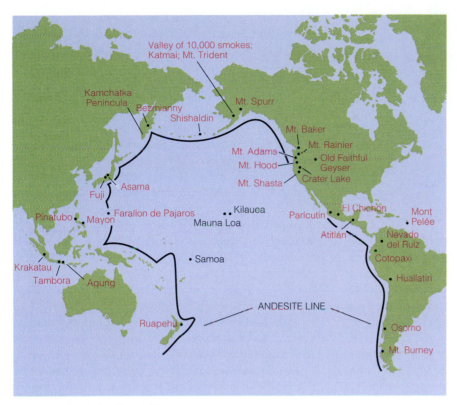

▲ **F I G U R E 4.21**

The Ring of Fire around the Pacific Ocean basin is formed by andesitic volcanoes. The ring, called the Andesite Line by geologists, coincides with subduction zones where lithosphere capped with oceanic crust is being subducted into the asthenosphere. Volcanoes within the Pacific Ocean basin, such as Mauna Loa, erupt basaltic magma but not andesitic magma.

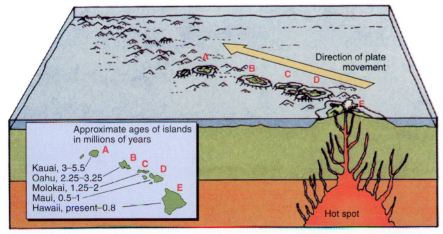

▲ **F I G U R E 4.22**

The Hawaiian Island chain of volcanoes formed above a deep-seated source of basaltic magma within the mantle, called a hot spot. The magma source has remained stationary for at least 70 million years while the Pacific Plate has moved over it, carrying with it the old volcanic landforms built over the hot spot. The Hawaiian volcanoes are progressively older with increasing distance from the hot spot. The currently active volcanoes, Mauna Loa, Kilauea, and Loihi (a submarine volcano), are located on top of the hot spot. Eventually they, too, will be carried off with the moving plate, and a new volcanic edifice will be built over the hot spot.

VOLCANIC HAZARDS

Since A.D. 1800 there have been 18 volcanic eruptions in which a thousand or more people died (Table 4.4). Yet millions of people live in areas where there are active volcanoes and must come to terms with the associated hazards and risks. Like other natural phenomena, volcanic hazards have *primary, secondary,* and *tertiary* effects. In this section we discuss those effects in detail.

Primary Effects of Volcanism

Lava Flows

Most types of volcanoes produce at least some lava, but extensive lava flows are most characteristic of the quieter types of volcanoes such as those found in Hawaii. Because lava flows are closely controlled by topography, it is often possible to predict the general direction and course of a flow. Some lavas can travel downhill at remarkably high velocities. Basaltic lava, for example, can travel as fast as 64 km/h down a steep slope. Such fluidity is rare, however, and rates of flow are more commonly measured in meters per hour or even meters per day. As can be seen in Fig.

4.23, which shows basaltic lava destroying a house in Hawaii, lava flows are usually slow enough that people are not endangered. This means that in dealing with the hazardous effects of lava flows the main focus is on preventing property damage, not on saving lives.

Lava flows are one of the few aspects of volcanism that can be controlled, at least in part, by human intervention. Several methods of controlling lava flows have been tested. In 1935 bombing was tried, with limited success, during an eruption of Mauna Loa in order to spare the city of Hilo from excessive damage. It was attempted again at Mount Etna in Sicily in 1983 and 1992. The goal of bombing is to block or divert the advancing lava flow, either by altering the topography ahead of the flow or by creating a barrier by blocking the channel or causing a lava tube to collapse. The construction of artificial barriers is based on the same principle, that is, creating a blockage and diverting the flow from its natural course. There are numerous examples of preexisting walls that have withstood and even diverted oncoming lava flows. Walls and bulldozed rock barriers constructed for this purpose have been tested—again with somewhat limited success—in Hawaii, Iceland, and Japan.

Hydraulic chilling, first tested during an eruption of Kilauea in 1960, involves spraying water on an advancing lava

T A B L E   4.4 • **Volcanic Disasters Since A.D. 1800 in Which a Thousand or More People Lost Their Lives**

Volcano	Country	Year	Primary Cause of Fatalities			
			Pyroclastic Eruption	Mudflow	Tsunami	Famine
Mayon	Philippines	1814	1,200			
Tambora	Indonesia	1815	12,000			80,000
Galunggung	Indonesia	1822	1,500	4,000		
Mayon	Philippines	1825		1,500		
Awu	Indonesia	1826		3,000		
Cotopaxi	Ecuador	1877		1,000		
Krakatau	Indonesia	1883			36,417	
Awu	Indonesia	1892		1,532		
Soufrière	St. Vincent	1902	1,565			
Mt. Pelée	Martinique	1902	29,000			
Santa Maria	Guatemala	1902	6,000			
Taal	Philippines	1911	1,332			
Kelud	Indonesia	1919		5,110		
Merapi	Indonesia	1930	1,300			
Lamington	Paupa-New Guinea	1951	2,942			
Agung	Indonesia	1963	1,900			
El Chichón	Mexico	1982	1,700			
Nevado del Ruíz	Colombia	1985		23,000		

Source: From a Report by the Task Group for the International Decade of Natural Disaster Reduction, published in *Bull. Volcano. Soc. Japan,* Series 2, Vol. 35, No. 1 (1990): 80–95.

▲ F I G U R E 4.23
An advancing tongue of basaltic lava setting fire to a house in Kalapana, Hawaii, during an eruption of Kilauea in June 1989. Flames at the edge of the flow are due to burning lawn grass.

flow so that the front of the flow solidifies. During a 1973 eruption of Heimay Island, off the coast of Iceland, fire boats sprayed seawater on advancing lava flows. This action is credited with having saved the harbor of the fishing village of Westmannaeyjar.

Violent Eruptions and Pyroclastic Activity

Many of the primary hazards of volcanism are directly related to the effects of violent eruptions, particularly pyroclastic activity. Unlike slowly moving lava flows, hot, rapidly moving pyroclastic flows and laterally directed blasts may overwhelm people before they can run away. The most destructive pyroclastic flow this century (in terms of loss of life) occurred on the Caribbean island of Martinique in 1902. In that eruption an avalanche of hot ash rushed down the flanks of Mont Pelée at a speed of more than 160 km/h, overwhelming the city of St. Pierre and instantly killing 29,000 people.

Pyroclastic debris sometimes contains blocks the size of cars. In the 1968 eruption of Arenál Volcano in Costa Rica, large falling blocks crashed through the roofs of houses 3 km away. However, much of the damage from pyroclastic eruptions is caused by the widespread fall of ash. For example, in A.D. 79 many citizens in the nearby towns of Pompeii and Herculaneaum were killed in the eruption of Mount Vesuvius. Most of the victims apparently were either buried and suffocated by falling ash or crushed by buildings that collapsed under the weight of the ash (Fig. 4.24). The prehistoric eruption of Mount Mazama covered

▲ F I G U R E 4.24
Casts of the bodies of five citizens of Pompeii, Italy, who were killed during the eruption of Mount Vesuvius in 79 A.D. Death was caused by poisonous gases; then the bodies were buried by pyroclastic material. Over the centuries the bodies decayed away, but their shapes were imprinted in the tephra blanket. When excavators discovered the imprints, they carefully recorded them with plaster casts.

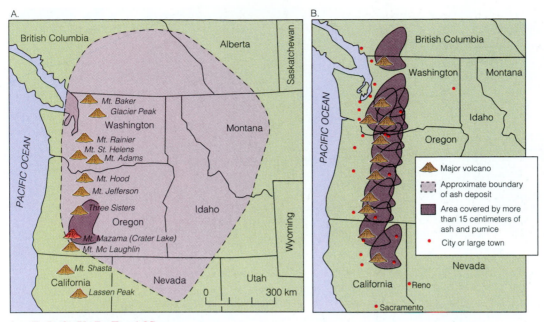

▲ F I G U R E 4.25
Area affected by ash fall from the prehistoric eruption of Mount Mazama. B. Area of Mount Mazama ash fall shown in A, transposed onto other dormant and active volcanoes in the region.

an area of 13,000 km² with a layer of ash more than 15 cm thick. The Cascade Range of volcanoes, which includes Mount Mazama and Mount St. Helens, stretches from California to southern British Columbia; imagine what the effect would be if another of the Cascade volcanoes were to undergo a similar eruption today (Fig. 4.25)! Even a relatively minor ash eruption can cause serious problems for residents of the area and wreak havoc on airplanes flying in the vicinity.

Poisonous Gas Emissions

Many volcanoes emit gases more or less continuously through fumaroles and geysers. Although water vapor is the main gas emitted by volcanoes, other kinds of gases are often present, many of them potentially harmful to people, animals, or vegetation. Some are toxic, such as, carbon monoxide (CO). Some are acidic (e.g., hydrochloric acid, HCl, and hydrofluoric acid, HF); in other cases the emitted gases mix with water vapor to form acidic solutions (e.g., sulfuric acid, H_2SO_4).

Perhaps the best known examples of the destructive capabilities of volcanic gases occurred in Cameroon in West Africa. Lake Nyos and Lake Monoun are part of a series of summit crater lakes in small, young basaltic volcanoes in northwestern Cameroon. In August 1984 a release of car-

bon dioxide (CO_2) gas from Lake Monoun caused the deaths of 37 people. In 1986 a much more serious episode occurred at Lake Nyos. As one research team reported:

On 21 August at about 21:30 a series of rumbling sounds lasting perhaps 15 to 20 seconds caused people in the immediate area of the lake to come out of their homes. One observer reported hearing a bubbling sound, and after walking to a vantage point he saw a white cloud rise from the lake and a large water surge. Many people smelled the odor of rotten eggs or gunpowder, experienced a warm sensation, and rapidly lost consciousness. Survivors of the incident, who awakened from 6 to 36 hours later, felt weak and confused. Many found that their oil lamps had gone out, although they still contained oil, and that their animals and family members were dead. The bird, insect, and small mammal populations in the area were not seen for at least 48 hours after the event. The plant life was essentially unaffected.

At least 1700 people and 3000 cattle lost their lives in this episode. Scientists concluded that the cause of death was asphyxiation by CO_2 gas. Interestingly, the "rotten egg" smell reported by some of the survivors was later attributed to "olfactory hallucinations," a common side effect of exposure to nonlethal doses of CO_2. As discussed in Box 4.1, it is likely that the gas was slowly released from the underlying magma chamber, saturating both the lake water and bottom sediments with CO_2.

Secondary and Tertiary Effects of Volcanism

Mudflows and Debris Avalanches

Tephra can be dangerous long after an eruption has ceased. Rain or meltwater from summit snow can loosen tephra piled on a steep volcanic slope and start a deadly mudflow. Mudflows can also occur when a glowing avalanche enters a river channel. A volcanic mudflow is technically referred to as a **lahar.** A related phenomenon (actually a type of lahar) is the volcanic *debris avalanche,* in which many different materials—mud, blocks of pyroclastic material, trees, and so on—are mixed together.

Lahars and debris avalanches are common features of volcanic eruptions and can have devastating consequences. For example, in 1985, a small eruption of the volcano Nevado del Ruíz in Colombia melted part of the icecap on the mountain's summit. Mudflows were formed when the volcanic ash mixed with the meltwater. The massive lahars moved swiftly down river valleys on the flanks of the volcano, killing at least 23,000 people and causing more than US$212 million in property damage. It was the second worst volcanic disaster in the twentieth century.

Because they follow topography, lahars are relatively predictable. Sometimes they can be diverted by barriers or tunnels, and they are subject to the same types of control techniques as are lava flows. Fig. 4.26A shows a lahar hazard map that was hastily produced in the weeks preceding the eruption of Mount Pinatubo in 1991. A map of the actual lahars resulting from the eruption (Fig. 4.26B) illustrates the remarkable success of the team in predicting the paths taken by the hot volcanic mudflows. Even so, lahars were a major source of property damage and loss of life resulting from that eruption.

Flooding

It is common for volcanic eruptions to be accompanied or preceded by flooding, which may in turn cause mudflows. In some cases flooding may be caused by the rupture of a summit crater lake. In Iceland, for example, volcanoes buried under permanent ice caps cause subsurface accumulation of meltwater, which eventually escapes in a huge gush of water known as a *glacier burst.* Flooding can also result when rivers are blocked by lava flows or lahars. The resulting erosion and sedimentation may cause long-term disruption of downstream waterways; it can take years for a river system to clear itself of ash deposits from a major pyroclastic eruption. Some river courses in the vicinity of Mount Pinatubo were permanently altered after the 1991

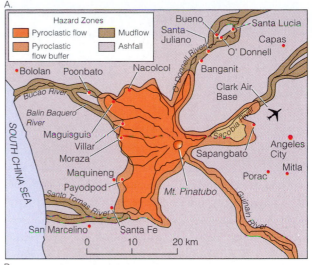

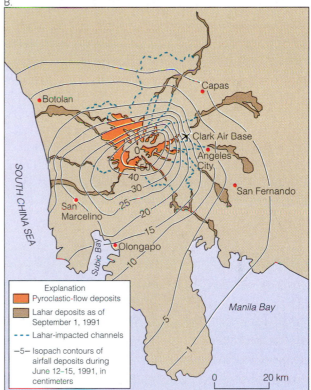

▲ **F I G U R E   4.26**

A. Hazard map produced by geologists immediately before the catastrophic eruption of Mount Pinatubo in 1991. B. Map of actual lahars resulting from the eruption shows the success of the monitoring team in predicting the volcanic hazards.

BOX 4.1

●

THE HUMAN PERSPECTIVE

LAKES OF DEATH IN CAMEROON

*T*wo small lakes in a remote part of Cameroon, a small country in central Africa (Fig. B1.1), made international news in the mid-1980s when lethal clouds of carbon dioxide (CO_2) gas from deep beneath the surface of the lakes escaped into the surrounding atmosphere, killing animal and human populations far downwind. The first gas discharge, which occurred at Lake Monoun in 1984, asphyxiated 37 people. The second, which occurred at Lake Nyos in 1986, released a highly concentrated cloud of CO_2 that killed more than 1700 people. The two events have similarities other than location: both occurred at night during the rainy season; both involved volcanic crater lakes; and both are likely to recur without some type of technologic intervention.

Immediately after the disasters scientists began monitoring the lakes in an attempt to understand what had triggered these catastrophic incidents. They identified several factors that caused instability in the lakes and they produced models of the conditions that probably preceded the degassing. The researchers found that Lake Nyos had a huge reservoir of CO_2 stored deep beneath unstable, den-sity-layered or *stratified* lake waters. When it was disturbed by some event—wind at the surface, a landslide, an earthquake, or a minor eruption—the stratified column of water turned over, allowing approximately 100 million m^3 of CO_2 to bubble to the surface in just 2 hours (Figs. B1.2 and B1.3).

It soon became apparent that despite the release of huge amounts of CO_2 during the overturns, vast amounts of gas remain dissolved at great depths in the lakes, and more is being added each year from natural sources. On the basis of 6 years of records of dissolved gases, temperature, density, and recharge rates in each lake, scientists predicted that Lake Monoun was in danger of a violent degassing event within 10 years. Lake Nyos, which is much larger and deeper, would build up its CO_2 to dangerous concentrations within about 20 years.

Gas-rich volcanic crater lakes are known to occur in other localities as well, including Japan, Zaire, and Indonesia. In theory, the hazards associated with such lakes can be alleviated by controlled degassing. This technique involves installing a subsurface network of pipes to reduce the con-

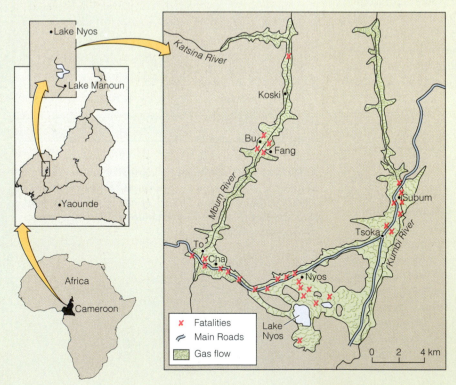

◄ F I G U R E B1.1
The path taken by the cloud of carbon dioxide emerging from Lake Nyos. Inset map shows the location of Cameroon and Lakes Monoun and Nyos.

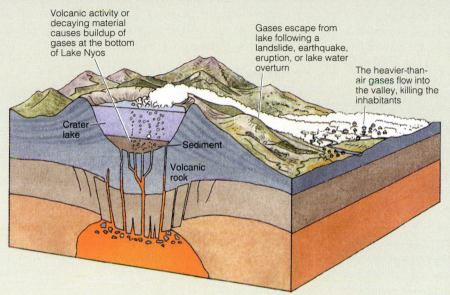

Volcanic activity or decaying material causes buildup of gases at the bottom of Lake Nyos

Gases escape from lake following a landslide, earthquake, eruption, or lake water overturn

The heavier-than-air gases flow into the valley, killing the inhabitants

Crater lake

Sediment

Volcanic rock

◀ **FIGURE B1.2**
A lethal cloud of CO_2 bubbled up from the bottom of Lake Nyos after the stratification of the lake water was disturbed. The gas, which is invisible and heavier than air, flowed over the natural dam surrounding the lake and down the adjacent valley, asphyxiating humans and animals in its path.

centrations of gas in deep water. Gas-charged water is pumped to the surface, where it releases gases to the air in nontoxic amounts. In Lake Monoun a prototype system of this type was tested successfully in 1992. It used three pipes, each 14 cm in diameter, placed deep beneath the surface of the water. Calculations showed that this procedure would restore CO_2 concentrations to stable levels in 2 to 3 years.

However, Lake Nyos presents a more complicated situation because of its greater size and depth. The challenge is

to control the flow of partially degassed water without upsetting the stratification of the lake water. One proposed solution is to divert water to a storage reservoir and then reinject it into the lake at the level of its natural buoyancy. The goal would be to establish a balance between the degassed water flowing into the lake and the bottom water removed from the lake. But unless the system is precisely calculated and perfectly designed, there is a danger that it will initiate the very situation the scientists are trying to prevent.

A.

B.

▲ **FIGURE B1.3**
Lake Nyos in Cameroon, Africa. A. The lake, muddy-brown in color after the gas release, sits in a depression at the top of a volcano. One night in 1986 this lake discharged a mass of lethal carbon dioxide. B. Cattle killed by asphyxiation as a result of the 1986 gas discharge from Lake Nyos.

eruption, primarily by lahars that filled and blocked major drainage channels and diverted the rivers.

Tsunamis

Violent undersea eruptions can cause giant sea waves called *tsunamis.* Tsunamis set off by the eruption of Krakatau in 1883 killed more than 36,000 dwellers on the coast of Java and other Indonesian islands. Tsunamis are discussed more fully in Chapter 5.

Volcanic Tremors and Earthquakes

Seismicity is another common accompaniment of volcanic activity. Eruptions are commonly preceded by local earthquakes, which may be caused by the cracking and splitting open of fissures as the magma chamber inflates. At both Mount St. Helens (1980) and Mount Pinatubo (1991), hundreds of small earthquakes were recorded daily before the main eruption sequence, providing information that was used in predicting the onset of the eruption. Sometimes seismic activity comes in phases; sometimes it ceases when the eruption begins. The seismic activity may last only a few days or weeks or may continue for months or even years. The seismic prelude to the 79 A.D. eruption of Mount Vesuvius lasted 16 years!

Another type of seismicity that accompanies volcanic eruptions, *volcanic tremor* or *harmonic tremor,* consists of a more or less continuous, low-frequency, rhythmic ground motion. It may be associated with actual movement of the magma (e.g., boiling, convection, or drag of the magma against the chamber walls). Volcanic activity is associated with tectonic seismicity because the distribution of both volcanoes and earthquakes is controlled by the locations of active plate boundaries. It has even been suggested—though not confirmed—that large earthquakes may contribute to the onset of major volcanic eruptions.

Atmospheric Effects

Among the most significant long-term hazards associated with volcanism are atmospheric and climatic effects. Climatic effects result primarily from the ejection of ash and extremely fine particles or droplets called *aerosols* high into the stratosphere during major eruptions. Some eruption columns reach such great heights that high-level winds transport fine debris and sulfur-rich gas around the world. By blocking incoming solar energy, such atmospheric pollution can lower average temperatures at the land surface and cause spectacular sunsets as the Sun's rays are refracted by the airborne particles and aerosols. As mentioned earlier, the 1991 eruption of Mount Pinatubo blasted more than 8 km^3 of fine pyroclastic material and sulfur-rich gas high into the atmosphere, causing significant global cooling for as long as two years. And in 1815, three days of total darkness followed the major pyroclastic eruption of Tambora in Indonesia; the darkness extended as much as 500 km from the volcano. The following year was called "the year without a summer;" average global temperatures fell more than 1°C below normal, and crop failures were widespread.

Volcanic material ejected into the atmosphere can also cause toxic or acidic fallout. Hydrofluoric acid (HF) in volcanic fallout has been known to cause fluoride poisoning in livestock. Salty and acidic precipitation can damage crops, contaminate soil, and corrode materials.

Famine and Disease

As we will see in the next section, periodic light ash falls can contribute significantly to soil fertility. However, a major tephra eruption may wreak such havoc on agricultural land and livestock that famine results. The effects can be exacerbated by long-term climatic changes and by the dislocation of people and interruption of basic services associated with other aspects of the eruption. Most of the deaths that resulted from the 1991 eruption of Mount Pinatubo were caused not by the effects of the eruption itself but by disease, lack of water and sanitation, and related problems in temporary camps for the homeless.

BENEFICIAL ASPECTS OF VOLCANISM

Although we tend to focus on the hazards associated with volcanism, volcanoes have actually done much more good than harm to human beings. We mentioned earlier that the origin and evolution of the atmosphere and oceans were (and still are) directly dependent on the outgassing of volatile materials through volcanoes. The origin and evolution of life itself has been dependent on this process. Volcanoes have also created many thousands of square kilometers of new land, including some of the most beautiful and awe-inspiring settings in the natural environment (Fig. 4.27).

Aside from the beauty of the surroundings, it is no accident that people tend to live near volcanoes. These are areas with very fertile soils and high agricultural productivity. In a sense, volcanoes represent the only mechanism through which the Earth can directly revitalize soils that have been depleted through extended agricultural use. Periodic ash falls, especially when they are rich in potassium, phosphorus, and other essential elements, are effective natural fertilizers.

Volcanism is also linked with the formation of mineral deposits. Many types of ore deposits are associated with ancient volcanic systems, particularly collapsed calderas, ring fracture complexes, and subsurface fracture systems related to siliceous volcanism. Circulating groundwater heated by a subvolcanic magma chamber was a central feature in the formation of many major ore deposits. Recent explorations of oceanic rift zones have shown that the submarine rifting environment, in which seawater is heated by contact with hot basaltic magmas, is another important setting for the formation of major ore deposits.

▲ **F I G U R E   4.27**
Seemingly tranquil, lush fields grow on rich volcanic soils around an active volcano in the Philippines. Although an active volcano can be deadly dangerous during an eruption, the volcanic ash deposited fertilizes and so enriches the soil that people return quickly once an eruption ceases.

Volcanic heat also plays a role in geothermal energy. Using water warmed by hot rocks, people can heat or cool their houses, generate electricity, and swim in geothermally heated pools. The places on the Earth where geothermal energy is most easily exploitable are places of recent volcanic activity. Most volcanic and magmatic activity is close to plate margins, and it is here, in such places as New Zealand, the Philippines, Japan, Italy, Iceland, and the western United States, that geothermal power is being used.

PREDICTING ERUPTIONS

Volcanic eruptions are not rare events. Some 600 volcanoes are considered to be **active**—that is, they have erupted recently, or at least within recorded history—and every year about 50 to 60 volcanoes erupt. Volcanoes that have not erupted within recorded history, are deeply eroded, and show no signs of future activity are referred to as **extinct.** In between these two categories are volcanoes that are said to be **dormant**—that is, they have not erupted in recent memory and show no signs of current activity, but they are not deeply eroded. Dormant volcanoes can become active with unnerving ease—Mount St. Helens had been dormant for 123 years before its 1980 eruption; Mount Pinatubo had been dormant for more than 400 years before its 1991 eruption; and Mount Vesuvius was widely considered extinct before its eruption in A.D. 79.

To some extent, volcanic hazards can be anticipated if experts are able to gather data before, during, and after eruptions. With sufficient information, the experts can advise civil authorities on when to implement hazard warnings and when to move endangered populations to areas of lower risk. There is much to be gained through successful forecasting and prediction of eruptions. In the context of volcanism, a *prediction* offers a fairly specific date for an expected event, whereas a *forecast* is a more general statement of likelihood. For example, in 1975 D. R. Crandell and colleagues wrote that Mount St. Helens "will erupt again, perhaps before the end of this century." The volcano erupted 5 years later. This is a good example of a successful long-term forecast.

Recently scientists have begun to achieve some success in the short-term prediction of volcanic eruptions. Since the 1980 eruption of Mount St. Helens, for example, volcanologists monitoring the volcano have managed to maintain a near-perfect record in predicting minor dome-building eruptions. Successes have also been achieved at Mount Pinatubo and at Augustine Volcano in Alaska, among others. An important component of prediction is the identification and intensive study of high-risk volcanoes. Unfortunately, this process cannot yet guarantee that major disasters will be avoided altogether.

Aside from identifying a volcano as active, dormant, or extinct, the first step in predicting an eruption is to study the volcano's past behavior. This effort has two main goals. The first is to ascertain the historic style of eruption. This is important in predicting the type of activity to be expected and the area that might be affected by an eruption. The second goal—a crucial step in predicting the actual time of the eruption—is to determine the volcano's eruption interval. Some volcanoes, notably those of Hawaii, display reasonably regular eruptive sequences with definable intervals. Mauna Loa tends to alternate between flank and summit eruptions; a flank eruption can usually be expected to occur within three years after a summit eruption. Other volcanoes show little or no periodicity. In the case of Vesuvius, for example, no pattern or periodicity is discernible in 2000 years of recorded history.

Volcano Monitoring

The aspect of prediction in which the most progress has been made is in the monitoring of volcanoes. Such monitoring facilitates the prediction and forecasting of eruptions in two ways: (1) it allows scientists to track the distribution and movement of magma in the volcanic "plumbing" system; and (2) it makes possible the detection of anomalies and the identification of *precursor phenomena*—events or processes that signal the onset or progression of activity within the volcano.

Monitoring the Movement of Magma

There are also some techniques whereby volcanologists can actually attempt to monitor the distribution and movement of a magma body within a volcano. Changes in the magma body may indicate that an eruption is imminent. Several techniques are used in such studies.

Seismic studies   Seismic waves are released both by earthquakes and by explosions. As discussed in Chapter 3, some types of seismic waves (S waves) cannot travel through liquids. Scientists can exploit this fact to study the distribution of magma underneath a volcano. The approach utilizes controlled explosives that are set off on one flank of the volcano, generating seismic waves. These are measured by seismographs set up on the other side of the mountain. If a body of liquid lies in the path of the seismic waves, the S waves will be blocked, creating a so-called *shadow zone* (Fig. 4.28). Through repetition of this technique, scientists can delineate the shape of the magma body and monitor any changes in its position.

Change in magnetic field   Rocks contain minerals that are naturally magnetic; notable among these is the mineral magnetite (Fe_3O_4), which is relatively common. The magnetic field of a rock containing magnetite can be measured with a magnetometer. If magnetite is heated past its *Curie point,* 575°C, its magnetic field will exhibit a sudden decrease in strength. Recall that the temperature of molten rock may range anywhere from 800 to 1200°C. Thus, if a magma body were to move into close proximity to a rock, the rock would likely be heated past its Curie point and the strength of its magnetic field would decrease. This is another technique whereby scientists can sometimes track the movement of magma upward through a volcanic system.

Change in electrical resistivity   The Earth has a natural (weak) electrical field. Changes in an underlying magma body should cause changes in the electrical resistivity of rocks. Such changes have been observed and measured and have been successfully correlated to magma movements that had not been detected by other methods.

Magma chamber modeling   Detailed seismic studies and ground deformational studies, often derived from satellite imagery, can be combined with knowledge about the tectonic setting of a volcano to yield sophisticated three-di-

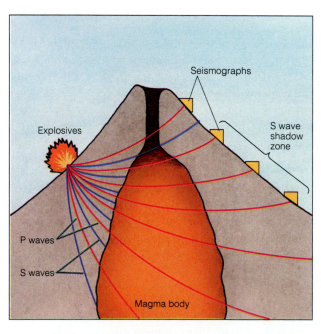

◀ **F I G U R E   4.28**
The use of controlled explosives to delineate an underlying magma body. Explosives set off on one flank of the volcano send seismic waves through the mountain. S waves, which cannot be transmitted through fluids, are blocked by any intervening body of fluid, creating an S-wave shadow zone on the other side of the mountain, where seismographs have been placed.

Seismographs

Explosives

S wave shadow zone

P waves

S waves

Magma body

◄ **F I G U R E   4.29**
False color (infrared) image of Vesuvius and the Bay of Naples, Italy. Vesuvius, an active volcano, is the circular structure, center right. Lavas erupted from Vesuvius over the past 300 years are little weathered rocks and show up bright red. Older lavas and volcanic ash, such as those that buried the Roman towns of Pompeii and Herculaneum in 79 A.D., show in yellows and oranges. The dark blue and purple region at the head of the bay is the city of Naples. Left of Naples, near the center of the image, is a cluster of smaller volcanoes called the Flegreian Fields.

mensional models of the subvolcanic "plumbing" system. This is a new approach that is at the forefront of the science of volcanology.

Physical Anomalies and Precursor Phenomena

To a great extent, the prediction of volcanic eruptions (especially short-term predictions) is based on the detection of physical anomalies, and the recognition that these are precursors or warning signs pointing to the onset of a major event. A physical anomaly is any physical occurrence that is out of the ordinary and is linked in some way to activity within the volcano. Following are some of the most common anomalies that typically precede volcanic eruptions.

Ground deformation   Major eruptions are sometimes preceded by a change in the shape or elevation of the ground, such as bulging, swelling, or doming related to the inflation of the underlying magma chamber. This type of change can be detected through the use of tiltmeters, geodolites, and electronic distance measurement instruments.

Change in the temperature of crater lakes, well water, fumaroles, or hot springs   In some cases, the temperature of water near a volcano rises, reaching a maximum at the onset of an eruption; in other cases, temperatures rise and then fall again prior to an eruption, or rise and fall with no eruption. Such physical anomalies are signs of activity in the volcano, even if there is no actual eruption.

Change in heat output at the surface   Ground temperature in the vicinity of a volcano may also change before a major eruption. Recently, thermal (infrared) remote sensing techniques have been used to detect subtle changes in ground temperature near active volcanoes. Infrared satellite imagery can also be used to produce temperature maps of lava flows and to determine the total volume of a flow (Fig. 4.29).

Change in the composition of gases   The composition of gases emitted from fumaroles has been observed to change before some eruptions. In general, the proportions of hydrochloric acid (HCl) and sulfur dioxide (SO_2) tend to increase relative to the proportion of water vapor, although other types of chemical changes have also been noted. Like changes in water temperature, changes in gas composition are not wholly reliable indicators, but they may contribute to a general scenario of an imminent event.

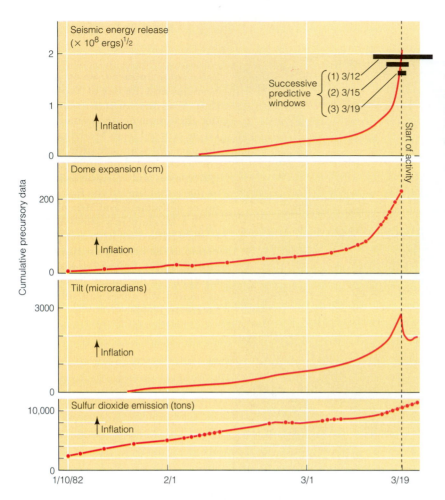

◄ **FIGURE 4.30**

The acceleration of precursor activity before the March–April 1982 eruption of Mount St. Helens. The increasing accuracy of volcanologists' predictions, issued on March 12, 15, and 19, are shown by the sizes of the bars at top.

Local seismic activity As discussed earlier in the chapter, several types of seismicity are associated with volcanic eruptions. Monitoring of precursory seismic activity has proved to be one of the most reliable tools for predicting eruptions. Seismic activity—especially local small earthquakes and harmonic tremor—often culminates in a major eruption. This occurred, for example, at Mount St. Helens and at Mount Pinatubo; in both of those cases monitoring of local seismic activity contributed to the issuance of successful predictions.

In most cases no single precursor indicator is sufficient to predict an eruption. Taken together, however, several physical or chemical anomalies may create an overall picture that provides a clear indication that an eruption is imminent (Fig. 4.30).

SUMMARY

1. Magma is molten rock, together with any suspended mineral grains and dissolved gases, that forms when temperatures rise sufficiently high for melting to happen in the Earth's crust or mantle. When magma reaches the surface, it does so through a volcano, a vent from which magma, solid rock debris, and gases are erupted. Magma that flows out and solidifies at the surface is called lava.

2. The most common types of magma are basaltic magma, which contains about 50 percent SiO_2; andesitic magma, containing about 60 percent SiO_2; and rhyolitic magma, containing about 70 percent SiO_2. The properties of magma—especially its composition (silica content and gas content) and viscosity—are important determinants of a volcano's eruptive style.

3. Nonexplosive eruptions are characterized by flows of fluid basaltic lava. Explosive eruptions are characterized by the eruption of large clouds of gas and pyroclastic material.

4. Some volcanoes erupt from fissures; others erupt from central vents. The most common landforms associated with central vent eruptions are shield volcanoes (gently

sloping volcanoes built up from successive lava flows) and stratovolcanoes (steep-sided composite volcanoes built up from alternating lava flows and deposits of pyroclastic material).

5. A close relationship exists between plate tectonics and the locations of different types of volcanoes. Rhyolitic volcanoes occur primarily on continental crust and are formed by the partial melting of continental rocks. Andesitic volcanoes form a Ring of Fire around the Pacific Ocean and are associated with the subduction and melting of crustal material. Basaltic magma, which is formed by the melting of mantle material, occurs at seafloor spreading centers and over deep-seated mantle hot spots.

6. Primary hazards associated with volcanism include lava flows, violent eruptions and pyroclastic activity, and emissions of poisonous gases. Secondary and tertiary hazards include mudflows and debris avalanches, flooding, tsunamis, local seismicity, atmospheric effects, and famine and disease.

7. Not all the effects of volcanism are negative. Volcanic eruptions have some beneficial aspects, including beautiful scenery, fertilization of agricultural soils, geothermal energy, and, in the long run, contributions to the chemical evolution of the atmosphere and the formation of mineral deposits.

8. Long-term forecasting of volcanic eruptions is based primarily on identification of a volcano's tectonic setting, past eruptive style, and eruptive interval.

9. Short-term prediction of volcanic eruptions and the issuance of warnings are based primarily on monitoring of volcanic activity. Monitoring has two main goals: to allow scientists to track the distribution and movement of magma in the volcanic "plumbing" system, and to permit the detection of physical anomalies and precursor phenomena—that is, events or processes that signal the onset or progression of activity within the volcano. Satellite imagery is an important aspect of modern volcano monitoring programs.

IMPORTANT TERMS TO REMEMBER

active volcano (p. 125)
andesite (p. 102)
basalt (p. 102)
caldera (p. 113)
crater (p. 113)
dormant volcano (p. 125)
eruption column (p. 108)
extinct volcano (p. 125)
fissure (p. 106)
fissure eruption (p. 114)

fumarole (p. 112)
geyser (p. 112)
glowing avalanche (p. 108)
lahar (p. 121)
lateral blast (p. 109)
lava (p. 101)
lava dome (p. 114)
magma (p. 101)
pyroclast (p. 107)
pyroclastic flow (p. 108)

pyroclastic rocks (p. 107)
resurgent dome (p. 114)
rhyolite (p. 102)
shield volcano (p. 110)
stratovolcano (p. 110)
tephra (p. 107)
tephra cone (p. 111)
thermal spring (p. 112)
viscosity (p. 103)
volcano (p. 101)

QUESTIONS AND ACTIVITIES

1. Imagine that you live in the vicinity of a shield volcano or a stratovolcano. What is the tectonic setting likely to be? What kinds of volcanic hazards do you think are most likely in each of these situations?

2. Are there any active volcanoes in your area? What about dormant or extinct volcanoes? You might wish to contact the local or regional Geological Survey for further information concerning ancient volcanic activity in your area.

3. Choose a volcanic event that interests you and investigate the impacts of the volcano on nearby residents. Some suggested events are the 1991 eruption of Mount Pinatubo or the 1980 eruption of Mount St. Helens.

4. Investigate a major eruption from long ago, such as the prehistoric eruption of Mount Mazama or the 1883 eruption of Krakatau. Try to project the events of the eruption into modern times. What hazards would be faced by residents if such an eruption were to occur today? On the basis of what you find out about the geology of the area, what is the likelihood that the event might recur in the near future?

5. Spend some time researching the role of volcanism and emissions of volcanic gas in the chemical evolution of the Earth's atmosphere. What kind of planet would we live on today if volcanism had been less active in this process?

TSUNAMIS

The bay was filling again; not slyly, as it had emptied, but in a great rushing wave, climbing the shores . . . Right over the chariot-road below me ran the salt sea, and climbed the plowland; spent itself, and paused, and went sucking back from the scoured land.

• **Mary Renault, The Bull From the Sea, 1962**

According to Greek legend, Theseus, the king of Athens, had a handsome son named Hippolytos. Theseus' wife, Phaedra, fell in love with Hippolytos. When he refused her advances, she hanged herself and left a letter accusing Hippolytos of having raped her. Theseus, convinced of Hippolytos' guilt by the fact of Phaedra's death, invoked a curse entrusted to him by his father Poseidon, the god of the sea. As Hippolytos drove his chariot along the rocky coast, a huge wave swept over the land, bearing on its crest a sea-bull. After the wave receded, Hippolytos' battered corpse was found and carried to Theseus, who had learned the truth about Phaedra's deception too late.

Recent archeologic research has established a factual context for many of the events in Greek mythology. The wave referred to in the legend of Phaedra and Hippolytos may have been the *tsunami,* or *seismic sea wave,* generated in 1600 B.C. by an eruption of the volcano Santorin. This island volcano, now called Thíra, is located in the Mediterranean Sea to the northeast of Crete. The cataclysmic eruption of 1600 B.C.—the most powerful volcanic eruption in recorded history—has been blamed for the weakening and ultimate downfall of the Minoan civilization of Crete. Sea waves generated by the eruption would have inundated the nearby shores of Crete to depths of tens of

meters, totally destroying agricultural land. Coastal lowlands around much of the eastern Mediterranean also would have flooded. The legendary floods of Deucalion, reputed to have submerged the mythical lost city of Atlantis, have also been tentatively identified with these waves.

The eruption of Santorin in the seventeenth century B.C. was probably very similar to another, much more recent cataclysmic volcanic eruption. During the night of August 27, 1883, the Indonesian volcano Krakatau erupted, with catastrophic results. The eruption was significant for many reasons, but its most disastrous effect was the generation of a series of huge sea waves that inundated nearby coastal areas. The waves reached heights of at least 40 m above normal sea level. Blocks of coral and other objects weighing up to 600 tons were carried far inland by the force of the waves. At least 36,417 people were killed, most of them by drowning, and 165 coastal villages were destroyed.

WHAT IS A TSUNAMI?

A **tsunami** is a very long ocean wave that is generated by a sudden displacement of the sea floor. The term is derived from a Japanese word meaning "harbor wave." Tsunamis are sometimes referred to as **seismic sea waves** because submarine and near-coast earthquakes are their primary cause. They are also popularly called "tidal waves," but this is a misnomer; tsunamis have nothing to do with tides.

Tsunamis can occur with little or no warning, bringing death and massive destruction to coastal communities. In this chapter we examine how they are generated and what

An earthquake offshore of Japan on the 13th of July, 1993, caused a tsunami that destroyed the shorefront of Okushiri, Hokkaido, Japan.

kinds of damage they can cause. We also look at efforts to develop an integrated and effective early warning system for tsunami-prone areas.

Physical Characteristics of Tsunamis

Figure 5.1 illustrates the terminology used to describe tsunamis. These terms are essentially the same as those used for other types of waves, including light waves and seismic waves. For all types of waves, the term **wavelength** is used to refer to the distance from one crest to the next. Normal ocean waves average about 100 m in wavelength. A tsunami, in contrast, can exceed 200 km in wavelength.

Tsunamis also travel very quickly compared to normal ocean waves. This is particularly true in open water because wave velocities increase with water depth. In the open ocean, where the water is deepest, tsunami velocities can reach 950 km/h or more. (Normal ocean waves travel at velocities closer to 90 km/h.) Because of the relationship between velocity and water depth, when the waves reach shallower coastal waters they slow down abruptly—in a sense, they "pile up" on themselves. This causes the height of the wave from the still water line—its **amplitude**—to increase dramatically. Thus, in the open ocean tsunami amplitudes rarely exceed 1 m, little more than a broad, gentle bulge. But once they reach the shore it is not unusual for tsunamis to crest at 5 or 10 m. In the most dramatic documented cases these waves have risen to levels of 20, 30, and even 40 m above normal sea level.

The behavior of a tsunami after it reaches the shore is also different from that of a normal ocean wave. Sometimes the level of water at the coast recedes noticeably just before the onset of the tsunami. This phenomenon, called *drawdown,* can prove deadly to onlookers who may be tempted to walk down to the water's edge to investigate, only to be trapped when the wave finally hits. In other cases, the first observed movement of water may be a rise. Because of its long wavelength, it can take a long time for the tsunami to crest and recede. The water level may rise, reach a peak, and then remain high for several minutes. It can also take a long time, as much as an hour, for the successive crests in a series

of tsunamis to reach the shore. This is because the **frequency** of the wave—the time interval between crests—depends on both velocity and wavelength. Thus, although tsunamis have high velocities, their long wavelengths result in a long time interval between crests.

The Influence of the Shoreline

The water level achieved by a tsunami once it hits the shore is called its **run-up** (usually expressed as height in meters above normal high tide). Run-ups resulting from a particular tsunami vary from place to place along the coast because the height of a wave is strongly influenced by the depth of the water, the profile of the sea floor, and the shape of the coastline. In some cases the shoreline may bend or *refract* the wave, drawing it into otherwise protected harbors or bays. When the energy of a long section of the wave is concentrated on a particular stretch of coastline as a result of sea floor topography or shoreline configuration, it is called a **wave trap**. If the onrushing wave becomes concentrated in a long, narrow bay or river mouth, it can form a wall of water called a **bore**.

The run-up of a tsunami is usually determined by a combination of eyewitness accounts and physical indicators such as water marks on the sides of buildings, the locations of seaweed and other debris transported by the wave, the height to which vegetation has been damaged or killed by saltwater, the landward limits of sand deposition, and the height to which buildings or trees have been damaged by transported objects.

Measurement of Tsunami Magnitudes

Various scales are in use for the measurement of tsunami magnitudes. These scales define the magnitude of a tsunami in terms of the logarithm of the maximum wave amplitude observed locally. (In this respect, tsunami magnitude scales are analogous to the Richter earthquake magnitude scale.) One such scale is the *Imamura–Iida scale,* in which tsunami magnitude is calculated as a function of the maximum wave height along the coast. This type of formula leads to different estimates of magnitude for a given tsunami because of the variability of run-ups along a given

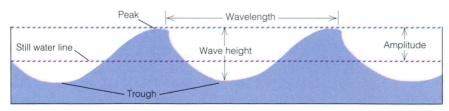

▲ **F I G U R E   5.1**
Terminology used in describing the characteristics of tsunamis. In most usages, *amplitude* refers to half of the distance from peak to trough. In the context of ocean waves it is used differently, referring to the height of the wave from the still water line. The peak-to-trough distance is called the wave height.

stretch of shoreline and the difficulties inherent in measuring the maximum run-up.

HOW TSUNAMIS ARE GENERATED

Any event that causes a significant displacement of the sea floor also causes the displacement of an equivalent volume of water. This, in turn, can generate tsunamis. Most tsunamis are generated by earthquakes, but they can also be caused by volcanic eruptions, submarine landslides, and human activity.

Earthquakes

Most tsunamis are produced by near-shore or offshore earthquakes. The primary causes of wave generation are the release of energy and the associated crustal deformation resulting from the earthquake. Any earthquake that generates a tsunami is called a **tsunamigenic earthquake**. Perhaps the most famous such earthquake occurred off the coast of Portugal in 1755. It produced a series of tsunamis with run-ups as high as 5 m above normal high tide that caused the deaths of 60,000 people in Lisbon (population 235,000). The waves were observed a few hours later as far away as the West Indies.

The magnitude of a tsunami is related to the magnitude of the earthquake that caused it; a large earthquake may be expected to generate a large tsunami. The correlation is not always that simple, however. For one thing, tsunamis are much more likely to result from vertical deformation of the crust than from horizontal deformation. No tsunami resulted from the great 1906 San Francisco earthquake, even though displacements of up to 6 m occurred along portions of the fault that are partly underwater. The San Andreas Fault, along which the quake occurred, is characterized by horizontal (strike-slip) motion, in which there is essentially no vertical displacement of the sea floor. By contrast, when the sea floor undergoes vertical deformation it acts like a huge paddle pushing the displaced volume of water outward from the zone of deformation. So earthquakes that occur along normal or reverse faults, in which the crustal displacement is primarily vertical, are much more efficient at displacing water and thus generating tsunamis.

Tsunami Earthquakes

Even along normal and reverse faults, very large earthquakes sometimes generate only moderate-size tsunamis or none at all. In other cases, earthquakes of small or moderate size generate unexpectedly large tsunamis. For this reason, Japanese tsunami experts have established a category of **tsunami earthquakes**, distinct from other tsunamigenic earthquakes. A tsunami earthquake generates a tsunami that is anomalously large with respect to the magnitude of the quake. Examples of tsunami earthquakes include the

▲ **F I G U R E 5.2**
Tsunami striking Hilo, Hawaii, in 1946. Toward the lower left-hand corner of the photo you can see the figure of a man; he was swept away by the wave. The photograph was taken by a sailor on the *Brigham Victory.*

1986 Sanriku (Honshu, Japan) earthquake, which caused a tsunami 24 m high in which 26,000 people drowned, and the 1946 Unimak Island (Aleutian Islands, Alaska) earthquake, which caused a tsunami that traveled at a velocity of 800 km/h, inundating Hilo, Hawaii, 4.5 hours later with a run-up 18 m higher than normal high tide. The Unimak tsunami killed 150 people (90 of them in Hilo) and caused $25 million worth of property damage (Fig. 5.2). Both the Sanriku and Unimak earthquakes were moderate-sized quakes that were unexpectedly efficient at generating tsunamis.

Tsunami earthquakes occur along steeply dipping fault surfaces with vertical motion, the most efficient setting for generating tsunamis. The plate tectonic setting is typically, though not always, an active plate boundary characterized by deep oceanic trenches. There are two main reasons that a moderate-size earthquake in these settings may generate a very large tsunami: sediment slumping and surface rupture.

Along some active plate margins, called *accreting margins,* large amounts of sediment are accumulating in the deep trench along the fault zone (Fig. 5.3A). This mass of sediment is called an *accretionary prism.* When an earthquake strikes, large volumes of sediments in the accretionary prism may collapse and slump into the trench, causing unusually large tsunamis. This may have been the cause of the Sanriku and Unimak tsunamis; both followed quakes in areas characterized by large-scale slump features in underwater sediments near the trench.

The second category of tsunami earthquake is represented by the September 2, 1992, quake off the coast of Nicaragua, an earthquake of moderate size (Richter magnitude 7) that generated an unusually large tsunami. The related tsunami had a maximum run-up of about 10 m. About 170 people were killed by the wave and 500 were injured; 13,000 were left homeless and 1500 homes were destroyed. Unlike the trenches associated with Sanriku and the Aleutian Islands, the oceanic trench off the coast of Nicaragua does not contain a large volume of sediment; that is, it occurs along a *nonaccreting margin,* in which sediments are not accumulating but are being subducted into the trench (Fig. 5.3B). It has been proposed that because of the absence of sediments in the trench, the slip along the fault was able to propagate all the way from the earthquake's focus to the ocean floor. This means that the actual sea floor ruptured, vertical displacement occurred, and a fault scarp was created, causing the "paddle" effect necessary to generate a sizable tsunami. (Along an accreting margin, by contrast, the presence of sediments would prevent the rupture from extending all the way to the ocean floor.)

For many years seismologists have used seismic wave forms to help them understand the complex tectonic movements of fault blocks that create earthquakes. This approach is called *seismic inversion.* Japanese scientists have applied this approach to the study of tsunami wave forms. To determine the type of crustal deformation that occurs during tsunami earthquakes, they look at historical records of tsunami arrival times and data from tidal gauges; then they work backward to recreate the wave form, its velocity and propagation pattern, and, ultimately, the characteristics of the crustal movement that have generated the wave. It is relatively easy to model the propagation of tsunamis because the factors that influence the velocities of water waves are better understood than those that influence the motion

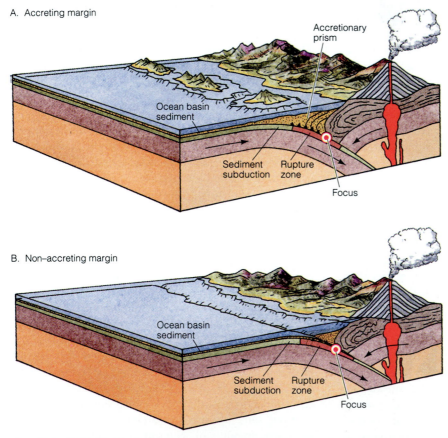

A. Accreting margin

Accretionary prism

Ocean basin sediment

Sediment subduction

Rupture zone

Focus

B. Non–accreting margin

Ocean basin sediment

Sediment subduction

Rupture zone

Focus

▲ **F I G U R E 5.3**
Diagram showing two types of active plate margins where tsunami earthquakes occur. A. An accreting margin, where slumping in the accretionary prism of sediments is the main tsunami-generating mechanism. B. A nonaccretionary margin, where surface rupture is the main tsunami-generating mechanism.

A.

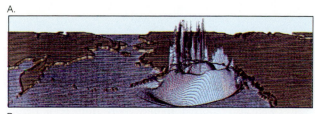

B.

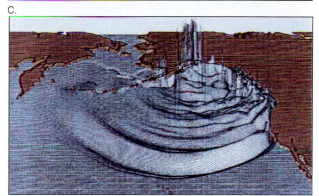

C.

◀ **F I G U R E   5.4**
Computer simulation of the tsunamis resulting from the Alaska Good Friday (March 28, 1964) earthquake. The crustal deformation that accompanied the earthquake included about 3 m of uplift and 1 to 2 m of subsidence over an area of about 500×300 km^2. Tsunami propagation after (A) 1 hour, (B) 3 hours, and (C) 4 hours. Vertical exaggeration is about 2×10^6; the elevation of the continents above the ocean in the figure corresponds to a tsunami height of about 5 cm. The tsunami was 30 m high at Valdez Harbor in Alaska. When it reached the shore in Hawaii, the tsunami was 5 m high.

Volcanic Eruptions

Volcanic eruptions can also be efficient generators of tsunamis, as shown by the Santorin and Krakatau eruptions mentioned at the beginning of the chapter. An explosive eruption can displace enormous volumes of rock; if the volcano is partly or mostly submerged, a corresponding volume of water will also be displaced, causing a tsunami. The collapse of steep walls of a volcanic structure and associated underwater debris or ash flows can contribute to the formation of a tsunami.

In the case of Krakatau, an enormous Plinian eruption in 1883 generated a series of at least three great tsunamis and many smaller waves. Most of the 36,417 people who died as a result of the eruption were swept away by the tsunamis; hot ash caused a number of injuries but only a few fatalities. Underwater ash flows may have generated some of the tsunamis. Krakatau is located west of Java in the Strait of Sunda, which was an important shipping corridor at the time (Fig. 5.5). Shipboard reports of the erup-

of seismic waves. A computer simulation of tsunami propagation after the 1964 Good Friday earthquake in Prince William Sound, Alaska, is shown in Fig. 5.4.

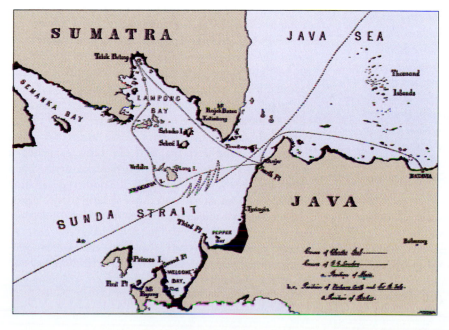

◀ **F I G U R E   5.5**
Map of Krakatau Island and the Strait of Sunda. The areas shaded black are those that were submerged by the tsunamis following the eruption of Krakatau. The dotted and dashed lines and small circles (a, b, c, d) indicate the paths and positions of ships from which eyewitness accounts were made.

tion, the resulting tsunamis, and the destruction in its aftermath were among the most important eyewitness accounts of this event. After the eruption the sea was generally very rough, with unpredictable currents, and cluttered with blocks of pumice, fallen trees, and corpses. Crews reported that their ships had been tossed and tumbled by the waves, and several large vessels were run aground. The government steamer *Berouw* was carried 2.5 km inland and stranded 24 m above sea level; its crew of 28 drowned.

An elderly Dutch pilot who guided ships through the Strait gave this account of the first of the tsunamis caused by the eruption:

> *Looking out to sea I noticed a dark black object through the gloom, traveling towards the shore. At first sight it seemed like a low range of hills rising out of the water, but I knew there was nothing of the kind in that part of the Sunda Strait. A second glance—and a very hurried one it was—convinced me that it was a lofty ridge of water many feet high, and worse still, that it would soon break upon the coast near the town. There was no time to give any warning, and so I turned and ran for my life . . . In a few minutes I heard the water with a loud roar break upon the shore. Everything was engulfed. Another glance around showed the houses being swept away and the trees thrown down on every side . . . I gave up all for lost, as I saw with dismay how high the wave still was. I was soon taken off my feet and borne inland by the force of the resistless mass. I remember nothing more until a violent blow aroused me. Some hard firm substance seemed within my reach, and clutching it I found I had gained a place of safety. The waters swept past me, and I found myself clinging to a cocoanut palm-tree. Most of the trees near the town were uprooted and thrown down for miles, but this one fortunately had escaped and myself with it.*
>
> *The huge wave rolled on, gradually decreasing in height and strength until the mountain slopes at the back of [the town of] Anjer were reached, and then, its fury spent, the waters gradually receded and flowed back into the sea. The sight of those receding waters haunts me still. As I clung to the palm-tree, wet and exhausted, there floated past the dead bodies of many a friend and neighbour. Only a mere handful of the population escaped. Houses and streets were completely destroyed, and scarcely a trace remains of where the once busy, thriving town originally stood. Unless you go yourself to see the ruin you will never believe how completely the place has been swept away.*

This wave was followed by at least two other tsunamis of equal or greater magnitude. The wave run-ups varied, reaching as high as 41 m in some places. One observer reported that "up to a height of 30 to 40 m, the sea had erased or knocked down every object, and covered it with a gray layer of mud and pumice-like material." The waves destroyed buildings, uprooted trees, and transported some astonishingly heavy objects; for example, a block of coral weighing approximately 600 tons was ripped from an offshore coral reef and swept 100 m inland. Many smaller waves were also reported. In one report, from Telok Betong Harbor (where the *Berouw* ran aground), the water level was observed to be 1 m below the level of the pier at one moment and 1 m above the level of the pier a moment later.

Landslides

Tsunamis can also be caused by coastal and submarine landslides. In most such cases the landslides themselves have been generated by earthquakes or volcanoes. Sediments can build up along the walls of a submarine canyon and, when shaken by an earthquake, collapse or slough off into the bottom of the canyon. Many of the most damaging tsunamis that have occurred along the West Coast of North America have been generated in this manner. For example, the Good Friday earthquake triggered at least 20 separate landslide tsunamis. About 80 of the 119 deaths linked to the quake were caused by those tsunamis.

Underwater Explosions

Tsunamis are occasionally caused by human activity. For example, tsunamis were generated by submarine nuclear testing at Bikini Atoll in the Marshall Islands in the 1940s and 1950s.

Tsunamis in Lakes, Bays, and Reservoirs

Tsunami-like waves can occur in smaller, enclosed bodies of water such as lakes, bays, and reservoirs. In most cases, these waves are generated by falls or slides of rock or soil. The slides can occur spontaneously or be triggered by earthquakes, explosions, or human activity. A famous example of a spontaneous landslide leading to a tsunami-like wave is the Vaiont Dam disaster, which occurred in Italy in 1965. Hillslope failure resulted in an avalanche of soil and rock that fell into the reservoir, creating an enormous wave that surged over the dam and flooded the valley below. Another well-known episode occurred in Rissa, Norway. Excavations adjacent to the coast of a lake initiated a major "quick clay" landslide; the slide, in turn, generated a tsunami-like wave that swept across the lake, destroying a village on the opposite shore. Still another devastating landslide-induced tsunami occurred in Lituya Bay, Alaska, following an earthquake of Richter magnitude 7 that occurred on July 9, 1958. The earthquake triggered an avalanche that fell into one corner of the enclosed bay, producing a wave that stretched completely across the bay and ran up the opposite side to a height of 60 m, stripping vegetation from the shoreline (Fig. 5.6).

◀ **F I G U R E   5 . 6**
Lituya Bay, Alaska, after a
tsunami caused by an avalanche
in 1958. The light areas around
the shoreline are regions where
forests were destroyed by the
wave.

Seiches

A related phenomenon occurring in enclosed bodies of water is a **seiche**, a periodic standing-wave oscillation of the water surface. Once initiated, a seiche travels back and forth at regular intervals determined by the depth, size, and bottom and shoreline configurations of the body of water. In other words, each body of water will have a characteristic period of oscillation for such waves—or sometimes several characteristic periods, depending on the complexity of the water body, the shoreline, and so on. The size of the stand-ing waves can range from a few centimeters to several meters, and the interval of repetition can range anywhere from a few minutes to a few hours.

Seiches are usually caused by unusual winds, currents, or changes in atmospheric conditions. They can also be initiated by earthquakes. For example, the Loma Prieta (San Francisco) earthquake of October 17, 1989, caused seiche oscillations in Monterey Bay (Fig. 5.7). The waves oscillated at an interval of about 9 minutes and were about 0.4 m high from crest to trough.

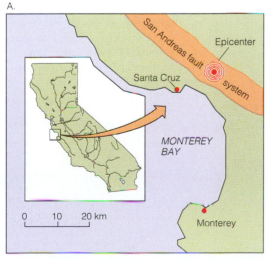

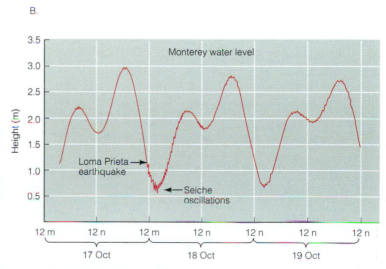

▲ **F I G U R E   5 . 7**
A. Location of Monterey Bay relative to the San Andreas fault and the epicenter of the Loma
Prieta earthquake. B. Records from Monterey Bay showing water-level oscillations resulting
from the Loma Prieta earthquake of 1989.

MITIGATION OF RISK AND HAZARDS

Tsunami Hazards

The main source of damage from a tsunami is the direct action of the wave on coastal structures. However, a variety of indirect mechanisms can also cause further damage. For example, flotation and drag can move houses, machinery, and railroad cars; debris such as trees, cars, and parts of destroyed structures can become projectiles; strong currents can erode foundations, leading to the collapse of buildings, bridges, and seawalls; and fires can result from the combustion of oil spilled by damaged ships and storage facilities.

The most serious hazard associated with tsunamis, of course, is loss of life. Over the past century, 94 destructive tsunamis have caused the deaths of a total of over 51,000 people. In 1992 alone, two tsunamis were responsible for the deaths of hundreds of people. In each of these cases (one caused by the earthquake in Nicaragua discussed earlier, the other in Riangkrok, Indonesia) the death toll was unusually high because the ferocity of the wave had not been adequately anticipated.

Prediction and Early Warning

The prediction of tsunamis centers on efforts to understand the mechanisms through which earthquakes generate tsunamis. Particular attention has been directed toward the so-called tsunami earthquakes, those that generate tsunamis that are unexpectedly large in comparison with the magnitude of the quake. As mentioned earlier, the magnitude of an earthquake is not the sole determinant of the magnitude of a tsunami; the degree, direction, and disposition of crustal deformation are also important. For this reason, in predicting whether an earthquake will generate a sizable tsunami, the Richter magnitude scale may not be the most useful measure of earthquake size. Some scientists believe that an earthquake's *seismic moment* may be a more appropriate measure. Seismic moment is a different measure of the size of an earthquake and takes into account the elastic properties of the Earth material, the fault area, and the average area over which dislocations occurred during the earthquake—all factors that are important in the generation of tsunamis.

Tsunamis and tsunamigenic earthquakes are a particular hazard for Pacific Ocean islands and locations around the Pacific rim. Hawaii is especially vulnerable to dangerous tsunamis (Fig. 5.8 and Table 5.1) because of its location in the path of waves generated at many seismically active points around the Pacific rim. Sirens and radio newscasts alert Hawaii's population to arriving tsunamis, and maps printed in telephone books show the coastal zones at greatest risk (Fig. 5.9); both measures have helped reduce the loss of life from tsunamis.

Regional Warning Systems

In spite of the rapid onset and quick travel times of tsunamis, there is often ample opportunity to warn residents of coastal areas, especially if they are located more than 750 km (approximately 1 hour in tsunami travel time)

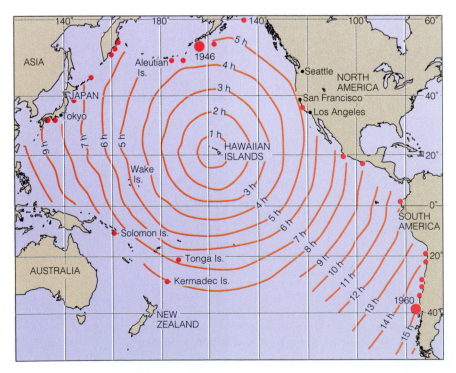

◀ **F I G U R E 5.8**
Map showing the time required for a tsunami to reach the island of Oahu, Hawaii. Small red dots mark the origins of tsunamis that have struck Hawaii. Large dots mark places where the disastrous tsunamis of 1946 and 1960 originated.

T A B L E 5.1 • **Major Tsunamis in Recorded History.**

Date	Source region	Visual run-up	Generated by	Comments
1600 B.C.	Santorin	?	Volcanic eruption	Devastation of Crete and Mediterranean coast
Nov. 1, 1755	Eastern Atlantic	5-10	Earthquake	Major damage in Lisbon, Portugal; tsunami reported from Europe to West Indies
Aug. 13, 1868	Peru–Chile	>10	Earthquake	Observed in New Zealand; damage in Hawaii
Aug. 27, 1883	Krakatau	40	Volcanic eruption	Over 30,000 drowned
June 15, 1896	Honshu	24	Earthquake	About 26,000 drowned
Mar. 2, 1933	Honshu	>20	Earthquake	3000 deaths from waves
April 1, 1946	Aleutian Islands	10	Earthquake	Over 150 drowned in Hilo, Hawai; $25 million in property damage
May 23, 1960	Chile	>10	Earthquake	909 dead and 834 missing along Chilean coast; 120 dead in Japan
Mar. 28, 1964	Alaska	6	Earthquake	In California, 119 deaths and $104 million damage
Dec. 2, 1992	Indonesia	26	Earthquake	137 people killed, village destroyed
Sept. 2, 1992	Nicaragua	10	Earthquake	170 deaths, 500 injured, 13,000 homeless

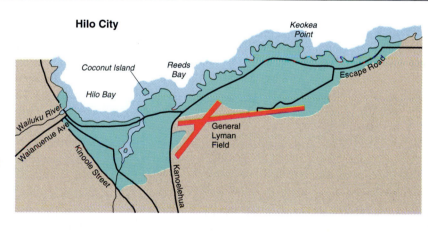

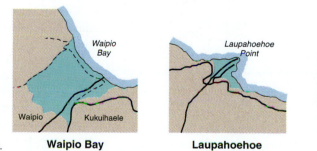
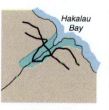

▲ **F I G U R E 5.9**
Page from a telephone book showing areas (in green) along the coast of the island of Hawaii that are susceptible to inundation by tsunamis.

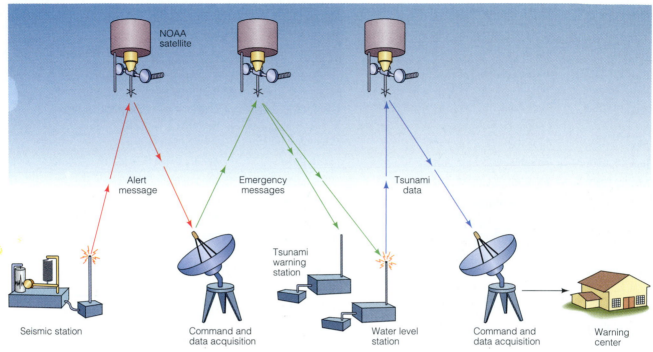

▲ **F I G U R E   5.10**
Early warning systems in the Pacific Ocean basin make use of information relayed from
seismic stations to tsunami warning stations via NOAA (National Oceanographic and
Atmospheric Administration) satellites.

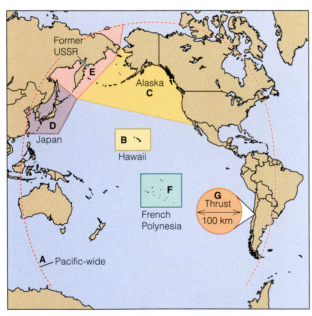

▲ **F I G U R E   5.11**
**Map of the Pacific Ocean basin showing: A. the
Pacific-wide early-warning system, which warns populations about 1 hour or more than 750 km from the
source of the tsunami; B. through F. regional warning
systems, which warn residents about 10 minutes or
100 to 750 km from the source; G. THRUST, a local
warning system designed to warn populations within
100 km of the source.**

from the source of the wave. Several different types of early
warning systems now exist in the Pacific basin. One such
system is operated by the National Oceanic and Atmospheric Administration (NOAA) Pacific Tsunami Warning
Center (PTWC) near Honolulu. It is an international
monitoring network consisting of about 30 seismic stations
and 78 tide stations located around the Pacific basin. When
an earthquake occurs, it is detected by the seismic stations.
If the quake meets certain criteria with respect to location
and magnitude, local tidal gauges are monitored for signs
of a tsunami. If a tsunami is detected, and if it is large
enough to be potentially hazardous, a Pacific-wide warning
is issued. Information is relayed from the seismic station to
the tsunami warning station via NOAA satellites (Fig.
5.10). Arrival times for the tsunami can be calculated for
different localities around the Pacific rim. The minimum
time required to gather this information and issue a warning is 1 hour; since tsunamis typically travel at 750 km/h,
this means that the Pacific-wide early-warning system is effective only for communities located more than 750 km
from the epicenter.

Regional warning systems are designed to provide warnings to areas in a 100- to 750-km radius (i.e., between
about 10 minutes and 1 hour) from the epicenter (Fig.
5.11). Regional systems have been established in areas that
are known to be particularly susceptible to earthquake-gen-
erated tsunamis, including Japan, Alaska, and French Poly-

BOX 5.1

•

THE HUMAN PERSPECTIVE

A TALE OF TWO TSUNAMIS

*T*wice in 15 months (July 1993 and October 1994) major earthquakes occurred near the coasts of Japan. Although the two quakes were of comparable magnitude (7.8 and 8.2, respectively), the first created a damaging tsunami of catastrophic proportions, whereas the second caused only minor tsunamis and a 6-hour media event. The differing impacts of the quakes may be attributed to differences in the topography of the ocean floor on Japan's east and west coasts as well as to the early-warning system installed by the government in 1994.

Tsunamis generated by earthquakes sometimes cause more deaths and damage in coastal areas than the earthquakes themselves. Japan is especially vulnerable to these devastating events because of its location in a tectonically active region and the high-density population centers located on its coasts (Fig. B1.1). The July 1993 quake, for example, triggered a 33-m seismic sea wave that struck the west coast of Okushiri Island within seconds of the first tremors, killing 239 people and destroying 558 houses.

Although the west coast of Japan historically has experienced fewer tsunamis than the east coast, current plate tectonic theory suggests that the west has a greater potential for quakes and tsunamis than was previously recognized. During the past 50 years a series of earthquakes has occurred in the Sea of Japan near Honshu and Hokkaido islands. The locations of these quakes is consistent with the theory that the North American and Eurasian plates meet just west of Honshu Island, whereas the Eurasian Plate is being subducted beneath western Japan. In addition, the islands in the area have coastlines with numerous bays, reflecting the shallow topography of the ocean bottom. Scientists predict that these characteristics of the seabed topography and the tectonic setting create conditions that may produce more damaging tsunamis in the future.

In contrast to the devastation on Okushiri Island caused by the 1993 tsunami, the tsunamis produced by an undersea earthquake east of Hokkaido in 1994 caused no casualties. The epicenter of the *M* 8.2 quake was located

about 160 km off the northern coast of Japan. Fatalities were averted through a new system that links the seismographs of the Japanese Meteorological Agency with the national media. Eight seconds after the first tremors were recorded at the Kushiro Observatory, warnings were automatically sent to radio and television stations. Those warnings were available to a nationwide network in screen flashes and voiceovers 2 minutes later, even before the jolt of the earthquake was felt in Tokyo. The local early-warning system had been operational for just 3 months, 1 year after the fatal Okushiri tsunami.

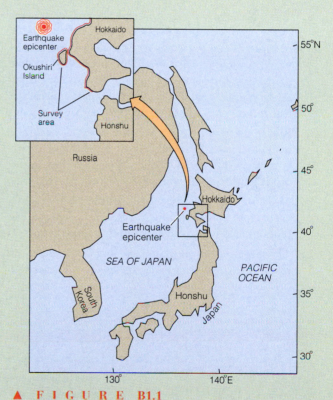

▲ F I G U R E   B1.1
Map of Japan, showing locations mentioned in the text.

nesia. The regional centers issue warnings on the basis of earthquake magnitude and location alone; warnings typically can be issued within 10 or 12 minutes of the earthquake's occurrence. In addition, local early-warning systems can be designed to provide warnings for populations less than 10 minutes (less than 100 km) from the source. An example of such a system is THRUST (Tsunami Hazards Reduction Utilizing Systems Technology), a pilot project in Valparaiso, Chile (Fig. 5.11).

The effectiveness of tsunami early-warning systems is impressive. For example, before the Japanese system was established, a total of over 6000 people had been killed by 14 tsunamis; since then, 20 tsunamis have killed a total of 215 people in Japan. In addition to the regional early-warning systems, some localities have established systems to issue warnings to the immediate coastal area within a few minutes of a large tsunami-generating event. The success of local early-warning systems depends heavily on emergency operations planning, as well as on access to timely information about earthquake occurrences and water levels. Other important factors are the ability of local authorities to assess the danger; the ability to disseminate information very quickly; and education of the public to respond appropriately in the event of a tsunami emergency.

SUMMARY

1. A tsunami is a very long ocean wave that is generated by a sudden displacement of the sea floor. Although they are sometimes mislabeled "tidal waves," they have nothing to do with tides.

2. The wavelengths of tsunamis typically exceed 200 km. In the open ocean, their velocities can reach 950 km/h or more. When the waves reach shallower coastal waters they slow down very abruptly and "pile up" on themselves. In open water, the amplitude of a tsunami rarely exceeds 1 m, but on shore it is not unusual for tsunamis to crest at 5 or 10 m above normal sea level.

3. Run-ups resulting from a tsunami vary from one place to another along the coast because the height of a wave is strongly influenced by the depth of the water, the profile of the sea floor, and the shape of the coastline.

4. Most tsunamis are generated by earthquakes, but volcanic eruptions, submarine landslides, underwater explosions, and human activity can also cause them.

5. The magnitude of an earthquake is not the only factor that determines whether the earthquake will generate a tsunami. Some earthquakes are unexpectedly efficient at generating tsunamis.

6. Tsunami-like waves sometimes occur in enclosed bodies of water such as lakes, bays, and reservoirs. Seiches, or oscillating standing waves, are a related phenomenon.

7. The direct action of waves is the main source of damage and loss of life caused by tsunamis. Other sources of damage include strong currents and debris acting as projectiles.

8. The prediction of tsunamis focuses on identifying earthquakes that are likely to generate tsunamis and on estimating the travel times of tsunamis across ocean basins.

9. Regional warning systems around the Pacific rim have been quite effective at minimizing loss of life from tsunamis.

IMPORTANT TERMS TO REMEMBER

amplitude (p. 132)
bore (p. 132)
frequency (p. 132)
run-up (p. 132)

seiche (p. 137)
seismic sea wave (p. 131)
tsunami (p. 131)
tsunami earthquake (p. 133)

tsunamigenic earthquake (p. 133)
wave trap (p. 132)
wavelength (p. 132)

QUESTIONS AND ACTIVITIES

1. Not all tsunamis originate on active plate margins. Investigate the Newfoundland earthquake of 1929, which occurred on a *passive,* or trailing, continental margin yet caused significant tsunamis. What was the tsunami-generating mechanism in that case?

2. Imagine that you live in a coastal community in a region that is susceptible to tsunamis (maybe you really do). What kinds of action do you think your community should carry out to minimize the potential damage and loss of life from tsunamis? What kinds of planning exercises do you think your local or regional government should undertake to determine the risk to your community?

3. If you do live in a tsunami-prone area, how well prepared is your community? Is there an early-warning system? If so, find out how it works. What kinds of action could you take to minimize risk if you own property near the coast? What would you do if you learned that a tsunami was heading toward your coast?

4. As discussed in this chapter, Japan is particularly susceptible to destructive tsunamis. Assess the tsunami risk to the Japanese public by investigating (a) earthquake activity in Japan, (b) the distribution of population centers along Japan's coastlines, (c) the history of damaging tsunamis in Japan, and (d) the status of Japan's participation in local, regional, and Pacific-wide early-warning systems. Write a summary of your findings.

CHAPTER 6

LANDSLIDES AND MASS-WASTING

The heights of our land are thus leveled with the shores; our fertile plains are formed from the ruins of mountains.

• James Hutton

*I*n the high Andes of South America, steep, unstable slopes rise above densely populated valleys. Here many active volcanoes with rugged peaks have been built up along converging lithospheric plates. These conditions hold the potential for disaster in a landscape where major earthquakes and volcanic eruptions can cause the steep slopes to collapse.

In Colombia, a group of active volcanoes lies west of Bogotá. One of them, Nevado del Ruíz, has a history of eruptive activity extending back to at least 1595. In late 1984 the dormant volcano awakened and began belching clouds of steam and ash; this activity continued through the autumn of 1985. People in the city of Armero, located far downvalley from the volcano, grew alarmed. Local authorities reassured them, even though recent geologic studies of the volcano had disclosed a history of repeated large volcanic mudflows. In early November, when the volcano showed signs of increasing activity, geologists warned that in the event of an eruption the resulting mudflows could pose a danger for Armero. At 3 p.m. on November 13 a technical emergency committee urged that Armero be evacuated, but the warning went unheeded.

That night, as the local radio station played cheerful music and urged people to be calm, the volcano erupted. Torrents of water released from rapidly melting ice and snow near the summit sent huge waves of muddy debris

surging down the volcano's slopes into the surrounding valleys. The largest of several mudflows moved rapidly toward Armero. Just after 11 p.m., as most of the citizens of Armero were sleeping soundly, a turbulent wall of mud came rushing out of a canyon and inundated the city. At least 23,000 people were buried in sulfurous volcanic mud. Had the geologists' warning been heeded in time, the tragedy might have been avoided.

MASS-WASTING AND ITS HUMAN IMPACTS

The landscapes we see about us may appear fixed and unchanging, but if we were to make a time-lapse motion picture of almost any slope it would be clear that the slope is constantly changing. Much of the recorded motion would be a result of **mass-wasting**, the movement of Earth materials downslope as a result of the pull of gravity. Any perceptible downslope movement of bedrock, regolith, or a mixture of the two is commonly referred to as a **landslide**, but, as we shall see, many different types of movement, materials, and triggering events may be involved in downslope mass movements of Earth material.

As the human population increases and cities and roads expand across the landscape, mass-wasting processes become increasingly likely to affect people. Landslides occur throughout the world; virtually no area is immune. The impacts of mass-wasting, in terms of loss of life and damage to property, can be devastating (Table 6.1). In the United States alone, landslides cause about $1.5 billion in economic losses and 25 to 50 deaths in a typical year. In less developed countries, the toll in damage and loss of life can be considerably more significant because of population density, lack of stringent zoning laws, scarcity of information about landslide hazards, and inadequate preparedness.

Although it may not always be possible to predict or prevent the occurrence of mass-wasting events, a knowl-

Mudflow of water-saturated volcanic ash in the town of Armero, Colombia, 1985. The thick, muddy slurry was released by an eruption of an ice-clad Andean volcano. More than 23,000 people were killed and very few buildings were left standing.

T A B L E 6.1 • **Fatalities Resulting from Some Major Landslides During This Century**[a]

Year	Location[b]	Fatalities
1916	Italy, Austria	10,000
1920	China[e]	200,000
1945	Japan[f]	1,200
1949	USSR[e]	12,000–20,000
1954	Austria	200
1962	Peru	4,000–5,000
1963	Italy	2,000
1970	Peru[e]	70,000
1985	Columbia[v]	23,000
1987	Ecuador[e]	1,000

[a]Source: National Research Council (1987).

[b]Landslides related to earthquakes ([e]), floods ([f]), and volcanic eruptions ([v]).

edge of the processes and their relationship to local geology can lead to intelligent planning that will help reduce losses of life and property. In this chapter we look at the factors that control slope stability, the different types of mass movements, and the impacts of mass-wasting on humans.

TYPES OF MASS-WASTING PROCESSES

All mass-wasting processes take place on slopes. There are many kinds of slope movements, but there is no simple way to classify such movements. The composition and texture of the material involved, the amount of water or air in the mixture, and the steepness of the slope all influence the type and velocity of movement. In effect, there is a progression from the flow of clear stream water to sediment-laden stream water to an array of mass-wasting processes ranging from those in which water promotes downslope movement to those in which water plays no direct or significant role.

For the purposes of this discussion, we will divide mass-wasting processes into two basic categories: (1) those involving the sudden failure of a slope, which results in the downslope transfer of relatively coherent masses of rock or rock debris by slumping, falling, or sliding; and (2) those involving the downslope flow of mixtures of sediment, water, and air. In the latter category, which involves internal motion of flowing masses of debris, processes are distinguished on the basis of their velocity and the amount of water in the flowing mixture. We also briefly examine some processes and deposits that are representative of mass-wasting in cold regions and on the ocean floor. This approach to the classification of mass movements is outlined in Table 6.2.

Slope Failures

The constant pull of gravity makes all hillslopes and mountain cliffs susceptible to failure. When failure occurs, material is transferred downslope until a stable slope condition is reestablished. Some of the most common types of slope failure are illustrated in Fig. 6.1.

Slumps

A **slump** is a type of slope failure involving *rotational* movement of rock or regolith, that is, downward and outward movement along a curved, concave-up surface (Fig. 6.1). Slumps can range from small displacements covering only one or two square meters to large complexes that cover hundreds or even thousands of square meters.

Slumps frequently result from artificial modification of the landscape. They are common along roads and highways

T A B L E 6.2 • **Outline for Classification of Types of Mass-Wasting Processes**

Slope failures	Sediment flows	Mass-wasting in cold climates	Subaqueous mass-wasting
Slumps	Slurry flows	Frost heaving	Slumps
Falls	Solifluction	Gelifluction	Slides
Rockfall	Debris flows		Flows
Debris fall	Mudflows		
Slides	Granular flows		
Block glide	Creep		
Rockslide	Earthflows		
Debris slide	Grain flows		
	Debris avalanches		

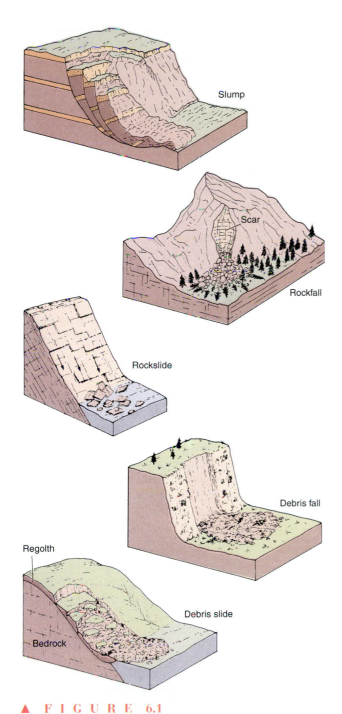

▲ **F I G U R E   6.1**
Examples of slope failures giving rise to slumps, falls, and slides.

▲ **F I G U R E   6.2**
Slump in agricultural land, Guatemala. Curvature of the slip surface has caused the toe of the slump block to rotate outward. The slump is believed to have been caused by ground saturation as a result of crop irrigation.

where bordering slopes have been oversteepened by construction activity. We can also see them along river banks or seacoasts where currents or waves have undercut the base of a slope. In some places slumping recurs seasonally and is associated with seepage of water into the ground during the rainy season (Fig. 6.2). Some slumps may be related to changing climatic conditions.

Falls

If you ask mountain climbers about the greatest dangers associated with their sport, they are likely to place falling rock near the top of the list. A **fall** is a sudden, vertical movement of Earth material, for example, from an overhanging cliff. *Rockfall*, the free fall of detached bodies of bedrock from a cliff or steep slope, is common in precipitous mountainous terrain, where rockfall debris forms conspicuous deposits at the base of steep slopes (Fig. 6.1).

As a rock falls, its speed increases. If we know the distance of the fall *(h)*, we can calculate the velocity *(v)* on impact as

$$v = \sqrt{2gh}$$

where *g* is the acceleration due to gravity. What this formula tells us is that a rock of a given size will be traveling at a much higher velocity if it falls from a point high on a steep mountain face than if it falls from a low cliff.

A rockfall may involve the dislodgement and fall of a single fragment or the sudden collapse of a huge mass of rock that plunges hundreds of meters, gathering speed until it breaks into smaller pieces on impact. The pieces continue to bounce, roll, and slide downslope before friction and decreasing slope angle bring them to a halt. Sometimes not only rock but overlying sediment and plants are dislodged. The resulting *debris fall* is similar to a rockfall, but it consists of a mixture of rock and weathered regolith as well as vegetation (Fig. 6.1).

Slides

Slides, like slumps and falls, involve the rapid displacement of masses of rock or sediment. In slides the movement is *translational,* that is, uniform movement in one direction with no rotation. *Translational slides* are also called *block glides* because they involve the movement of relatively coherent blocks of material along well-defined, inclined surfaces such as faults, foliation planes (in metamorphic rocks), or layering (in sedimentary rocks or alternating sequences of rock types). A *rockslide* is the sudden downslope movement of detached masses of bedrock (or of debris, in the case of a *debris slide*) (Fig. 6.1). Like falls, rockslides and debris slides are common in high mountains where steep slopes abound.

Flows

When a sufficiently large force is applied, any deformable material will begin to flow. In mass-wasting, the force is gravity and the material consists of dense mixtures of sediment and water (or sediment, water, and air). Mass-wasting processes that involve the movement of such mixtures are called **flows**. The way a sediment flows depends on the relative proportions of solids, water, and air in the mixture and on the physical and chemical properties of the sediment. All streams carry at least some sediment, but if the sediment becomes so concentrated that the water can no longer transport it, a sediment-laden stream becomes a very fluid sediment flow. In such a case the water helps promote flow, but the pull of gravity remains the primary reason for movement.

In Fig. 6.3 flows are subdivided into two classes—*slurry flows* and *granular flows*—based on water content (i.e., concentration of sediment). Slurry flows are water saturated mixtures, whereas granular flows are not water saturated. Each of these two classes is further subdivided on the basis of the velocity of the flow, which can range from very slow (millimeters or centimeters per year) to very fast (kilometers per hour). In this classification of flows, the boundaries between processes are approximate and depend on the size distribution of the grains, the concentration of the sediment, and other factors.

Slurry Flows

A *slurry flow* is a moving mass of water-saturated sediment. In slurry flows, the sediment mixture is often so dense that large boulders can be suspended in it. Boulders that are too large to remain in suspension may be rolled along by the flow. When the flow ceases, fine and coarse particles remain mixed, resulting in an unsorted sediment.

Very slow downslope movement of water-saturated soil and regolith is known as *solifluction*. As can be seen in Fig. 6.3, solifluction lies at the lower end of the velocity scale

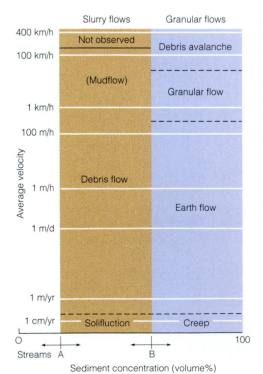

◄ **F I G U R E 6.3**

Classification of sediment flows on the basis of their average velocity and sediment concentration. The transition from a sediment-laden stream to a slurry flow occurs when the concentration of sediment becomes so high that the stream no longer acts as a transporting agent; instead, gravity becomes the primary force causing the saturated sediment to flow. As the percentage of water decreases further, a transition from slurry flow to granular flow takes place. Now the sediment may contain water and/or air. The boundaries between muddy streams and slurry flows (A) and between slurry and granular flows (B) are not assigned sediment-concentration percentages because they can shift to the left or right depending on the physical and compositional characteristics of the mixture. Different types of slurry and granular flows are recognized on the basis of their mean velocity.

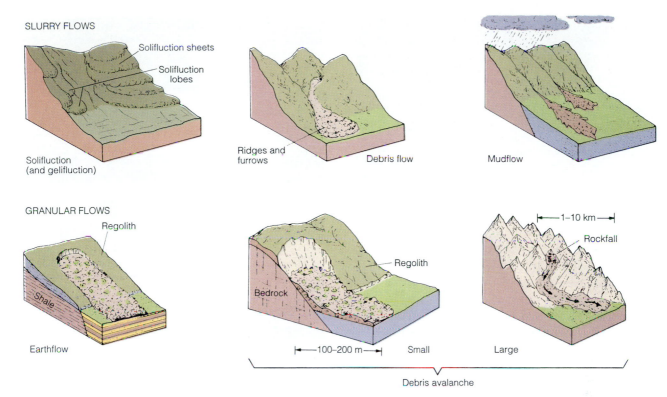

▲ F I G U R E 6.4
Examples of slurry flows and granular flows.

for flowing sediment–water mixtures. The rates of movement of such mixtures are generally so slow as to be detectable only by measurements made over several seasons. Solifluction occurs on hill slopes where sediment remains saturated with water for long intervals. It results in distinctive surface features, including lobes and sheets of debris that sometimes override one another (Figs. 6.4 and 6.5).

A *debris flow* involves the downslope movement of regolith whose consistency is coarser than that of sand, at rates ranging from 1 m/year to as much as 100 m/h (Fig. 6.3). In some cases a debris flow begins with a slump or debris slide, whose lower part then continues to flow downslope (Figs. 6.4 and 6.6). Once mobilized, a typical debris flow moves along a stream channel and may then spread

◄ F I G U R E 6.5
A meter-thick solifluction lobe has slowly moved downslope and covers glacial deposits on the floor of the Orgière Valley in the Italian Alps.

▲ **F I G U R E   6.6**
Debris slides that turned into debris flows on a steep mountainside in southern Puerto Rico have stripped away vegetation and inundated two houses at the base of the slope.

out to form a poorly sorted deposit. Debris flow deposits commonly have a tongue-like front and a very irregular surface, often with concentric ridges and depressions. They are frequently associated with intervals of extremely heavy rainfall that lead to oversaturation of the ground.

A debris flow that has a water content sufficient to make it highly fluid is commonly called a *mudflow*. In Fig. 6.3, the velocity of mudflows lies at the upper range of the velocity scale for debris flows (more than about 1 km/h). Most mudflows are highly mobile and tend to travel rapidly along valley floors (Fig. 6.4). The consistency of mudflow sediment can range from freshly poured concrete to a souplike mixture only slightly denser than very muddy water. After a heavy rain in a mountain canyon, a mudflow can start as a muddy stream that continues to pick up loose sediment until its front becomes a moving dam of mud and rubble, extending to each wall of the canyon and urged along by the force of the flowing water behind it (Fig. 6.7). When it reaches open country the moving dam collapses, floodwater pours around and over it, and mud mixed with boulders is spread out in a wide, thin sheet.

On active volcanoes in wet climates, layers of tephra and volcanic debris commonly cover the surface and are easily mobilized as mudflows called *lahars*. When closely associated with an actual eruption, lahars can be very hot. A particularly large mudflow that originated on the slopes of Mount Rainier about 5700 years ago traveled at least 72 km. The sediment spread out as a broad lobe as much as 25 m thick; its volume is estimated to be well over a billion cubic meters. Throughout much of its history Mount St.

A.

B.

C.

▲ **F I G U R E   6.7**
Passage of a muddy debris flow along a canyon near Farmington, Utah, in June 1983. A. The boulder-laden front of a muddy debris flow advances from left to right along a stream channel in the wake of an earlier surge of muddy debris. B. The steep front, about 2 m high and advancing at 1.3 m/s, acts as a moving dam, holding back the flow of muddy sediment upstream. C. The main slurry, having a sediment concentration of about 80 percent and now moving at about 3 m/s, is thick enough to carry cobbles and boulders in suspension.

Helens has produced mudflows, the most recent of which occurred during the huge eruption of May 1980 (Fig. 6.8). And as discussed in Chapter 4, lahars that resulted from the mixing of ash with typhoon rains caused extensive damage during the 1991 eruption of Mount Pinatubo in the Philippines.

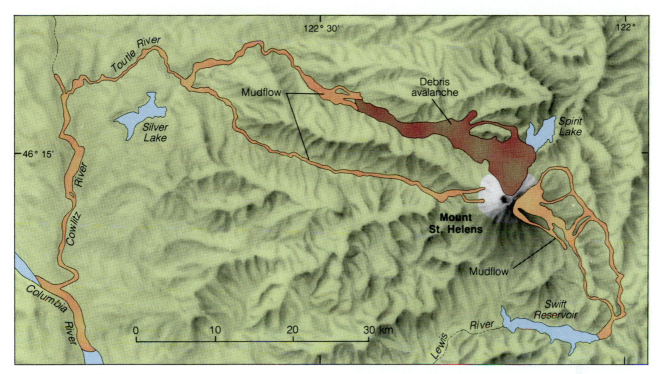

▲ **F I G U R E 6.8**
During the 1980 eruption of Mount St. Helens in Washington, volcanic mudflows were channeled down valleys west and east of the mountain. Some mudflows reached the Columbia River after traveling more than 90 km. Flow velocities were as high as 40 m/s and averaged 7 m/s.

Granular Flows

A *granular flow* is a mixture of sediment, air, and water but, unlike a slurry flow, it is not saturated with water; instead, the weight of the flowing sediment is supported by contact or collision between grains. The sediment of granular flows may be largely dry, with air filling the pores, or it may con-

tain water but include a range of grain sizes and shapes that allows the water to escape easily.

Creep is an imperceptibly slow granular flow. Most of us have seen evidence of creep in curved tree trunks or old fences, telephone poles, or gravestones leaning at an angle on hill slopes (Fig. 6.9). Steeply inclined rock layers may be

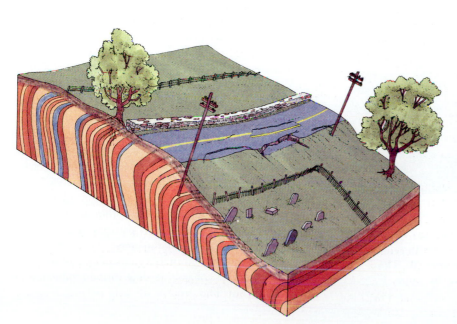

◄ **F I G U R E 6.9**
Effects of creep on surface features and bedrock. Steeply inclined rock layers have been dragged over near the surface by creep so they appear folded. Telephone poles and fence posts affected by creep are tilted, stone fences are deformed, roadbeds are locally displaced, and gravestones are tilted or have fallen.

bent over in the downslope direction just below the surface of the ground, another sign of creep. A number of factors contribute to creep, including the growth and decay of plants (which can bind sediment particles together or wedge them apart); the activities of animals (such as bur-

rowing or trampling); and heating, cooling, wetting, and drying (all of which cause changes in the volume of mineral particles). However, as with all types of mass-wasting, gravity is the main downslope force.

Although creep occurs too slowly to be seen, careful measurement of the downslope displacement of objects at the surface enables scientists to record the rates involved (Fig. 6.10). As might be expected, rates of creep tend to be higher on steep slopes than on gentle slopes. Measurements in Colorado, for example, document a creep rate of 9.5 mm/year on a slope of 39° but a rate of only 1.5 mm/year on a 19° slope. Creep rates also tend to increase as the amount of moisture in the soil increases. However, in wet climates the density of vegetation also increases, and roots, which bind the soil together, tend to inhibit creep. Despite the slow rates of movement involved, creep affects all hill-slopes covered with regolith, and its cumulative effect is therefore very great.

Earthflows are among the more common types of mass-wasting. An *earthflow* is a downslope granular flow that is more rapid than creep (Fig. 6.3). Earthflows may continue for several days, months, or even years. Even after their initial motion ceases, they may be highly susceptible to renewed movement. Earthflows occur where the ground is saturated intermittently, and they are frequently associated with intervals of excessive rainfall. An earthflow typically heads in a steep **scarp**, a cliff that is formed where displaced material has moved away from undisturbed ground upslope (Fig. 6.11).

If you have ever walked along the crest of a sand dune and stepped too close to the steep slope that faces away from the wind, your footsteps likely started a cascade of sand flowing down the dune face. This is an example of still another type of mass-wasting, called *grain flow,* which involves the movement of a dry or nearly dry granular sediment with air filling the pore spaces.

A *debris avalanche* is a type of granular flow that travels at high velocity (tens to hundreds of kilometers per hour) and can be extremely destructive (Fig. 6.4 and Table 6.3). Large debris avalanches are rare but spectacular events involving huge masses of falling rock and debris that break up, pulverize on impact, and then continue to travel downslope, often for great distances. Because they are infrequent and extremely difficult to study while they are occurring, there are few observational data about the processes involved. It has been suggested that the debris actually rides on a layer of compressed air. If this is true, debris avalanches behave somewhat like a commercial hovercraft that travels across land or water on air compressed by a large propeller. Alternatively, air trapped and compressed within the moving debris may reduce friction between particles and cause the mass to behave in a highly fluid manner.

The slopes of steep, unstable stratovolcanoes are especially susceptible to collapse, leading to the production of debris avalanches. The deposits of such avalanches can be

A.

B.

▲ **F I G U R E   6.10**
Colored targets placed in a straight line across a hill-slope in Greenland (A) had been moved differentially by creep when photographed a year later (B). The maximum recorded movement along the slope averaged 12 cm/year.

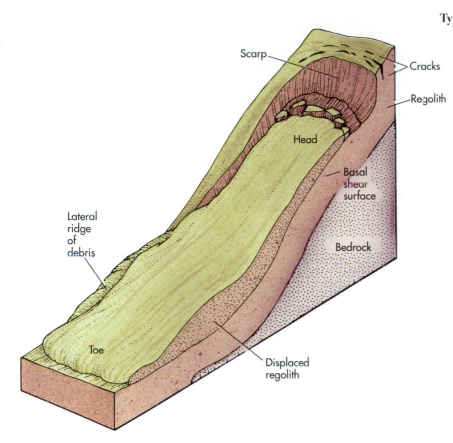

Scarp
Cracks
Regolith
Head
Basal shear surface
Bedrock
Lateral ridge of debris
Toe
Displaced regolith

▲ **F I G U R E 6.11**
In this section through an idealized earthflow, regolith moves downslope across a basal shear surface, leaving a scarp at the head, where the sediment separated from the slope above. A bulging toe protrudes beyond ridges of sediment that have piled up along the lower margins of the earthflow.

difficult to recognize because of their huge dimensions. For example, a broad valley that extends some 40 km north of Mount Shasta in northern California contains a complex of hills and mounds of volcanic rock that resulted from the collapse of a flank of the volcano about 300,000 years ago. The volume of rock involved was at least 26 km³—almost 10 times the volume of the huge debris avalanche associated with the 1980 eruption of Mount St. Helens. Whereas the St. Helens deposits cover an area of 60 km², the Shasta debris avalanche overwhelmed at least 450 km².

Mass-Wasting in Cold Climates

Mass-wasting is especially prevalent at high latitudes and high altitudes, where average temperatures are very low. In such regions much of the landscape is underlain by perennially frozen ground, and frost action is an important geologic process. When water freezes, its volume increases. Ice forming in saturated regolith therefore pushes up the ground surface in a process called **frost heaving**. Frost heaving strongly influences the downslope creep of sedi-

T A B L E 6.3 • **Characteristics of Some Large Debris Avalanches**

Locality	Date	Volume (million m³)	Vertical Movement (m)	Horizontal Movement (km)	Calculated Velocity (km/h)
Huascarán, Peru	1971	10	4000	14.5	400
Sherman Glacier, Alaska	1964	30	600	5.0	185
Mount Rainier, Washington	1963	11	1890	6.9	150
Madison, Wyoming	1959	30	400	1.6	175
Elm, Switzerland	1881	10	560	2.0	160
Triolet Glacier, Italy	1717	20	1860	7.2	≥125

ment in cold climates. When freezing occurs, the ground surface is lifted essentially at right angles to the slope. As the ground thaws, each particle tends to drop vertically, pulled downward by gravity. The net result of repeated episodes of freezing and thawing, during which a particle experiences a succession of upward and downward movements, is slow but progressive downslope creep.

In cold regions that are underlain by frozen ground year round, a thin surface layer thaws in summer and refreezes in winter. During the summer the thawed layer becomes saturated with meltwater and is very unstable, especially on hillsides. As gravity pulls the thawed sediment slowly downslope, distinctive lobes and sheets of debris are produced. This process, which is similar to solifluction in temperate and tropical climates, is known as *gelifluction*. Although measured rates of movement are generally less than 10 cm/year, gelifluction is so widespread on high-latitude landscapes that it constitutes a highly important agent of mass transport. Hill slopes in arctic Alaska and Canada, for example, are often mantled by sheets or lobes of gelifluted regolith.

▲ **F I G U R E 6.12**
In the Wrangell Mountains of southern Alaska, a jumbled mass of angular rock debris supplied from a steep cliff moves slowly downslope as a rock glacier.

A *rock glacier,* another characteristic feature of many cold, relatively dry mountain regions, is a tongue or lobe of ice-cemented rock debris that moves slowly downslope in a manner similar to the movement of a glacier (Fig. 6.12). Rock glaciers generally originate below steep cliffs, which provide a source of rock debris. Active rock glaciers may reach a thickness of 50 m or more and advance at rates of up to about 5 m/year. They are especially common in high interior mountain ranges such as the Swiss Alps, the Argentine Andes, and the Rocky Mountains.

Subaqueous Mass-Wasting

As geologists have extended the search for petroleum to offshore regions, their explorations have shown that mass-wasting is an extremely common and widespread means of sediment transport on the sea floor. Mass-wasting also has been documented in lakes. As on land, the potential for gravity-induced movement of rock and sediment exists wherever there are *subaqueous* (underwater) slopes.

Extensive studies of the offshore slopes of eastern North America using a variety of modern techniques (deep ocean drilling, sonar, echograms, and submersible vessels, among others) have shown that vast areas of the sea floor are disrupted by submarine slumps, slides, and flows. Some large slide complexes cover areas of more than 40,000 km² and reach depths as great as 5400 m. A subaqueous slope failure can give rise to a *turbidity current,* a type of sediment flow that travels down submarine canyons and deposits sediments off the continental shelf. Such flows have been known to cause extensive damage to underwater communications cables.

Major marine deltas also frequently display surface features and sediments that can be attributed to slope failures. In such subaqueous environments, failure can occur even on slopes as low as 1°. The slope failures generally display three distinct zones: a source region where subsidence (sinking or downward settling) and slumping take place; a central channel where sediment is transported; and a zone where sediment is deposited, often in the form of overlapping lobes of debris. These and other features are common on the submarine slopes of the Mississippi Delta, among the best studied deltas in the world (Fig. 6.13). In places, more than 30 m of sediment have been deposited by sediment flows in the last 100 years.

FACTORS THAT INFLUENCE SLOPE STABILITY

Under natural conditions, a slope evolves toward an angle that allows the quantity of regolith reaching any point from upslope to be balanced by the quantity that is moving downslope from that point. Such a slope is said to be in a balanced, or *steady-state,* condition. The slope may appear

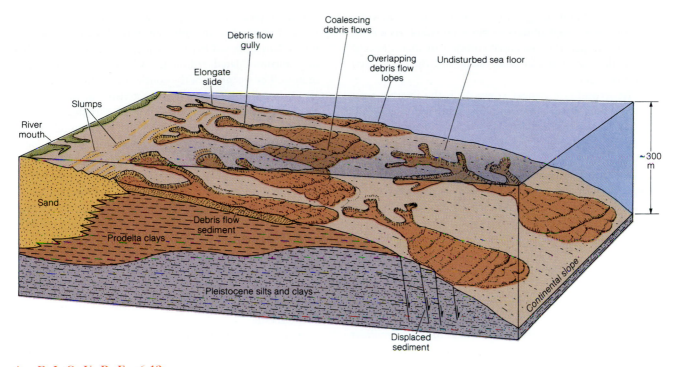

Coalescing
debris flows

▲ **F I G U R E 6.13**
**Block diagram showing various mass-wasting features on the submarine surface of
the Mississippi Delta.**

stable and show little evidence of geologic activity. Yet if we
examine the regolith beneath the surface we will probably
find some rock particles derived from bedrock located farther upslope. We can deduce, therefore, that the particles
have moved downslope. As you have just learned, the particle's downward journey can be very slow or very fast, but in
either case the movement is controlled primarily by gravity.

Many factors affect slope stability. A change in any one or
a combination of these factors can alter the steady-state condition of the slope, decreasing its stability and sometimes
leading to slope failure. In some cases, the change may take
the form of a relatively sudden triggering event, whether natural (such as an earthquake) or human-generated (such as an
explosion). In other cases, the slope may have been slowly
changing over time; again, the cause can be natural (such as a
long period of intense precipitation) or a result of human activities (such as the construction of a dam).

The main factors that influence slope stability are (1)
the force of gravity, and therefore the gradient of the slope;
(2) water, and therefore the hydrologic characteristics of the
slope; (3) the presence of troublesome Earth materials; and
(4) the occurrence of a triggering event. In this section we
examine each of these factors in detail.

Gravity and Slope Gradient

Two opposing forces determine whether a body of rock or
debris located on a slope will move or remain stationary.

These forces are *shear stress* and *shear strength*. The first of
these forces, **shear stress**, causes movement of the body
parallel to the slope. The primary factor influencing shear
stress is the pull of gravity, which is related to the slope's
gradient, or steepness. On a horizontal surface, gravity
holds objects in place by pulling on them in a direction
perpendicular to the surface (Fig. 6.14). On any slope,
however, gravity consists of two component forces. The
perpendicular component (g_p in Fig. 6.14) acts at right angles to the slope and tends to hold objects in place. The

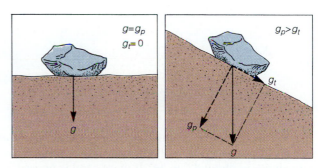

▲ **F I G U R E 6.14**
**Effects of gravity on a rock lying on a hillslope.
Gravity acts vertically and can be resolved into two
components, one perpendicular (g_p) and the other
parallel (g_t) to the surface.**

tangential component (g_t in Fig. 6.14) acts along and down the slope and causes objects to move downhill. As a slope becomes steeper, the tangential component increases relative to the perpendicular component and the shear stress becomes larger.

The second force, **shear strength**, is the internal resistance of the body to movement. Shear strength is governed by factors inherent in the body of rock or regolith, such as friction and cohesion between particles and the binding action of plant roots. As long as shear strength exceeds shear stress, the rock or debris will not move. However, as these two forces approach a balance, the likelihood of movement increases. This relationship is expressed in a ratio known as the **safety factor** (denoted *Fs*):

$$Fs = \frac{shear\ strength}{shear\ stress}$$

When the safety factor is less than 1 (i.e., shear strength is less than shear stress), slope failure is imminent.

The relationship between shear stress and slope gradient (the steeper the slope, the greater the shear stress) means that conditions favoring mass movement tend to increase as slope angle increases. Steep slopes, of course, are most common in mountainous areas, so it is not surprising that mass-wasting is most frequent in high mountains.

Water

Water is almost always present within rocks and regolith near the Earth's surface, and it plays a variety of important roles in mass-wasting of both solid rock and regolith. Unconsolidated (loose, uncemented) sediments behave in different ways depending on whether they are dry or wet, as anyone knows who has constructed a sand castle at the beach. Dry sand is unstable and difficult or impossible to mold. When poured from a bucket, dry sand (or any other dry, unconsolidated sediment) will form a cone-shaped mound. The steepness of the cone's sides, called the *angle of repose,* is determined by the characteristics of the material, primarily the size and angularity of the particles. Sand, for example, will always pile up with slopes of about 32° to 34° (Fig. 6.15). When a little water is added, the sand gains strength; its angle of repose is greater, so it can be shaped into vertical walls. The water and sand grains are drawn together by *surface tension,* a property of liquids that causes the exposed surface to contract to the smallest possible area. This force tends to hold the wet sand together as a cohesive mass. However, the addition of *too* much water saturates the sand; the spaces fill with water, and the sand grains lose contact with one another. The mixture turns into a slurry that easily flows away, as the sand castle builder sees with dismay when the rising tide destroys the elaborate work of an afternoon.

Moist or weakly cemented fine-grained sediments, such as fine silt and clay, may be so cohesive that they can stand in near-vertical cliffs. But if the silt or clay becomes saturated with water and the internal fluid pressure rises, the fine-grained sediment may also become unstable and begin to flow like the water-saturated sand castle.

The movement of some large masses of rock has been attributed to water pressure in voids in the rock. If the voids along a surface separating two rock masses are filled with water, and the water is under pressure, a buoying effect may result. In other words, the water pressure may be high enough to support the weight of the overlying rock mass, thereby reducing friction along the points of contact. The result can be a sudden failure. An analogous situation can make driving in a heavy rainstorm extremely dangerous.

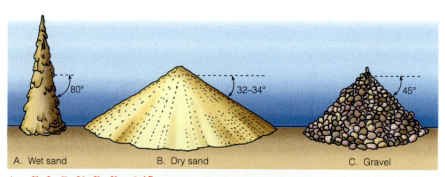

▲ F I G U R E 6.15
Angle of repose in unconsolidated materials. A. Wet sand can be piled up steeply, but (B) dry sand will always come to rest in a cone-shaped mound with slopes of about 32° to 34°. If more sand is poured onto the pile, it will simply roll down the slopes. C. Coarser material, such as gravel, will have a steeper angle of repose. Moisture content and the angularity of particles can also affect a material's angle of repose.

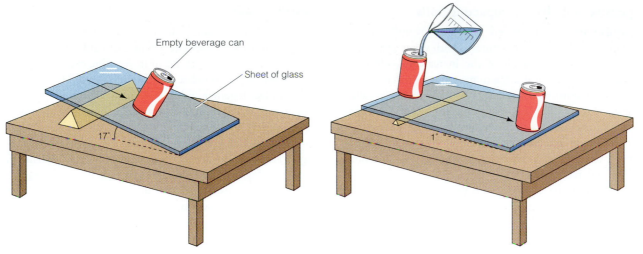

Empty beverage can

Sheet of glass

17°

1°

▲ F I G U R E 6.16

An experiment illustrating how water can reduce friction at the base of a mass resting on a slope. A. An empty beverage can placed on a wet sheet of glass will begin to slide down the surface when the angle reaches about 17°. B. If a small hole is made in the bottom of the can and water is poured in through the top, the can will begin to slide down the slope at a much lower angle. In an analogous way, high water pressure at the base of a large rock mass may promote downslope movement of the rock.

When water is compressed beneath the wheels of a moving car, the increasing fluid pressure can cause the tires to "float" off the roadway, a condition known as hydroplaning.

You can demonstrate this principle by conducting a simple experiment (a variation on a classic experiment first described by geologists M. King Hubbert and William Rubey in 1959). An empty beverage can is placed in an upright position on the wetted surface of a sheet of glass (Fig. 6.16A). If the glass is slowly tilted, the can will not begin to slide until a certain critical angle is reached. For the particular substances used in this demonstration (metal and wet glass), sliding begins at an angle of approximately 17°. Next, punch a small hole in the bottom of the can. Place it on the glass sheet and slowly pour some water into the open top (Fig. 6.16B). The can will begin to slide at a much gentler angle because the conditions at the base of the can have changed. In an analogous way, high water pressure at the base of a large mass of rock may promote downslope movement of the rock.

These examples show that water can be instrumental in reducing shear strength and thereby promoting the movement of rock and sediment downslope under the pull of gravity. It does so by reducing the natural cohesiveness between grains or by reducing friction at the base of a mass of rock through increased water pressure. As we will see shortly, water can act in other ways that contribute to slope instability, such as undercutting the base of the slope or altering the chemical composition of the sediment.

Troublesome Earth Materials

Some Earth materials are particularly susceptible to the types of changes and disturbances that can lead to slope failure. Such materials, sometimes referred to as *problem soils,* are often involved in mass-wasting.

Liquefaction

As discussed earlier, when water is added gradually to a dry soil, the material first becomes *plastic,* or moldable. If enough water is added, the particles lose contact with one another and the material turns into a loose slurry, losing its shear strength in the process. The transformation of a soil from a solid to a liquid state, usually (but not always) as a result of increased water content, is called **liquefaction**. The point at which this transition occurs, called the *liquid limit,* varies from one soil to another. Some materials, particularly some clay-bearing soils, have very high liquid limits and may remain plastic over a broad range of water contents. These soils can be particularly troublesome because by the time the liquid limit is exceeded, the moisture content of the soil is so high that the material behaves in an extremely fluid manner.

Expansive and Hydrocompacting Soils

Expansive soils, also referred to as *shrink–swell soils,* expand greatly when they are saturated with water and shrink when they dry out. Much of the increase in volume is caused by the chemical attraction of water molecules between the submicroscopic layers of clay minerals called *smectites* (Fig. 6.17A). Expansion resulting from increased water content can drastically reduce the shear strength of an Earth material, often contributing to downslope movement. It can also cause extensive damage to structures built on that material (Fig. 6.17B).

When expansive clays dry out, they undergo a decrease in volume. The process of shrinkage and/or collapse resulting from water loss is referred to as *compaction* (or, more precisely, *hydrocompaction,* since soil compaction can be caused by other processes besides water loss). In addition to expansive clays, there are other types of soils that can exhibit substantial decreases in volume when they dry out. Extreme compaction associated with drying is a particular problem in water-saturated, organic-rich soils such as peat. Soil compaction often results in *subsidence,* a type of mass movement involving the lowering or collapse of the ground surface. (Subsidence is discussed in greater detail in Chapter 7.)

Sensitive Soils

In some clay-rich soils, the particles are arranged in an open, porous structure like a house of cards (Fig. 6.18A). Such an arrangement can occur in very fine marine clays, in which salt acts to stabilize the "house of cards" by "gluing" the particles together end to end. Eventually, fresh groundwater may move through the area, changing the chemical composition of the clay and washing away the salt. Without the stabilizing effect of the salt, the clay particles collapse and take on a new, more compact arrangement, as shown in Fig. 6.18B. The transition from open to compact arrangement causes a sudden and dramatic loss of shear strength—in other words, liquefaction—which can propagate with astonishing speed throughout the entire mass of clay. Liquefaction or compaction that results from a disturbance of the internal structure of a soil is referred to as *remolding.* Materials that lose shear strength as a result of remolding are called **sensitive soils**. Those that are the most susceptible to remolding and liquefaction are called *quick clays.* Some other types of soils lose their shear strength suddenly when disturbed but gradually strengthen and resume their original properties when left undisturbed; these are referred to as *thixotropic clays.*

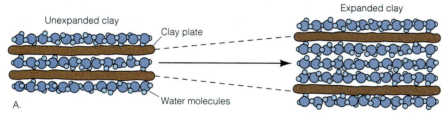

◄ **F I G U R E 6.17**
A. Expansion of a single smectite grain as a result of the addition of water between clay layers. B. The type of structural damage that can result from the expansion of soil beneath a building.

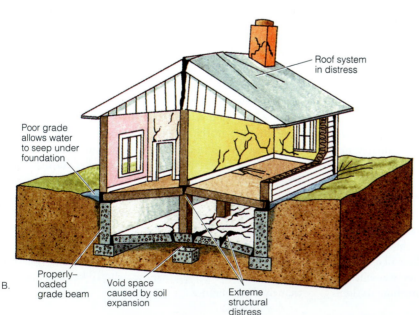

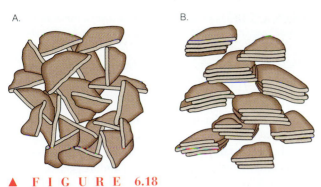

▲ **F I G U R E   6.18**
Sensitive clays. A. The "house-of-cards" structure of microscopic plates in a sensitive clay. B. Structural change in a sensitive clay as a result of compaction, remolding, or thixotropism.

The most extensive deposits of quick clay are found in Canada (in the St. Lawrence River Valley) and Scandinavia (in Norway and Sweden). In both cases the clays were deposited in shallow, quiet marine environments at the edges of glaciers. Both areas have been subjected to damaging mass-wasting events resulting from the remolding and liquefaction of the quick clays. A well-documented example is the quick clay slide that occurred in Rissa, Norway, in 1978. This event began when a farmer made a small excavation for a new barn, piling the excavated soil along the shore of a nearby lake. The extra load on the quick clay at the edge of the lake initiated a small slide, which quickly propagated more than a kilometer up the valley behind the farmhouse. What had been solid ground completely liquefied. Entire buildings were carried intact toward the lake by the fluidized mass, traveling at rates of 20 km/h or more. After the slide, the Norwegian Geotechnical Institute carried out extensive studies and testing in the area. Some remaining quick clay was removed, and the rest of the slopes in the area were regraded and stabilized.

Triggering Events

As in the example of the Rissa quick clay slide, slope failures are often triggered by some extraordinary activity or occurrence. It is also very common for a combination of conditions to lead to slope failure: a moderate-size earthquake might not generate landslides under normal conditions, but it could do so in an area underlain by sensitive soils; a slope steepened by construction might not fail under normal conditions, but it might during a period of exceptionally heavy precipitation. Among the most common types of triggering events are earthquakes, volcanic eruptions, slope modifications, and changes in the hydrologic characteristics of an area (including the effects of prolonged or exceptionally intense rainfall).

Earthquakes and Other Shocks

An abrupt shock, such as an explosion, an earthquake, an electrical storm, or even a truck passing by, can increase shear stress and contribute to slope failure. Intense shaking can cause a buildup of water pressure in the pore spaces of a sediment, leading to liquefaction. In other words, liquefaction is not always related to an increase in water content; sometimes shaking causes the pore water *already* present in the sediment to coalesce so that the sediment grains lose contact with one another. The result is fluidization of the sediment and abrupt failure. Any structure built on such sediments or in their path may be demolished (Fig. 6.19).

▲ **F I G U R E   6.19**
Chaotically tilted trees and houses in suburban Anchorage, Alaska show how violent shaking of the ground during the great 1964 earthquake caused sudden liquefaction of underlying clays and widespread slumping.

Earthquakes frequently generate landslides. In 1970 a large earthquake in Peru triggered a debris avalanche that roared more than 3.5 km down the steep, rocky slopes of Mount Huascarán, reaching speeds of 400 km/hr. The villages of Yungay and Ranrahirca were destroyed and as many as 20,000 people killed (Fig. 6.20). An eyewitness to the event (a geophysicist, by chance) made the following report:

I heard a great roar coming from Huascarán. Looking up, I saw what appeared to be a great cloud of dust and it looked as though a large mass of rock and ice was breaking loose from the north peak. . . The crest of the wave [of rock and ice] had a curl, like a huge breaker coming in from the ocean. I estimated the wave to be at least 80 m high. I observed hundreds of people in Yungay running in all directions and many of them towards

Cemetery Hill. All the while, there was a continuous loud roar and rumble. I reached the upper level of the cemetery near the top just as the debris flow struck the base of the hill and I was probably only 10 seconds ahead of it . . . It was the most horrible thing I have ever experienced and I will never forget it.

Some earthquakes release so much energy that many landslides of different types and sizes are triggered simultaneously. For example, in 1929 a major earthquake in northwestern South Island, New Zealand, triggered at least 1850 landslides larger than 2500 m^2 within a 1200 km^2 area near the quake's center. As mentioned in Chapter 3, the great 1964 earthquake in Alaska also generated many landslides, including a spectacular one in Turnagain Heights, a residential section of Anchorage. In that slide, a section of coastline 2.5 km long broke off up to 0.5 km in-

▲ F I G U R E 6.20
View of Mt. Huascarán, Peru and the debris avalanche that destroyed the villages of Yungay (remains are lower right) and Ranrahirca, in the May 1970 earthquake.

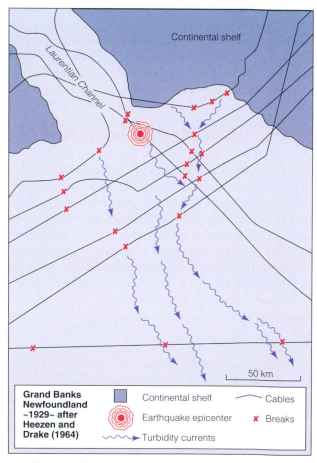

Continental shelf

Laurentian Channel

50 km

Grand Banks Newfoundland ~1929~ after Heezen and Drake (1964)

Continental shelf
Earthquake epicenter
Turbidity currents
Cables
✕ Breaks

▲ F I G U R E 6.21
Submarine slope failures off Grand Banks, Newfoundland, 1929. Turbidity currents generated by an earthquake traveled at least 470 km, breaking telephone cables on the ocean floor.

BOX 6.1
·
THE HUMAN PERSPECTIVE

VALLEY FEVER LINKED TO LANDSLIDES

After the initial devastation of the 1994 Northridge (Los Angeles) earthquake, residents in a widespread area surrounding Los Angeles continued to experience repercussions for several months. In Ventura County, for example, 166 cases of acute coccidioidomycosis, commonly known as valley fever, were reported from January through March—a dramatic increase over the 53 cases reported in the entire previous year.

Valley fever is contracted by inhaling airborne spores of *Coccidioides immitis,* a fungus that lives in the top few inches of surface soils. The illness is not passed from person to person. In January 1994, the Northridge earthquake triggered thousands of landslides in the Santa Susana Mountains, which form the northern edge of the San Fernando and Simi Valleys. The landslides generated huge clouds of dust that were blown by prevailing northeasterly winds across Simi Valley and eastern Ventura County.

The Santa Susana Mountains consist of very young, weak sedimentary rock that is extremely susceptible to sliding during earthquakes. The highest concentration of valley fever cases were found directly downwind from the area of the most concentrated landslides. Little change in the number of valley fever cases was reported in the more highly populated San Fernando Valley, however, where the earthquake was centered. Therefore, researchers from the U.S. Geological Survey and the Centers for Disease Control deduced that landslides triggered the Simi Valley/Ventura County valley fever outbreak.

---

land, and slid toward the ocean in great blocks. The movement of the blocks was facilitated by liquefaction, in which sandy layers underlying the blocks lost their shear strength and were reduced to a slurry as a result of the shaking. The landslide began approximately 2 minutes after the earthquake and lasted for 5 minutes, during which time the sliding blocks moved as far as 300 m and created an entirely new coastline and local landscape.

Major submarine turbidity currents and slumps off eastern North America are also known to have been caused by strong earthquakes. One such quake occurred in Charleston, South Carolina, in 1886, and another in Cape Ann, Massachusetts, in 1755. In 1929 an earthquake off Grand Banks, Newfoundland, was followed by a succession of breaks in underwater telephone cables (Fig. 6.21). In piecing together the evidence from the sequence and locations of the cable breaks, it was determined that the quake had generated a turbidity current at least 150 km wide in which an enormous volume of sediment was moved. The distance from the source area to the farthest cable break was 470 km, and the maximum velocity of the submarine slide was 93 km/h.

Volcanic Eruptions

Volcanic eruptions are another mechanism for triggering mass-wasting events. Large stratovolcanoes consist of inherently unstable accumulations of interlayered lava flows, rubble, and pyroclastic material that form steep slopes. The slopes of high, ice-clad volcanoes may be further steepened by glacial erosion. Large volumes of water, released when summit glaciers and snowfields melt during an eruption of hot lavas or pyroclastic debris, can combine with unconsolidated deposits to form rapidly moving lahars. As in the case of Armero, Colombia, these highly fluid mudflows can travel great distances and at such high velocities that they constitute one of the major hazards associated with volcanic eruptions.

Slope Modifications and Undercutting

Landslides often result when natural slopes are modified, either by natural processes or by human activities. Translational slides can occur, for example, where roads have been cut into regolith or unstable rock, creating an artificial slope that exceeds the angle of repose or exposes natural planes of weakness (Fig. 6.22). Such landslides are especially common along mountainous and coastal cliffs where roads have been carved into deformed sedimentary or metamorphic rocks.

Overloading—placing a building or a mass of excavated material at the top of a slope, for example—can also contribute to slope failure because of the added weight as well as the steepening effect of the load. In 1966, for example, ex-

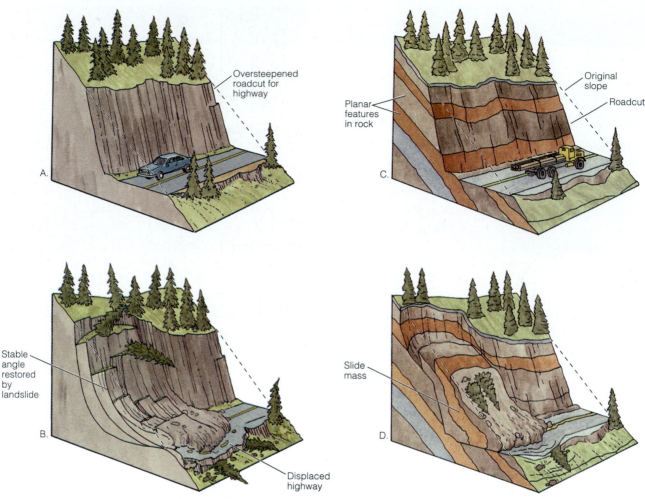

▲ F I G U R E 6.22
Modification of a slope during road construction can lead to slope failure. A. The natural
angle of repose of the material in the slope is exceeded. B. The oversteepened slope fails,
and a landslide buries the road. In the process, the natural angle of repose is reestablished.
C. Roadcuts can expose natural planes of weakness, such as bedding planes in sedimentary
rock or foliation in metamorphic rock, that can contribute to slope failure (D).

cessive rain in Rio de Janeiro caused an oversteepened slope in a road cut to fail. The mass of material involved in this small slide overloaded the top of the underlying slope, triggering a large landslide that destroyed several houses and two apartment buildings and killed 132 people (Fig. 6.23).

Oversteepened artificial slopes also fail occasionally. Perhaps the most famous example occurred in 1966 in the coal-mining town of Aberfan in Wales. The debris left over from the mining process—mostly very fine clay, in this case—was routinely piled up in large artificial hills called *tips.* One morning in October, a major slope failure occurred on one of the tips (Fig. 6.24). It was later determined that drainage within the tip had been inadequate, resulting in saturation and eventually liquefaction of the lower portion. The upper part of the tip moved as a coher-

ent mass, carried along on the liquefied material below. When the mass of fluidized material came to rest, it had lost most of its water content and returned to its original solid state. Part of the town was engulfed by the landslide, including the elementary school where 116 children and 5 teachers were killed; 144 people were killed in all.

Slumps and other types of landslides can also be triggered by the natural steepening of slopes as a result of the undercutting action of a stream along its bank or waves along a coast. This is a contributing factor in the continuing movement of a large slide block at Portugese Bend, California (Box 6.2). Coastal landslides are often associated with major storms that direct their energy against rocky headlands or the bases of cliffs composed of unconsolidated sediments (Fig. 6.25).

▲ F I G U R E 6.23
An oversteepened roadcut caused an initial failure of
this slope in Rio de Janeiro, Brazil. The weight of the
initial slide overloaded the slope and caused a major
landslide. One hundred and thirty-two people are
known to have been killed; others probably remain
buried.

▲ F I G U R E 6.24
The failure of coal tip No. 7 at Aberfan, Wales, in
October 1966.

Changes in Hydrologic Characteristics

Changes in the characteristics of subsurface water or
drainage in an area often contribute to landslides. For example, heavy or persistent rains may saturate the ground and
make it unstable. Such was the case in 1925 when prolonged rains, coupled with melting snow, started a large debris flow in the Gros Ventre River basin of western
Wyoming. The water saturated a porous sandstone overlying an impermeable rock unit that sloped toward the valley
floor. This water-saturated condition was an ideal trigger for
slope failure. An estimated 37 million m³ of rock, regolith,
and organic debris moved rapidly downslope and created a
natural dam that ponded the river. Two years later the natural dam failed, causing a flood that resulted in several
deaths. Today, more than 70 years after the debris flow

▲ F I G U R E 6.25
Steep cliffs along the coast of Hawaii are undercut
by pounding surf. When a cliff collapses, the resulting
landslide debris is rapidly reworked by waves and
currents, and the process begins anew.

BOX 6.2
•
THE HUMAN PERSPECTIVE

THE PORTUGUESE BEND LANDSLIDE

*R*esidents of Los Angeles face a variety of geologic hazards, including steep, unstable slopes; frequent earthquakes; intense storms separated by extensive periods of drought; and coastal flooding. Major slope failures have been particularly damaging to urban developments on the slopes surrounding Los Angeles. An example is the large and costly landslide at Portugese Bend on the Palos Verdes Peninsula, which has become a classic example for hazard mitigation studies.

The topography of the Palos Verdes Peninsula is cliffed and hummocky, with poorly drained depressions. Slopes are steepened and made unstable by marine erosion at the base of the cliffs, which plunge steeply to the Pacific Ocean. Portugese Bend itself consists of an enormous block of sedimentary rock interlayered with volcanic ash deposits. The volcanic ash has weathered to a clay mineral, bentonite, which readily absorbs water and thus becomes susceptible to downslope movement. The bentonite layers provide a convenient slide surface for the block of sedimentary rock.

The region had been landslide prone for a long time—probably hundreds or thousands of years—before its development as real estate. Evidence of a prehistoric landslide mass at Portugese Bend was documented on geologic maps published in the 1940s, suggesting that the area might best have been left undeveloped. The ancient slide block was dormant until 1956, when unusually heavy precipitation following years of urban development initiated intermit-

tent movement of the mass. The situation was aggravated by earthquakes and by high tides that contributed to erosion. Septic tank drainage, which passed directly into the hillside, and construction of new roads, which added weight to the top of the slide block, compounded the problem.

By the 1980s, 150 homes in the Portugese Bend area had been damaged or destroyed by the continuing movement of the slide block, and $10 million in losses were assessed (Fig. B2.1). The County of Los Angeles was held accountable in court for at least some of the private property losses and had to pay the land owners more than $1 million.

During the 1970s a smaller slide-prone area to the north of Portugese Bend, called Abalone Cove, became active. A management plan was devised to stabilize its slopes by installing dewatering wells. The plan was so successful that in 1985 a similar plan was created for Portugese Bend. In the latter case, engineers recommended surface grading to inhibit infiltration, as well as a network of drainage wells to pump out excess groundwater. New development in the Portugese Bend area has stopped, and stricter building codes have been adopted. These efforts have alleviated the slide hazard in Portugese Bend. The people who have chosen to remain in the active slide area have adjusted to the constant slow movement of their homes; for example, they periodically employ contractors to level buildings, and all utilities are placed at the surface of the property.

B.

A.

▲ F I G U R E B2.1

Damage to homes and road as a result of landslides near Los Angeles, California. A. Houses on Point Fermin sliding into the sea, 1969. B. Fractures resulting from landslides offset lines of a tennis court, Abalone Cove, 1983.

◄ **F I G U R E 6.26**
Debris flow in the Gros Ventre River basin, Wyoming. The flow, which occurred in 1925, left massive scars on the landscape. The source of the flow was the bare area, upper center; the toe of the flow is in the foreground, so the direction of the flow is toward the viewer.

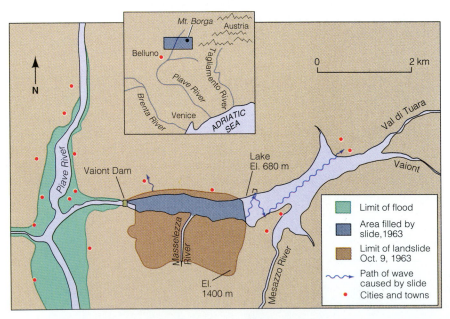

◄ **F I G U R E 6.27**
Sketch map of the Vaiont Dam and Reservoir showing the extent of the 1963 landslide. When the slope failed, a large mass slid into the reservoir, displacing water and causing a giant wave to overflow the dam. Both downstream and upstream areas suffered flooding and damage as a result.

▲ **F I G U R E   6.28**
Photo of Scarborough Bluffs, Toronto, along the shore of Lake Ontario. Along some portions of the shoreline, the cliffs are retreating at a rate of 5m/year as a result of slope failure.

began, the scar at the head of the slide is still quite visible, as is the distinctive chaotic topography downslope (Fig. 6.26).

The filling of a large reservoir can also cause changes in subsurface water conditions. Sometimes the increased water pressure in the pores of the underlying rock combines with other destabilizing factors to produce mass-wasting. Such factors caused the world's worst dam disaster, which occurred in Italy in 1963. A huge mass—almost 250,000,000 m³—of rock and debris slid into the reservoir behind the Vaiont Dam (Fig. 6.27). The material filled the reservoir and created a wave 100 m high, which overflowed the dam and swept both up and down the valley, killing almost 3000 people. There were several driving forces in this event: the slopes were composed of inherently unstable blocks of limestone, with open fractures dipping toward the reservoir; the water impounded in the reservoir had increased pore water pressure in the rock walls of the valley; and a long period of above-average precipitation had also contributed to high water pressures in the valley walls. The landslide itself followed a period of very slow creep that extended over 3 years. The rate of creep had increased dramatically in the days prior to the event, but engineers were caught off guard when they discovered, just 1 day before the landslide occurred, that a huge portion of the valley wall had been moving downslope as a single mass.

Urban and suburban development on hillslopes can lead to hydrologic changes that ultimately contribute to slope failure. For example, in Toronto, Ontario (Fig. 6.28), houses have been constructed along the edge of the Scarborough Bluffs, a series of dramatic cliffs along the shore of Lake Ontario. The cliffs, which are made of fine glacial silts and clays, are continuously steepened by wave action at their base. The sediments are fractured, and subsurface drainage tends to be focused along zones of weakness. Periodic slope failures in these zones have created a series of deeply incised ravines. The problems inherent in the natural drainage system have been amplified by clifftop deforestation, an increased proportion of impermeable (paved) surfaces on the clifftops, and focused drainage from septic tank systems, swimming pools, and other household sources. Along some portions of the shoreline, the cliffs are retreating at a rate of 5 m/year as a result of slope failures.

ASSESSING AND MITIGATING MASS-WASTING HAZARDS

Landslides and other forms of mass-wasting are ubiquitous, and they cause extensive damage and loss of life each year.

◄ F I G U R E 6.29
A new apartment building at the base of a steep mountain slope in the Italian Alps was struck by a large boulder falling from the cliffs above. This relatively small rockfall, which occurred just one day before the new owners were to move in, demolished the bedroom and most of the living room.

With careful analysis and planning, together with appropriate stabilization techniques, the impacts of mass-wasting processes on humans can often be reduced or eliminated.

Prediction and Hazard Assessment

Assessments of the hazards posed by potential mass-wasting events are based on reconstruction of similar past events in order to evaluate their magnitude and frequency; mapping and testing of soil and rock properties; and analysis of slopes to determine their susceptibility to destabilizing processes. Such information can be used in determining how often an event of a certain magnitude is likely to recur in a given locality.

Maps showing areas that could be affected by mass-wasting events are important tools for land-use planners. For example, large debris avalanches and small rockfalls are ever-present hazards in the northern Italian Alps (Fig. 6.29). Field studies have shown that large debris avalanches have repeatedly blanketed valley floors with rocky debris during the last 3000 years. From this evidence, a map has been constructed showing areas that could be affected by future rockfalls with various trajectories and distribution patterns. A number of small communities are at risk from small or medium-size rock avalanches (traveling 3 to 5 km), and several large communities, including one village with a population of several thousand, could be affected by large debris avalanches (traveling up to 7 km) such as those recorded in the deposits on the valley floors.

Valleys in the Cascade Range of Washington and Oregon contain deposits created by large mudflows that repeatedly spread from high volcanoes during the last 10,000 years. Lahar hazard maps have been prepared from information about the number and extent of such deposits, the current status of volcanic activity, and the topography of the slopes. Hazard maps prepared before the 1980 eruption of Mount St. Helens and the 1991 eruption of Mount Pinatubo (Fig. 4.26) proved prophetic, for the mudflows generated by those eruptions had distributions very similar to those predicted on the basis of the geologic studies.

Eliminating or restricting human activities in areas where slides are likely to occur may be the best way to mitigate such hazards. For example, land that is susceptible to mild failures might be suitable for some types of development (e.g., recreation or parkland) but not others (e.g., intensive agriculture or housing). In Fig. II.1, landslide hazard mapping in California provided a basis for land-use recommendations. Scientific understanding of the geology of an area and its potential hazards is combined with building codes and zoning laws in setting limits on the types of activities permitted in the area.

Early-warning systems can also help reduce loss of life and, in some cases, property damage caused by landslides. The U.S. Geological Survey has developed a system for forecasting landslides in coordination with National Weather Service forecasts in some regions. The system combines analyses of rainfall data and forecasts with the delineation of areas known to be susceptible to landslides.

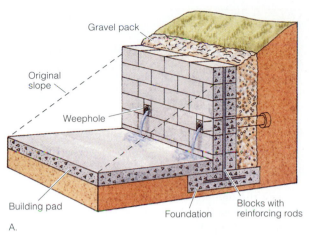

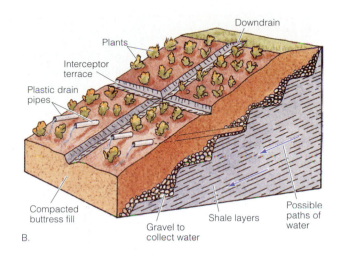

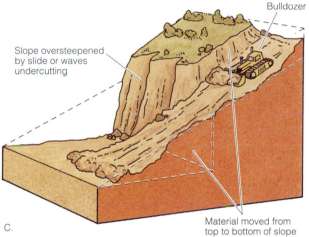

▲ F I G U R E 6.30
**Engineering techniques to stabilize slopes and prevent
failure. A. Rock bolts and retaining wall. B. Drainage
pipes. C. Slope regrading.**

Prevention and Mitigation

In addition to assessment, prediction, and early warning,
some engineering techniques can be used to mitigate or
even prevent landslides. These include retaining devices;
drainage pipes; grading; and diversion walls (Fig. 6.30).
One of the most common approaches is the use of concrete
block walls, poured or sprayed concrete, rock bolts, or
gabions (rocks contained in wire mesh cages) to strengthen
slopes (Fig. 6.30A). Slopes that are subject to creep can be
stabilized by draining or pumping water from saturated
sediment (Fig. 6.30B); this is accomplished by the insertion
of permanent drainage pipes, often in combination with a
wall. Oversteepened hill slopes can be prevented from
slumping if they are regraded to angles equal to or less than
the natural angle of repose (Fig. 6.30C). Sometimes the
slope itself cannot be stabilized but downslope structures
can be protected by the construction of diversion walls. In
some mountain valleys subject to mudflows from active
volcanoes, reservoirs can be quickly emptied so that dams
will halt mudflows before they reach population centers.

SUMMARY

1. Mass-wasting (the movement of Earth materials downslope as a result of the pull of gravity) has devastating impacts in terms of both property damage and loss of life.

2. Two basic types of movement may be involved in mass-wasting: the sudden failure of a slope, resulting in the downslope transfer of relatively coherent masses of rock or rock debris by slumping, falling, or sliding; and the downslope flow of mixtures of sediment, water, and air.

3. Slumps involve rotational movement of blocks of material; falls involve vertical free-fall of material; and slides involve translational movement along a well-defined plane.

4. Flows can be divided into slurry (water-saturated) and granular (unsaturated) flows. The water content of a flow may range from heavily sediment-laden stream flow to dry grain flow in a sand dune. Flows also vary in velocity, from imperceptibly slow (creep, solifluction) to extremely rapid (mudflows, debris avalanches).

5. Mass-wasting is especially prevalent at high latitudes and high altitudes, where average temperatures are very low and frost-heaving is an important force. Subaqueous (underwater) mass-wasting is also very common, especially on continental slopes and in major marine deltas.

6. The main factors that influence slope stability are (a) the force of gravity, and therefore the gradient of the slope; (b) water, and therefore the hydrologic characteristics of the slope; (c) the presence of troublesome Earth materials; and (d) the occurrence of a triggering event.

7. In any body of rock or rock debris located on a slope, two opposing forces—shear stress and shear strength—

determine whether the body will move or remain stationary. As long as shear strength exceeds shear stress, the rock or debris will not move. However, as these two forces approach a balance, the likelihood of movement increases.

8. Water plays a variety of important roles in mass-wasting of both solid rock and regolith. Water can decrease slope stability by reducing the natural cohesiveness between grains or reducing friction at the base of a rock mass through increased water pressure. Water can also contribute to slope instability through wave action undercutting the base of the slope, or by altering the chemical composition of sediment.

9. Some types of Earth materials are considered "troublesome" because they are particularly susceptible to failure under certain conditions. These include sensitive soils, expansive clays, and hydrocompacting clays.

10. Among the most common triggering events are earthquakes, volcanic eruptions, slope modifications, and changes in the hydrologic characteristics of an area (including the effects of prolonged or exceptionally intense rainfall).

11. Scientific understanding of the types of rock and characteristics of slopes in a given area can be combined with weather analysis to delineate landslide hazards and, sometimes, to issue predictions and early warnings of major landslides.

12. With careful planning, building regulations, and zoning laws, and the use of appropriate stabilization techniques, the impacts of mass-wasting processes on humans can often be reduced. Commonly used techniques include retaining devices (concrete, rock bolts, gabions), drainage pipes, grading, and diversion walls.

IMPORTANT TERMS TO REMEMBER

expansive soils (p. 158)
fall (p. 147)
flow (p. 148)
frost heaving (p. 153)
gradient (p. 155)

landslide (p. 145)
liquefaction (p. 157)
mass-wasting (p. 145)
safety factor (p. 156)
scarp (p. 152)

sensitive soils (p. 158)
shear strength (p. 156)
shear stress (p. 155)
slide (p. 148)
slump (p. 146)

QUESTIONS AND ACTIVITIES

1. Is the region where you live particularly susceptible to landslides? Have there been any major mass-wasting events there in recent history? If so, were the causes primarily natural (e.g., undercutting of a coastal cliff by waves, or slope failure in a mountainous region)? Or were human activities involved (e.g., alterations in natural drainage patterns or slope steepening by roadcuts)?

2. Has systematic landslide hazard mapping been carried out in your area? If so, which organization(s) was responsible? Has the mapping been coordinated with planning for new building regulations and zoning laws? Is an emergency response plan or early-warning system in effect for areas at greatest risk?

3. Investigate a major mass-wasting event. Some of the most interesting are the Gros Ventre, Vaiont Dam, and Turnagain Heights events mentioned in this chapter, but you may also want to find out about a major historic event that occurred in an area near where you live.

4. Make a table that classifies major mass-wasting events according to the conditions or triggering event primarily responsible for slope failure. Start with the examples given in this chapter and see how many you can add using other sources of information. For how many of the events in your table was more than one condition or triggering event responsible for slope failure?

5. Almost all parts of the world—even relatively flat areas—are susceptible to mass-wasting by creep. Go on a walking or driving tour of your neighborhood and see how many signs of creep you can discover on hillslopes. Look for bent tree trunks, curved fences, lobes of soil on grassy slopes, tilted gravestones, and other types of evidence.

6. As you walk or drive around your town, keep an eye out for the structures used to stabilize slopes or protect property from mass-wasting. Are the slopes in your area heavily engineered, or have they been left more or less in their natural state? Where you find such structures as retaining walls or drainage pipes, do they appear to have stabilized the slope as intended?

CHAPTER 7

SUBSIDENCE

*In nature things move violently to their place,
and calmly in their place.*

• **Sir Francis Bacon**

At 7:00 P.M. on May 8, 1981, a small tree in a vacant lot in Winter Park, Florida, suddenly disappeared. Within 10 hours a steep-sided crater 20 m in diameter had developed around the central depression where the tree originally stood. The crater continued to grow and deepen throughout the night, and by noon on the following day it was 100 m wide and about 30 m deep and was beginning to fill with water. By the time it stabilized, the depression had swallowed up parts of a house, six commercial buildings, and the municipal swimming pool, as well as several automobiles. The total cost of the damages was estimated at over $2 million.

The Winter Park event was the largest in a series of collapses that occurred in the Orlando area in May 1981. Much of Florida is underlain by carbonate rocks, which can be dissolved by slightly acidic groundwater. Dissolution of this rock has produced extensive cave systems in Florida and other areas in the southern United States. The caves are enlarged whenever groundwater levels are high, but the pressure of the water in the caverns helps support the weight of the overlying rocks. The collapses in the Orlando area occurred during a time of drought, when groundwater levels were particularly low. This left underground spaces and passageways unsupported, facilitating the collapse of overlying rocks into the spaces below.

Sudden, dramatic collapses like the Winter Park occurrence are unusual. However, a wide variety of factors can contribute to changes in the surface of the land, whether abrupt and dramatic or gradual and almost imperceptible. The process of sinking or collapse of the land surface, *subsidence,* is the subject of this chapter.

◄

Sinkhole in Winter Park, near Orlando, Florida. The crater appeared at 7 pm, May 8, 1981 and grew to be 100 m wide within about 20 hours.

SURFACE SUBSIDENCE AND COLLAPSE

Subsidence is the sinking or collapse of a portion of the land surface. The movement involved in subsidence is essentially vertical; little or no horizontal motion is involved. It may take the form of a sudden, dramatic collapse or a slow, almost imperceptible lowering. In some cases subsidence is a localized phenomenon; in others it is regional in extent.

Like mass-wasting, subsidence differs from other types of erosion, transportation, and deposition of Earth materials in that it does not require a transporting medium such as water or ice. However, water can facilitate ground collapse in a variety of ways, and it usually plays an important role in the process of subsidence. Also like mass-wasting, subsidence is driven by gravity. Subsidence differs from mass-wasting, however, in that it is controlled by the physical properties of the rocks and sediments underlying the collapse area and is not a function primarily of slope stability.

The mechanisms of collapse, and sometimes the conditions existing before the collapse, result from natural physical processes. In many cases, however, the conditions leading up to a subsidence event are exacerbated or even created by human actions. In this chapter we will examine some of the conditions that lead to subsidence, how those conditions can be caused or aggravated by human activities, what types of damage may result, and how subsidence-related hazards can be addressed.

CARBONATE DISSOLUTION AND KARST TOPOGRAPHY

In regions underlain by rocks that are highly susceptible to chemical weathering, groundwater creates underground caverns and distinctive landscapes that are among the most interesting and picturesque on our planet. These extensive cave systems, however, can set the stage for catastrophic events such as the one in Winter Park. To appreciate the mechanisms through which underground caverns can lead to surface subsidence, we must first consider the processes of *carbonate dissolution* and *cave formation.*

Dissolution

As soon as rainwater infiltrates the ground, it begins to react with minerals in the regolith and bedrock, causing chemical weathering. An important part of that process is **dissolution,** in which minerals and rock materials pass directly into solution.

Among the rocks of the Earth's crust, the *carbonate* rocks (those consisting primarily of minerals based on the CO_3^{2-} anion, such as calcite and dolomite) are most readily attacked by dissolution. Limestone, dolostone, and marble are the most common carbonate rocks; they underlie millions of square kilometers of the Earth's surface. Carbonate minerals are nearly insoluble in pure water but are readily dissolved by carbonic acid (H_2CO_3), a common constituent of rainwater. The weathering attack occurs mainly along fractures and other partings and openings in the carbonate bedrock, often with impressive results. When granite is weathered chemically, quartz and other resistant minerals are little affected and remain as part of the weathered regolith. However, when limestone weathers, nearly all of its volume may be dissolved away in slowly moving groundwater.

In carbonate terrains, the rate of dissolution can exceed the average rate of erosion of the surface by streams, mass-wasting, and other processes. Through periodic measurement of the amount of dissolution observed on small, precisely weighed limestone tablets placed at open sites in various areas, geologists have obtained estimates of the average rate at which limestone landscapes are being lowered by dissolution. In temperate regions with high rainfall, a high water table, and nearly continuous vegetation cover, carbonate landscapes are being lowered at average rates of up to 10 mm/1000 year. In dry regions with scanty rainfall, low water tables, and discontinuous vegetation, dissolution rates are lower.

Caves

Caves are large underground open spaces. The earliest evidence we have of human dwellings comes from limestone caves in Europe and Asia that provided shelter for paleolithic peoples during the Pleistocene glacial ages. The walls of those caves served as rocky canvases for prehistoric artists, whose polychrome paintings provide us with superb renditions of the prey of ice-age big-game hunters.

Caves come in many shapes and sizes. Although most caves are small, some are very large. A large cave or system of interconnected cave chambers is often called a **cavern.** The Carlsbad Caverns in southeastern New Mexico (Fig. 7.1) include a chamber that is 1200 m long, 190 m wide,

▲ **F I G U R E 7.1**
Caves and caverns are underground openings that are created when carbonate rocks undergo dissolution. This spectacular room is part of the Carlsbad Caverns region of New Mexico.

and 100 m high. Mammoth Cave in Kentucky consists of interconnected caverns with an aggregate length of at least 500 km. The recently discovered Good Luck Cave on the island of Borneo includes a chamber so large that into it could be fitted not only the world's largest previously known chamber (in Carlsbad Caverns) but also the largest chamber in Europe (in Gouffre St. Pierre Martin, France) and the largest chamber in Britain (Gaping Ghyll).

Cave Formation

Caves are formed through a chemical process in which carbonate rock is dissolved by circulating groundwater. The process begins with dissolution by percolating groundwater along a system of interconnected open fractures and bedding planes. A cave passage then develops along the most favorable flow route. Carbonate formations (called *speleothems*) are deposited like icicles on the cave walls while a stream occupies the floor. After the stream has stopped flowing, similar formations are deposited on the floor. Although geologists disagree as to the exact conditions under which cave formation occurs, available evidence favors the idea that most caves are excavated near the top of a seasonally fluctuating water table.

The rate of cave formation is related to the rate of dissolution. In areas where the water is acidic, the rate of disso-

▲ F I G U R E 7.2

The Karst region in the southeastern corner of the former Yugoslavia, adjacent to the border with Albania. The Adriatic Sea is in the lower left-hand corner. The curiously patterned landscape is produced largely by solution of underlying limestones.

lution increases with increasing velocity of flow. As a passage grows and the flow becomes more rapid and turbulent, the rate of dissolution increases. The development of a continuous passage by slowly percolating waters has been estimated to take up to 10,000 years, and the further enlarge-

ment of the passage by more rapidly flowing water to create a fully developed cave system may take an additional 10,000 to 1 million years.

Karst Topography

In regions where rock is exceptionally soluble, the topography is characterized by many small, closed basins and a disrupted drainage pattern. Streams disappear into the ground and eventually reappear elsewhere as large springs. Such terrain is called **karst topography** after the Karst region of the former Yugoslavia (Fig. 7.2). Karst is most typical of limestone landscapes but can also develop in areas underlain by dolomite, gypsum, and salt.

Several factors control the development of karst landscapes: The topography must permit the flow of groundwater through soluble rock under the pull of gravity. Precipitation must be adequate to supply the groundwater system. Soil and plant cover must supply an adequate amount of carbon dioxide (to make carbonic acid from rainwater), and temperatures must be high enough to promote dissolution. Although karst terrain occurs in a wide range of latitudes and at varied altitudes, it is most fully developed in moist temperate or tropical regions underlain by thick and widespread soluble rocks.

The most common type of karst terrain is **sinkhole karst,** a landscape dotted with closely spaced circular collapse basins of various sizes and shapes (Fig. 7.3). Such landscapes are seen in southern Indiana, south-central Tennessee, and Jamaica, among other places.

▼ F I G U R E 7.3

Sinkholes in limestone, near Roswell, New Mexico, are characteristic of sinkhole karst topography.

Sinkholes

In contrast to a cave, a **sinkhole** is a large dissolution cavity that is open to the sky. Some sinkholes are caves whose roofs have collapsed; these are sometimes called *collapse sinkholes* or *solution sinkholes.* Others are formed at the surface in places where rainwater is freshly charged with carbon dioxide and hence is most effective as a solvent.

Some sinkholes are funnel shaped; others have high, vertical sides. Sinkholes on the Yucatan Peninsula in Mexico, locally called *cenotes* (a word of Mayan origin), are steep sided and contain water because their floors lie below the water table (Fig. 7.4). The cenotes were the primary source of water for the ancient Maya and formerly supported a considerable population. A large cenote at the ruined city of Chichen Itza was dedicated to the rain god. Remains of more than 40 human sacrifices, mostly young children, have been recovered from the cenote, together with jade, gold, and copper offerings.

Sinkhole Formation

In many carbonate landscapes, such as those in Florida, new sinkholes are forming constantly. In one small area of about 25 km² more than 1000 collapses have occurred in recent years. In this case, lowering of the water table as a result of drought and excessive pumping of water wells has led to extensive collapse of cave roofs.

Some sinkholes are formed catastrophically. An account of one such event, which took place in rural Alabama, describes how a resident was startled by a rumble that shook his house. He then distinctly heard the sound of trees snapping and breaking. A short time later, hunters walking through nearby woods discovered a 50-m-deep sinkhole that was 140 m long and 115 m wide.

The sinkhole in Winter Park, Florida, described at the beginning of the chapter, was formed by the collapse of surface materials into a preexisting cavity in underlying carbonate rocks. In this case, the cavern was connected to near-surface rocks and unconsolidated overburden by an open, chimneylike passageway called an **aven** (Fig. 7.5A). Sinkhole formation can be initiated either by the sudden wholesale collapse of the bedrock "roof" into the underground void, a process called **stoping,** or by much more gradual downward movement of unconsolidated material into the aven. The latter process, called **raveling,** eventually leaves the roof materials unsupported; surface fractures begin to develop and the roof eventually collapses (Fig. 7.5B), filling the aven with sediment and debris. This type of process is thought to have led to the formation of the Winter Park sinkhole. The lowering of the water table in the region as a result of drought probably contributed to the sediment raveling. Groundwater percolating downward through the passageway provided by the aven carried sediment with it, facilitating the downward movement of material.

REMOVAL OF SOLIDS AND MINE-RELATED COLLAPSE

Sinkholes in karst terrains develop naturally in response to the dissolution of underlying carbonate rocks. However, as noted earlier, other types of rock are also susceptible to dissolution, and there are other ways in which the removal of solid materials can create spaces underground that may result in subsidence.

Removal of Salt

Because rock salt (composed mainly of the mineral halite, NaCl) can be dissolved by groundwater, karst terrains and sinkholes can develop in areas underlain by salt as well as in carbonate terrains. In 1980 a sinkhole 110 m wide and 34 m deep developed in this manner near Kermit, Texas. This depression, called the Wink Sink (Fig. 7.6), appeared without warning and grew to its full size within 48 hours. Groundwater had dissolved caverns in underlying salt beds in much the same manner as that described earlier for carbonate terrains.

One technique used for mining salt is to inject fluids and induce dissolution of the salt underground so that it can be withdrawn in a liquid state. This is called *solution mining.* When the salt-saturated solution is pumped out it leaves cavities in the rock, thereby weakening the support for overlying material. In 1974 solution mining of salt was directly responsible for the rapid development of a water-filled sinkhole 300 m in diameter in Hutchinson, Kansas (Fig. 7.7).

▲ **F I G U R E 7.4**
The sacred well at Chichen Itza, a ruined Mayan city on the Yucatan Peninsula of Mexico. This cenote, formed in flat-lying limestone rocks, contained a rich store of archaeological treasures that had been cast into the water along with human sacrifices centuries ago.

A. One quiet day in Winterpark. . .

B. Roof has collapsed, aven is filled with sediment

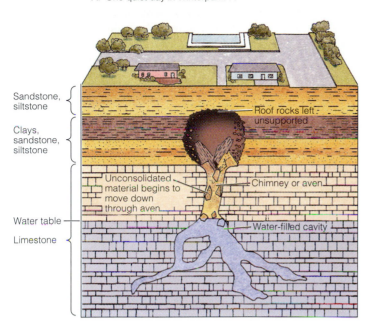

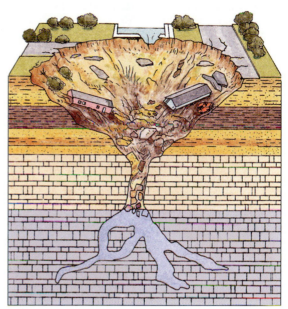

Sandstone, siltstone

Clays, sandstone, siltstone

Roof rocks left unsupported

Unconsolidated material begins to move down through aven

Chimney or aven

Water table

Water-filled cavity

Limestone

▲ F I G U R E 7.5

Formation of the sinkhole in Winter Park, Florida, as a result of the collapse of surface materials into a preexisting cavity in underlying carbonate rocks. A. The cavern was connected to near-surface rocks and unconsolidated overburden by an open, chimneylike passageway called an aven. B. Formation of the sink-hole was initiated by raveling, or gradual downward movement of unconsolidated material into the aven. This eventually left the roof materials unsupported; surface fractures began to develop and the roof even-tually collapsed, filling the aven with sediment and debris.

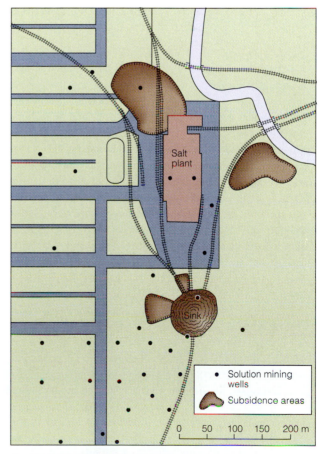

Salt plant

Sink

• Solution mining wells

Subsidence areas

0   50   100   150   200 m

▲ F I G U R E 7.7

Map of the sinkhole area near Hutchinson, Kansas, where solution mining of salt led to the development of collapse sinkholes.

▲ F I G U R E 7.6

The Wink Sink, Kermit, Texas. The road and cars provide scale.

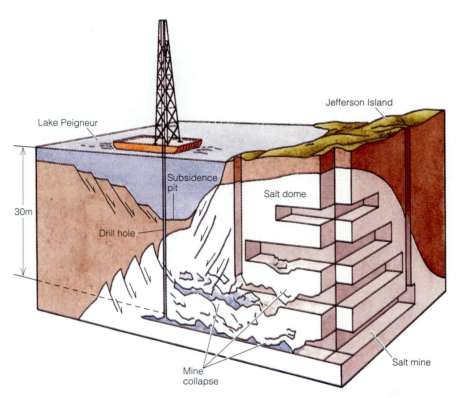

Diagram of the Jefferson Salt Dome under Lake Peigneur, Louisiana. In 1980 an oil-drilling rig accidentally punched a small hole through the lake bottom into a mining shaft. The hole grew rapidly as lake water drained into it, dissolving the salt and eventually causing the mine roof to collapse.

Perhaps the most astonishing collapse associated with a salt mine occurred at Lake Peigneur, a shallow lake in southern Louisiana. Under the lake was a still-active, multi-million-dollar salt mine in the Jefferson Island Salt Dome. On November 21, 1980, an oil-drilling rig accidentally punched a small hole through the lake bottom into a mining shaft (Fig. 7.8). The hole grew rapidly as lake water drained into it, dissolving the salt and eventually causing the roof of the mine to collapse. The entire lake drained into the mine in a whirlpool, carrying with it ten barges, a tugboat, and an oil-drilling rig. More than 25 hectares of land were lost, including a private home and substantial portions of a botanical garden. Incredibly, nine of the barges popped to the surface 2 days later as the depression began to stabilize and refill with water!

Salt and Oil

Salt is often associated with oil because of the shallow marine (i.e., saltwater) setting in which many oil deposits were formed. The Wink Sink, for example, occurred in an oil field, although the collapse itself was caused by the natural dissolution of salt layers. In the Lake Peigneur event, an oil-drilling operation was responsible for the hole that triggered the collapse, although it was the dissolution of salt that ultimately led to the collapse of the mine roof. The oil company was held legally responsible for the event and was required to pay damages to the salt mining company as well as to affected landowners.

Because of the frequent association of salt and oil, oil drilling commonly brings large quantities of salty fluid, or *brine,* to the surface along with the oil. The brine is often reinjected into waste disposal wells, which are drilled to depths of several hundred meters and may pass through thick salt layers. If the disposal well casing leaks, the brine may escape and dissolve large underground cavities in the salt layers (Fig. 7.9). Such an occurrence was responsible for the formation of an 80-m-wide sinkhole in Macksville, Kansas, in the summer of 1988. The underground cavern had probably existed for some time prior to the collapse; the disposal well had been plugged since 1984. The cavern that ultimately caused the Macksville sinkhole was probably the size of the Houston Astrodome! Other underground voids—some with no evidence of surface subsidence—have been detected in nearby areas, leading to concerns that a similar event could occur in the future.

Coal Mining

Removal of solid material from underground can lead to subsidence even when the rock is not soluble in water. Areas in which coal is mined are particularly susceptible to such events. Coal is commonly mined underground using the *"room-and-pillar"* method, in which large masses of coal are removed, creating the "rooms," and columns of coal are left behind to form the supporting "pillars." In the past, many mining companies removed as much coal as possible,

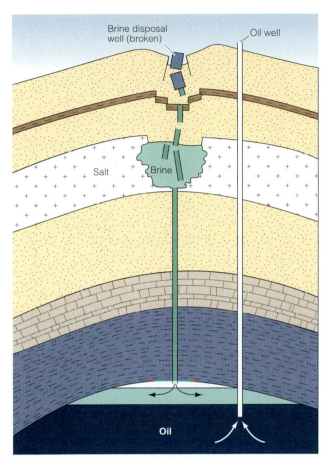

▲ **F I G U R E 7.9**
Brine (salty water) is a waste product generated during oil drilling. The brine is often reinjected into disposal wells. If the well casings break, the brine can leak out and dissolve large openings in salt layers. This can ultimately lead to subsidence or collapse, as occurred in Macksville, Kansas.

even from seams relatively near the surface. In some cases, subsidence occurred because too much coal was removed, and the pillars left behind could not support the weight of the overburden. In other cases, the mine was excavated too close to the surface and the roof material between the pillars collapsed.

Some relatively recent episodes of subsidence can be attributed to mining done 50 years ago or more. Grasslands near the town of Sheridan, Wyoming, are scarred and pitted with features resulting from subsidence of the ground surface into underground rooms from which coal was mined. Most mining activities associated with the subsidence ended decades ago. The pattern of the subsidence features reveals the layout of the underground rooms and pillars of the former mines (Fig. 7.10). Some of the features are broad, shallow depressions; others are deep, steep-sided pits.

Another hazard commonly associated with coal mining is underground fires, which are caused by spontaneous ignition of coal dust or methane gas. When such fires burn out of control they remove even more coal, increasing the likelihood of surface collapse. Extensive coal fires have burned for more than a quarter of a century in old mines underneath the towns of Laurel Run and Centralia, Pennsylvania, with subsequent ground fracturing, subsidence, and release of toxic gases. The fires, which have migrated several kilometers since they were ignited, are rekindled by oxygen passing through fractures and cracks in the ground surface.

A study investigating the land-use impacts of coal mining estimated that of the 8 million acres that have been undermined for coal in the United States, about 1.85 million—nearly 25 percent—have subsided. Loss of productive farmland is the main problem associated with subsidence in rural areas. In some cases groundwater supplies

◄ **F I G U R E 7.10**
Subsidence above abandoned coal mine workings, 15 km north of Sheridan, Wyoming. The Dietz Mine was active from the 1890s to the 1920s. The subsidence pattern reflects the design of support pillars and mine openings.

and underground drainage systems may also be adversely affected. Urbanization in previously mined areas can cause further problems, not only because of the potential damage to structures but also because the associated surface loading can increase the possibility of subsidence.

SUBSIDENCE CAUSED BY FLUID WITHDRAWAL

Withdrawal of fluids from underground can lead to surface subsidence, but the mechanisms involved generally differ from those described earlier. Sometimes fluid withdrawal and subsidence occur naturally. During a drought, for example, rows of trees or plants with very deep roots (such as alfalfa) may deplete soil moisture to such an extent that fractures as wide as 1 m are formed. Surface soils may bridge the subsurface fractures, hiding them from view; when it rains, the soil bridge may collapse, producing a gaping hole.

Subsidence resulting from the withdrawal of fluids can also be caused or facilitated by human activities. The fluids most often involved are groundwater, oil, natural gas, and associated brines, as well as mixtures of steam and water used for geothermal energy. In contrast to the sudden, localized events associated with mine-related collapses and sinkhole karsts, subsidence associated with fluid withdrawal is often gradual but may be regional in extent.

The subsidence that occurs when fluids are withdrawn is caused primarily by **compaction,** that is, a decrease in the thickness of a layer of sediment or rock. The weight of overlying materials causes the mineral particles to pack to-gether more tightly when the supporting pressure of their pore water is withdrawn. The amount of subsidence is proportional to the volume of fluid withdrawn and the thickness of the compressible layers. Crystalline (i.e., igneous or metamorphic) rock formations subside least; highly compressible materials, such as clays, or organic-rich sediments, such as peat, subside most.

Water

Subsidence is often caused by removal of subsurface water, especially if the rate of removal is faster than the rate of replenishment. *Hydrocompaction*—compaction caused by the drying or dewatering of sediments—and associated subsidence can occur whenever water is removed from layers of compressible materials, as in the draining of a marsh or wetland. (Hydrocompaction in clays and organic-rich sediments and its relationship to slope failure is discussed in Chapter 6.)

When groundwater is pumped out of an **aquifer** (a water-bearing rock or sediment unit), the pumping creates a depression in the surface of the water table in the shape of an inverted cone (Fig. 7.11). When many pumped wells are close together, as in some urban areas or irrigated farmlands, these features—called **cones of depression**—may begin to overlap. This can cause a regional depression in the level of the water table. In addition to creating water supply problems, regional lowering of the water table commonly leads to ground subsidence in the area overlying the aquifer.

Sometimes it is possible (although costly) to detect and reverse the process of subsidence by pumping water into the aquifer to raise the fluid pressure. However, the real

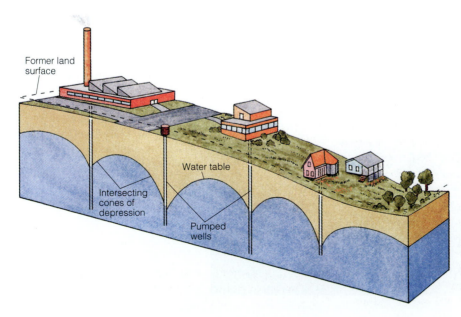

◀ F I G U R E 7.11
Overlapping cones of depression in an area of closely spaced pumped wells, leading to regional subsidence and depression of the surface of the aquifer (the water table).

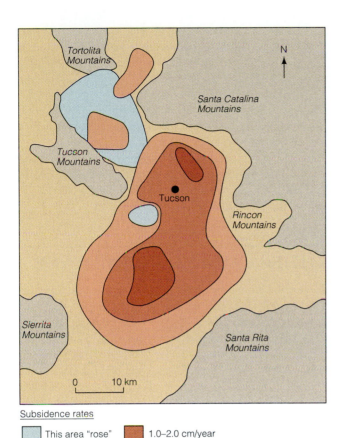

▲ F I G U R E 7.12
Subsidence rates in the Tucson Basin. Studies show that the weight of the basin, combined with overpumping of the aquifer, is damaging its water storage capacity.

Subsidence rates

- This area "rose"
- 0–1.0 cm/year
- 1.0–2.0 cm/year
- > 2.0 cm/year

problems begin when an aquifer becomes so compacted that it passes from *elastic* to *inelastic* subsidence. As noted in Chapter 3, inelastic refers to an irreversible change. In other words, the inelastic compaction of sediment or clay particles is permanent (not reversible, as would be an elastic compaction). Even if water were pumped back into the aquifer, it would not regain its original thickness. In such cases, the decrease in *porosity* (the amount of open or *pore* space in the sediment or rock) caused by compaction means that the water-bearing capacity of the sediment or rock has become permanently impaired.

The rate of ground subsidence relative to the decline in the level of the water table can be used to determine whether compaction is elastic or inelastic: Large water loss with only slight subsidence indicates elastic subsidence, whereas small water loss with significant subsidence suggests inelastic subsidence. Records from the Tucson Basin in southern Arizona show that before 1981 the ground sur-

face dropped 3 mm for every meter of decline in the water table. More recent measurements show that the ground surface is now dropping 24 mm for every meter drop in water level—eight times the previous rate (Fig. 7.12). Researchers have suggested that the dramatic increase in rate of compaction may signal a transition from elastic to inelastic subsidence, with potentially serious implications for the aquifer's water-storage capacity.

Subsidence caused by excessive withdrawal of groundwater is of particular concern in the southern and southwestern United States, where many thousands of square kilometers of land are affected. In the Los Banos–Kettleman City area of California, 6200 km^2 of land have subsided more than 0.3 m, with a maximum subsidence of 8.5 m. In areas where subsidence is especially pronounced, cracks or fissures as much as 1 m wide and 1 km long have opened around the edges of the subsiding basins.

Oil and Gas

The withdrawal of oil, natural gas, and associated briny fluids can lead to subsidence through the same mechanisms as the withdrawal of water. A map of the Inglewood oil field in California (Fig. 7.13) shows the correlation between the area from which oil was withdrawn, beginning in 1924,

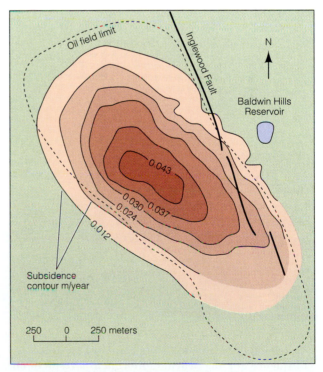

▲ F I G U R E 7.13
Map of the Inglewood oil field in Los Angeles showing the area from which oil was withdrawn and the area of surface subsidence.

and the resulting subsidence. As in the cases of groundwater withdrawal just described, the subsidence proceeded gradually and continued over an extended period. Decades after the subsidence began, fissures appeared around the edges of the subsidence basin. This would ultimately prove disastrous for the Baldwin Hills Reservoir, located on the northeast edge of the basin. In 1963 the Baldwin Hills Dam failed, releasing all of the water in the reservoir. The resulting flood killed five people and caused $12 million in damage. The conditions that led to the failure of the dam were caused primarily by the subsidence, which was associated with the reactivation of a preexisting fault underlying the dam. The injection of water (used, in this case, to flood the oil field, not to control the subsidence) may have played a minor role in reactivating the fault.

▲ **F I G U R E 7.14**
The friendly fire hydrant is high and dry. Ground subsidence at Long Beach, California as a result of oil removal. The ground sank but the pipe structures remained at their former height. Repressuring of the former oil reservoirs has restored the land surface closer to, but not equal to, its former level.

Subsidence associated with the Wilmington oil field near Long Beach, California, has cost over $100 million since it was first recognized in 1940. The area of the subsidence basin reached 50 km² at its maximum; near the center of the basin the vertical subsidence was as great as 9 m, causing localized flooding. In 1958 an aggressive program was undertaken with the goal of repressurizing the rocks from which the oil had been withdrawn. Large quantities of water were injected into the rocks to raise the fluid pressure. By 1962 the sinking was stopped and the area of the subsidence basin had been reduced to 8 km². A small amount of vertical rebound was observed at some sites (Fig. 7.14).

OTHER CAUSES OF SUBSIDENCE

In certain circumstances mechanisms other than those discussed so far can lead to subsidence. For example, in areas of former glaciation, depressions known as *kettles* are often observed. These were created where masses of debris-covered ice melted away, causing collapse of the overlying sediment. The resulting bowl-shaped depressions often contain a lake or swamp. A famous example of a kettle lake is Walden Pond in Massachusetts. Kettles can range up to tens of meters in depth and a kilometer or more in diameter.

Another special circumstance arises when flowing lava forms *tubes* or tunnels. As the lava cools and solidifies, it forms a lid of solid rock under which molten lava can continue to flow. If the source of molten lava is cut off, the flow of lava ceases, leaving a hollow tube or "skin" of solidified material. If the roof of the tube is thin enough, it will collapse; this produces a long, sinuous depression where the lava river once flowed.

Endogenous Subsidence

The types of subsidence discussed so far originate near the surface of the Earth, either as a result of natural weathering and groundwater processes or as a consequence of human activities such as mining or fluid extraction. Another broad category of subsidence mechanisms is associated with internal or *endogenous* Earth processes. These include such tectonic processes as folding, faulting, and plate motion, as well as sedimentation in certain tectonic environments. Except in the case of catastrophic episodes related to earthquakes or volcanism, the rate of subsidence associated with such processes rarely exceeds 10 mm/year.

Coastal areas, especially deltas, are particularly susceptible to the gradual subsidence that results from endogenous processes. Downwarping of the crust in such areas is normally offset by the influx of sediment carried by rivers. But the amount of sedimentation can decrease dramatically if a river is dammed, channeled, or diverted, resulting in higher rates of erosion. Natural isostatic subsidence combined with decreased sedimentation, increased erosion, and, in some cases, accelerated subsidence induced by the withdrawal of water or oil may create problems for some low-lying coastal areas in the future. For example, researchers at Woods Hole Oceanographic Institution have speculated that by the year 2100 local sea level in the Nile Delta could be as much as 3.3 m higher than it is now, meaning that Egypt could lose up to 26 percent of its habitable land as well as invaluable coastal assets such as fisheries and mangrove forests. Potential increases in sea level because of global warming over the next few decades could further exacerbate the situation.

SINKING CITIES[*]

In urban settings, factors conducive to subsidence often occur together. Given the high density of people, buildings, and utilities, this can create a significant potential for urban subsidence, which may cause extensive and costly damage (Table 7.1).

All of the cities that have experienced major problems with subsidence are located on unconsolidated sediments consisting of silt, clay, peat, and sand associated with river floodplains (e.g., New Orleans, London, Bangkok), coastal

T A B L E   7.1 • Subsiding Lands[a]

	Maximum subsidence (m)	Area affected (km²)
Coastal		
London	0.30	295
Venice	0.22	150
Po Delta	1.40	700
Taipei	1.90	130
Shanghai	2.63	121
Bangkok	1.00	800
Tokyo	4.50	3,000
Osaka	3.00	500
Niigata	2.50	8,300
Nagoya	2.37	1,300
San Jose	3.90	800
Houston	2.70	12,100
New Orleans	2.00	175
Long Beach/Los Angeles	9.00	50
Savannah	0.20	35
Inland		
Mexico City	8.50	225
Denver	0.30	320
San Joaquin Valley	8.80	13,500
Baton Rouge	0.30	650

[a]From Dolan, Robert, and Goodell, H. Grant, "Sinking Cities," *American Scientist*, Vol. 74, Jan.–Feb. 1986.

[*]This section is based on "Sinking Cities," *American Scientist*, Jan.–Feb. 1986, vol. 74, pp. 38–47, by Robert Dolan and H. Grant Goodell.

▲ **F I G U R E   7.15**
Land subsidence and the relative rise of the water table forces the citizens of New Orleans, Louisiana, to bury their dead in above-ground cemeteries.

marshes and/or delta complexes (e.g., Venice, Houston/Galveston, Tokyo, Shanghai), or lake beds (e.g., Mexico City). Subsidence initially occurs in response to the sheer weight of urban development. Unconsolidated sediments settle, and fluids are squeezed out of pore spaces. As discussed earlier, in such highly compressible sediments as clay and peat the particles may rearrange themselves permanently, flattening and compacting in response to loading, with the result that the subsidence becomes inelastic.

In addition, many cities in which subsidence is a problem (with the notable exception of Mexico City) are located in coastal areas that are prone to natural subsidence. In New Orleans, regional long-term subsidence from the compaction of delta sediments and isostatic downwarping ranges between 12 and 24 cm per 100 years, with local subsidence caused by groundwater withdrawal as high as 3 m in 50 years. The situation has been aggravated by the construction of levees and channels that direct the sediment-laden waters of the Mississippi River into deep water offshore. As a result of this combination of natural and human activities, the Mississippi Delta is sediment-starved, land is being lost rather than created, and 45 percent of the urbanized area of New Orleans is at or below sea level (Fig. 7.15).

Many large cities also contain concentrations of buildings that were constructed before the emergence of zoning laws based on an understanding of subsidence and soil mechanics. Capuchinas Church in Mexico City, constructed

BOX 7.1

•

THE HUMAN PERSPECTIVE

SUBSIDENCE IN MEXICO CITY

*N*atural subsidence has occurred in the Mexico Basin since before the Aztec period. In 1325 A.D. the Aztecs established Mexico City (originally named Tenochtitlan). At that time the Mexico Basin was a network of lakes and islands surrounded by volcanic mountains, and the city was established on several of the islands. Large structures built by the Aztecs, and by the Spanish several hundred years later, placed excessive loads on the fine-grained volcanic ash and lake sediments of the Mexico Basin aquifers. Also during this period the lakes were drained and canals were built. These actions initiated the first subsidence episodes directly related to human activity in the Mexico Basin.

Water confined within the layers of the aquifer provides pressure that supports the overburden of sediment by pushing apart the individual grains of subsurface material. This pressure can also support buildings and other structures at the surface. However, in this century a critical subsidence problem has developed as a result of the exploitation of groundwater reserves to meet the needs of Mexico City's rapidly growing population. When groundwater is removed at rates that exceed the natural rate of replenishment, subsurface material is compacted vertically and the land subsides. The subsidence creates problems such as uneven settling of large buildings; stranding of sidewalks, well casings, and other structures above the land surface; and weakening of foundations (a particular concern in this earthquake-prone region). Even after the withdrawal of groundwater is stopped, the subsidence may continue because of the weight of the surface material and structures. Furthermore, the compacting of sediments caused by the subsidence may be irreversible.

Most of the critical problems of subsidence in Mexico City originated in the early 1900s, when the population began placing tremendous demands on freshwater resources. In the 1850s there were 140 free-flowing artesian wells in the Mexico Basin aquifers. By 1922, the water table in the regional aquifers had declined more than 35 meters and land subsidence resulting from groundwater extraction in Mexico City totaled 1.25 m. In 1940, more than 100 deep-water production wells were dug to reach additional water supplies, greatly accelerating the rate and severity of subsidence. Subsidence reached depths of 8 m in some places and occurred at rates of between 15 and 46 cm/year. By 1954 most of the wells in the metropolitan area were closed to alleviate subsidence. However, as the population continued to grow, new large-capacity wells were installed at the periphery of the Mexico Basin aquifer system. During this time the rate of subsidence was accelerated in outlying areas even as it was reduced in the central city.

Chalco Plain, an area south of Mexico City, has been used since 1960 to provide additional water for the central metropolitan area. As a consequence of the pumping of water from wells, this region has subsided 6 m in places. If extraction of groundwater continues, the rate of subsidence will exceed that of the Valley of Mexico. Subsidence in Chalco Plain has already caused contamination of groundwater and flooding of agricultural and urban land. It remains to be seen whether land use planners for the world's largest city can manage their limited water resources without perpetuating the problems associated with subsidence.

in the eighteenth century, was seriously out of plumb as a result of differential settling until extensive restorations were undertaken. The Leaning Tower of Pisa (Fig. 7.16) is probably the best known example of a sinking building. Immediately after its construction was begun in 1174, the tower began to sink as the foundations settled into the soft sediment of the Arno floodplain. In the 1960s the rate of tilting increased as a result of the withdrawal of groundwa-

ter from aquifers underlying the city. Recent efforts to secure the tower have centered on the injection of concrete to stabilize the foundations.

The most severe problem facing urban areas, however, is the accelerated subsidence caused by intensive withdrawal of oil and/or water from underlying reservoirs. In the past 50 years, groundwater pumping in response to the pressures of industrialization and rapid population growth has

◄ F I G U R E 7.16
The Leaning Tower of Pisa is a campanile adjacent to the Cathedral. It leans because it settled unevenly into the soft sediment of the River Arno floodplain. Ground water withdrawal in modern times has increased the angle of tilt.

greatly accelerated rates of urban subsidence. In Venice, Italy, for example, subsidence is a critical problem (Fig. 7.17). Venice is built on a series of interconnected marshy islands in a saltwater lagoon on the coast of the Adriatic Sea. Subsidence has made the city and its historic buildings extremely vulnerable to storm surges and flooding associated with high tides. In the 1950s the withdrawal of large quantities of groundwater to meet industrial needs led to a 10-fold increase in the rate of subsidence, accompanied by a sharp decrease in well water pressures. When groundwater pumping was stopped in 1969, subsidence stopped and water pressures increased; there was actually a slight (2 cm) rebound in elevation in some parts of the city, although most of the subsidence became permanent because of inelastic compaction in underlying clay layers.

London is another city where a combination of factors has led to significant and damaging subsidence. The gradual regional tilting of southeastern England, which has occurred at a rate of about 30 cm per century, combined with

◄ F I G U R E 7.17
Venice, one of the world's most beautiful and fascinating cities, is slowly sinking. Periodic flooding is the result. Here, people are crossing St. Mark's Square during a time of flooding.

▲ **F I G U R E   7.18**
Flood barriers on the River Thames, London, looking upstream. The barriers are designed to prevent high tides and storm surges from flooding central London. The dome-shaped pedestals are fixed. The cylindrical structures between the domes are barriers that can be rotated up, to prevent flow, or down, to allow flow.

the dewatering and compaction of clay layers and the pumping of groundwater from an underlying chalk aquifer, has resulted in an increase in tide height (measured at London Bridge) of 60 cm over the past 100 years. This has greatly increased the potential for urban flooding. It is estimated that 40 stations and 127 km of subway track in the London underground rail system are in flood-prone areas. In response to the growing danger, the equivalent of $1 billion has been spent to construct the Thames Barrier, a series of 10 floodgates designed to prevent floods and storm surges in the lower Thames from reaching central London (Fig. 7.18).

Urban flooding is a common side effect of subsidence in cities located on river floodplains, coastal areas, or deltas. Bangkok, for example, is located on the floodplain of the Chao Phraya River, 25 km north of the Gulf of Thailand. Intensive withdrawal of groundwater to support the rapidly growing urban population has resulted in subsidence of up to 1 m, which has occurred at rates as high as 14 cm/year in the southeastern part of the city. Subsidence and urban development have interfered with the drainage of water into the Chao Phraya via a series of canals constructed for this purpose. As a result, the city is now regularly inundated with sewage-choked floodwaters after rainstorms. Differential subsidence in Bangkok has also damaged buildings, cracked roads and sidewalks, and broken utility pipes.

PREDICTING AND MITIGATING SUBSIDENCE HAZARDS

When removal of solid material is the primary cause of subsidence, prediction of subsidence and assessment of the related hazards are generally based on mapping of susceptible areas. In former coal-mining areas, for example, maps showing the distribution of previously mined areas are combined with information about subsidence in similar areas, the depth of the mine workings, and the locations where mine fires have occurred or are in progress. This results in a map showing the areas where the risk of mine-related subsidence is greatest. Prediction of subsidence in carbonate terrains in Florida has used aerial photography to identify areas of disrupted drainage. This information can be combined with studies of historic and prehistoric sinkhole occurrences to produce a map of subsidence potential (Fig. 7.19).

Where fluid withdrawal is the main cause of subsidence, it is important to know the rate of fluid withdrawal and the characteristics of the underlying rock layers, particularly with respect to compressibility. For example, in a study of the Houston area, researchers compared historical records of the drawdown of water table surfaces with records of land surface subsidence. They found that subsidence was greatest in areas of maximum drawdown. Specific predic-

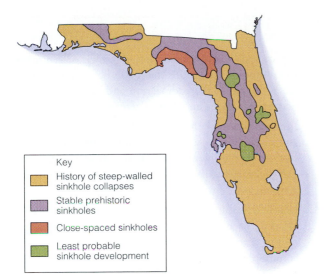

◄ F I G U R E 7.19
Map of Florida showing areas of differing potential for subsidence due to dissolution.

Key

History of steep-walled sinkhole collapses

Stable prehistoric sinkholes

Close-spaced sinkholes

Least probable sinkhole development

tions of future subsidence are based on an understanding of historical relationships between subsidence and water withdrawal, combined with models of the maximum compressibility of underlying materials (Fig. 7.20). Predictions of subsidence can be calculated for different water withdrawal scenarios.

So far, the most viable approach to the prevention, control, and mitigation of hazards associated with subsidence is to modify the human activities that contribute to its occurrence. In part, this can be accomplished through zoning reg-

ulations based on an understanding of existing geologic hazards. A complicating factor in the United States is that laws concerning the right to sue for damages from subsidence conflict with laws concerning the right to withdraw an underground resource (such as groundwater), even when such withdrawal may lead to subsidence in a neighboring property. Subsidence insurance and other commercial and government programs are increasingly available to property owners in areas where the risk of subsidence is high.

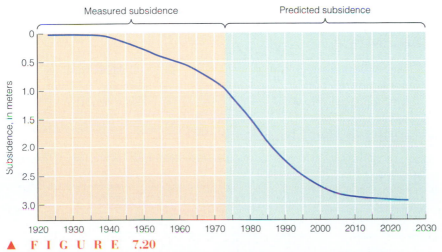

▲ F I G U R E 7.20
Measured rates of surface subsidence at Houston, Texas and predicted rates of subsidence. The predictions are based on measured land surface subsidence, measured drop in level of water table, and rates of fluid withdrawal.

SUMMARY

1. Subsidence—the vertical sinking or collapse of a portion of surface land—can be slow or sudden, localized or regional in extent. It can result from human actions as well as from natural processes.

2. In regions underlain by rocks, such as carbonates, that are highly susceptible to chemical weathering, extensive systems of caves and caverns may form. Cave formation is mainly a chemical process involving dissolution of carbonate rock by circulating groundwater; the rate of cave formation is related to the rate of dissolution.

3. In regions where rocks are exceptionally soluble, karst topography may form. This peculiar terrain is characterized by many small, closed basins and a disrupted drainage pattern.

4. Sinkholes—large dissolution cavities open to the sky—are a common feature of karst terrains. Some sinkholes are formed when the ground surface collapses into existing underground cavities, whereas others are formed directly at the surface as a result of interaction with acidic rainwater. Sinkholes can result from sudden collapses or from the gradual raveling of sediment through a chimneylike passageway called an aven.

5. The use of solution mining to mine salt sometimes causes subsidence. Near-surface coal mining also is commonly associated with subsidence.

6. The withdrawal of fluids—especially water, oil, or gas—from underground rock or sediment units can also lead to subsidence. This type of subsidence is usually caused by the compaction of a rock or sediment when fluid no longer fills the spaces between the particles. Sometimes the compaction and associated subsidence are inelastic and, hence, irreversible.

7. Internal tectonic processes such as folding, faulting, and plate motion, as well as sedimentation in certain tectonic environments, can lead to subsidence. Coastal deltas are particularly susceptible to isostatic downwarping and subsidence.

8. In urban settings a variety of factors conducive to subsidence (such as excessive groundwater withdrawal, mining activity, and the weight of urban development) commonly occur together. Given the density of people, buildings, and utilities, this can create a high potential for extensive and costly damage. Cities located on unconsolidated silt, clay, peat, and sand associated with river floodplains, coastal marshes and/or delta complexes, or lake beds are most likely to experience problems of subsidence.

9. When the removal of solid material is the primary cause of subsidence, prediction of subsidence and assessment of related hazards are generally based on mapping of susceptible areas. Knowledge of the characteristics of underlying rock and rates of fluid withdrawal are needed when fluid withdrawal and/or compaction of compressible layers is the primary cause of subsidence.

IMPORTANT TERMS TO REMEMBER

aquifer (p. 180)
aven (p. 176)
cave (p. 174)
cavern (p. 174)
compaction (p. 180)

cone of depression (p. 180)
dissolution (p. 174)
karst topography (p. 175)
raveling (p. 176)
sinkhole (p. 176)

sinkhole karst (p. 175)
stoping (p. 176)
subsidence (p. 173)

QUESTIONS AND ACTIVITIES

1. If you live (or vacation) in a region underlain by carbonate rocks, make a point of visiting caves or caverns in the area. Many spectacular caverns in North America and elsewhere in the world are open to visitors and offer guided tours. Try to observe the processes of cave formation in action. Where can you see carbonate being dissolved? Where is it being precipitated to form speleothems? A word of warning: don't go caving ("spelunking") without an experienced guide; it can be very dangerous.

2. If you visit a cavern or live in an area underlain by caverns, see if you can spot the signs of karst topography on the surface. Can you see sinkholes or small, pothole-shaped lakes? Get a map of the area—a geologic map, if possible—and try to determine the drainage patterns of streams and rivers.

3. What do you think would be the implications of karst topography and underground cave and drainage systems in the event of a chemical spill? How would the presence of karst topography affect your assessment of the suitability of an area for a hazardous waste disposal site or a landfill? (You may wish to revisit this question after reading Chapters 16 and 17.)

4. Select an urban subsidence problem to investigate in detail. There are lots of interesting examples to choose from—Bangkok, Venice, New Orleans, and Mexico City, among many others. Try to find out the history of subsidence in the city you have chosen. To what extent was the area naturally predisposed to subsidence, and to what extent have human actions caused or contributed to the problem? What is the estimated cost (both economic and environmental) to the city? What solutions to the problem are being considered or undertaken?

FLOODS

Rain added to a river that is rank
Perforce will force it overflow the bank.

• **William Shakespeare**

The Mississippi River has a habit of shifting its channel from time to time in order to avoid flowing around large bends. In seeking a new and shorter route, the river cuts across a bend at its narrowest point, thereby abandoning a piece of the old channel. In the nineteenth century riverboat pilots spoke of such an event as a "cutoff." Former steamboat pilot Mark Twain, an observant amateur geologist, speculated about the future history of the Mississippi River in *Life on the Mississippi* (1883):

The Mississippi between Cairo and New Orleans was 1215 miles long 176 years ago. It was 1180 after the cutoff of 1722. It was 1040 after the American Bend cutoff. It has lost 67 miles since. Consequently, its length is only 973 miles at present.

Now, if I wanted to be one of those ponderous scientific people, and "let on" to prove what had occurred in the remote past by what had occurred in a given time in the recent past, or what will occur in the far future by what has occurred in late years, what an opportunity here! Geology never had such a chance, nor such exact data to argue from! Please observe:

In the space of 176 years the lower Mississippi has shortened itself 242 miles. That is an average of a trifle over one mile and a third per year. Therefore, any calm person, who is not blind or idiotic, can see that in the Old Oolitic Silurian Period, just over a million years ago next November, the Lower Mississippi River was upwards of 1,300,000 miles long, and stuck out over the Gulf of

Mexico like a fishing rod. And by the same token any person can see that 742 years from now the Lower Mississippi will be only a mile and three-quarters long, and Cairo and New Orleans will have joined their streets together, and be plodding comfortably along under a single mayor and a mutual board of aldermen. There is something fascinating about science. One gets such wholesale returns of conjecture out of such a trifling investment of fact.

Twain wrote of the river shortening itself through natural processes, but the Mississippi has also been shortened by human intervention. During the 1930s, for example, a massive flood protection and channel modification program resulted in the shortening of the river's course by 185 km between Cairo and New Orleans. In spite of those efforts, the Mississippi River asserted itself on a grand scale in the summer of 1993. The main portion of the "Great Flood" of 1993 lasted for 79 continuous days, surpassing the previous record of 77 days set by the flood of 1973. Fifty people were killed and 55,000 homes destroyed or damaged. At the height of the flooding, more than 40,000 km² of land bordering the Missouri and Mississippi rivers were under water. The impact of this record-shattering flood on the lives of people in the midwestern United States will not soon be forgotten.

Fleeing the flood. An estimated 55,000 homes were flooded when the Mississippi River overtopped its levees during the great flood of 1993. This farm family's home was in St. Charles County, Missouri.

THE WATER'S EDGE

From the beginnings of civilization, people have tended to live by the water's edge—beside streams, rivers, or lakes; in delta regions; or along the sea coast. Land adjacent to water has traditionally offered many advantages to settlers, advantages that at first were necessary for survival and later were conducive to development and industrialization (Fig. 8.1).

◄ F I G U R E 8.1
Like many of the world's great cities, Paris is situated along the banks of a major river. The Seine provides water for human and industrial use, serves as an avenue of transportation, and has great aesthetic and recreational value. However, people and buildings in the urbanized floodplain must be protected from the constant threat of flooding. This has been accomplished mainly through extensive structural modifications of the river's channel.

Those advantages include fertile soil; access to water, food, energy, and other resources; ready transportation routes; and the capacity for absorption and dispersal of wastes. River valleys, in particular, are favored localities for human settlement because valley floors are flat and easy to build on and the soils typically are deep and arable. But living by the water's edge has some disadvantages, of which the most obvious is the repeated threat of flooding.

A **flood** occurs when the level of a body of water rises until it overflows its natural or artificial confines and submerges land in the surrounding area. Because of the human tendency to develop and inhabit such land, floods can take a high toll in loss of life and damage to structures. This is especially true in less developed countries, where factors such as population density, absence of strict zoning regulations, and lack of effective response, control, and early-warning systems combine to increase the risks associated with flooding.

Bangladesh is more susceptible to flood disasters than any other nation in the world (Fig. 8.2). Half of that nation's land is less than 8 m above sea level, and as much as 30 percent of its total area has been covered during a single flood! High population density in an active delta with constantly shifting rivers combines with heavy monsoon rainfalls and cyclones to increase the hazard. In the spring of 1991 more than 200,000 people were killed in delta flooding associated with cyclone activity.

In industrialized countries loss of life because of flooding is typically much lower than in developing countries, but the amount of damage and disruption can be staggering. Governments and communities spend countless dollars and much effort perfecting and installing flood-control devices. Yet, as demonstrated by the flooding of the Mississippi River in the summer of 1993, such measures sometimes create a false sense of security.

Understanding the causes of flooding and the dynamics of the natural systems in which floods occur is the first step toward predicting and—to a certain extent—controlling floods. There are many different types of floods and many different causes of flooding. In this chapter we concentrate primarily on *river flooding* because of its impact on human activities and interests.

▲ **F I G U R E 8.2**
Cutting of forests in the foothills of the Himalaya, combined with high rainfall, has led to repeated instances of massive downstream flooding of the Brahmaputra River in Bangladesh.

CAUSES OF FLOODING

People who experience floods are frequently surprised and even outraged at what a rampaging torrent of water can do. Geologists, however, tend to view floods as normal and expectable events, for the geologic record shows that floods have been occurring as long as rain has been falling on the Earth.

The Role of Precipitation

In general, natural systems of water flow, cycling, and containment are *mass-balanced;* that is, water flowing into each part of the system is balanced by water flowing out of that part. In the hydrologic cycle as a whole, evaporation is bal-
anced by precipitation. But precipitation is not uniformly distributed over the surface of the Earth. The balance of inflow and outflow in one part of a surface water system may be temporarily disturbed by a period of exceptional precipitation. *Drought* can result when precipitation is exceptionally low, and flooding may occur when precipitation is unusually intense or prolonged. Regardless of their impact on humans, these processes help maintain the long-term balance of the hydrologic cycle as a whole. Flooding therefore is a necessary consequence of the uneven distribution of precipitation and a normal part of the hydrologic cycle. Most floods are caused or exacerbated by intense precipitation, usually in combination with other factors. When a period of exceptional precipitation is combined with such factors as melting snow, inadequate drainage, water-saturated ground, or unusually high tides, the potential for flooding may increase dramatically.

Coastal Flooding

A variety of processes can lead to flooding in coastal areas. Some are natural processes, such as tsunamis (Chapter 5), hurricanes, cyclones, and unusually high tides (Chapter 9). Other floods in coastal areas are caused by human activities. For example, urban development on unconsolidated, compressible sediments in deltas and coastal regions can lead to subsidence of the ground surface, with associated flooding. As noted in Chapter 7, urban flooding in Bangkok is an oft-cited example of this phenomenon.

The failure of protective seawalls during storms can also lead to flooding in low-lying coastal lands reclaimed from the sea. In January 1953 a combination of raging storms and exceptionally high tides led to failure of dike systems in The Netherlands, with tragic consequences. When the flooding was over, 100,000 hectares (6 percent of the country's cultivated land) had been inundated by salt water, affecting well over half of the population. About 5000 houses were destroyed, 60,000 animals drowned, and 1490 people killed. The total cost of the damages was estimated to be at least $250 million.

About 27 percent of the current land area of The Netherlands has been reclaimed from below sea level as a result of efforts going back a thousand years or more. But the costs—both economic and environmental—of reclaiming this land have been extremely high. The new land must be protected by a vast network of dikes, dams, canals, and drainage stations; the Dutch government spends more than $400 million a year on drainage alone. Other environmental problems, primarily sediment compaction and water pollution, have been linked to these structural interventions. These problems, along with concerns about a possible rise in sea level because of global warming, have led the Dutch government to undertake limited flooding of some of the reclaimed land, with the goal of eventually reverting it to

◀ F I G U R E 8.3
The flooding of low-lying land in the Netherlands, January 1995.

marshland. The plan may have been hastened in the spring of 1995 by extensive coastal flooding (Fig. 8.3) associated with a long period of above-average rainfall in the region.

Dam Failures

Dam failures are another cause of flooding related to human activities. At midnight on March 12, 1928, the St. Francis Dam in Saugus, California, failed, releasing 46,500,000 m^3 of water in 1 hour; 450 people lost their lives in the flood. Another spectacular dam failure occurred in Johnstown, Pennsylvania on May 31, 1889, when an old, poorly maintained earthen dam collapsed. The dam contained a reservoir with a maximum capacity of 13,568,060 m^3. A period of intense rainfall in April and May had caused the dam to deteriorate. When the structure collapsed, a wave estimated to be 12 m high roared toward Johnstown, killing 2200 people. In the Vaiont, Italy, dam disaster of 1963 (discussed in Chapter 6), the dam itself did not fail but landsliding in the walls of the reservoir caused a tsunami-like wave, with associated flooding both upstream and downstream from the dam.

Natural dams may also fail. It is common, especially in mountainous terrain, for debris from large landslides to block the narrow part of a valley, thereby creating a natural reservoir. Natural dams made of weathered, unconsolidated material are likely to collapse eventually, causing flooding in downstream areas. Lakes in the summit craters of volcanoes are highly susceptible to breaching. Valleys at high elevations can be temporarily blocked by ice; when spring melting occurs, the blockage may be breached and lake waters released.

A curious landscape in eastern Washington, called the Channeled Scablands (Fig. 8.4), provides evidence that a

prehistoric flood of catastrophic proportions occurred when a large glacially impounded lake was suddenly released. During the last glaciation, part of the ice sheet covering western Canada dammed a huge lake in the vicinity of Missoula, Montana. The lake, which contained between 2000 and 2500 km^3 of water when it was filled, remained in existence only as long as the ice dam was stable. When the glac-

▲ F I G U R E 8.4
The Channeled Scablands of the Columbia Plateau. Giant ripples formed by raging floodwaters as they swept around a bend of the Columbia River. Composed of coarse gravel, the ripples are up to several meters high and their crests as much as 100 m apart.

ier retreated or began to float in the rising lake water, the dam failed and the water was released from the basin, flowing across the Channeled Scablands region and down the Columbia River to the sea. This catastrophic flood created such landforms as huge current ripples, dry canyons with abrupt cliffs marking the sites of former waterfalls, and massive piles of gravel containing huge boulders.

RIVER SYSTEMS

A body of water that flows downslope along a clearly defined natural passageway, transporting particles and dissolved substances, is called a **stream.** A **river** is a stream of considerable volume with a well-defined passageway. This discussion is concerned primarily with river flooding. However, the processes described are generally applicable to stream systems of all sizes. By studying the dynamics of streams and rivers we can begin to understand some of the factors that contribute to flooding.

In addition to their immediate practical and aesthetic importance, rivers are vital geologic agents; they carry most of the water that goes from land to sea and hence are an essential part of the hydrologic cycle. Rivers and streams of all sizes transport billions of tons of sediment to the oceans each year. They also carry soluble salts released by weathering; these play an essential role in maintaining the saltiness of seawater. Rivers shape the surface of the Earth; most landscapes consist of river valleys separated by higher ground and have been produced by weathering, mass-wasting, and river erosion working in combination.

Channels

The passageway of a river is its **channel,** and the particles of sediment and dissolved matter it carries with it constitute the bulk of its **load.** The quantity of water passing by a given point on the bank within a given interval of time is the river's **discharge.** Discharge varies both along the channel and over time, mainly because of changes in precipitation.

A river's channel is an efficient conduit for running water. In response to variations in discharge and load, the channel continuously adjusts its shape and orientation. Therefore, a river and its channel are dynamic elements of the landscape. The dimensions and characteristics of a river's channel determine its ability to handle the flow of water. Developing an understanding of channel dynamics is therefore an important step toward understanding and predicting river flooding.

The size and shape of a channel's cross-section reflect the river conditions prevailing at that point. Very small streams may be as deep as they are wide, whereas very large rivers are usually many times wider than they are deep. Because the volume of water moving through a channel generally increases with increasing distance from the source, the ratio of channel width to depth is also likely to increase.

If we measure the vertical distance that a channel falls between two points along its course, we obtain a measure of a stream's **gradient.** The average gradient of a steep mountain stream, such as the Sacramento River in California, may reach 60 m/km or even more, whereas near the mouth of a very large river, such as the Missouri, the gradient may be as low as 0.1 m/km or even less (Fig. 8.5).

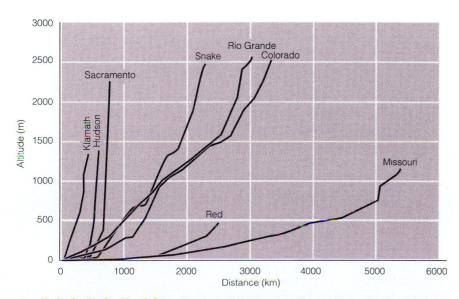

▲ **F I G U R E 8.5**
Long profiles of some rivers in the United States. The Klamath, Hudson, and Sacramento are relatively short, steep-gradient streams, whereas the Missouri is a long river with a low average gradient.

A river's *long profile* (a line drawn along the river's surface from its source to its mouth) is a curve that decreases in gradient with distance from the source (Fig. 8.5). However, the curve is not perfectly smooth because of irregularities in gradient along the channel. For example, a local change in gradient may occur at the point where a channel passes from a bed of resistant rock into one that is more erodible, or when a landslide or lava flow forms a temporary dam across the channel. A hydroelectric dam also introduces an irregularity in the long profile of a river channel and may create an extensive reservoir upstream.

Channel Patterns

From an airplane, it is easy to see that no two rivers are alike. The variety of channel patterns can be explained if we understand the relationships among river gradient, discharge, and sediment load. In this section we explore some of the most common types of river channels.

Straight Channels

Straight channel segments are rare. Generally they occur for only brief stretches before the channel begins to curve. Close examination of a straight segment of natural channel shows that it has some of the features of *sinuous* (curving) channels. For example, a line connecting the deepest parts of a straight channel typically wanders back and forth across the channel (Fig. 8.6A). This may initially be due to random variations in channel depth. However, when the deepest water lies at one side of a channel, a deposit of sediment (a *bar*) tends to accumulate on the opposite side, where the velocity of flow is lower. The sinuous flow of water within a channel thus causes a succession of bars to form on alternating sides of the channel.

Meandering Channels

In many rivers the channel follows a series of smooth bends like the switchbacks of a mountain road (Fig. 8.6B). Such a bend is called a *meander*, after the Menderes River (Latin *Meander*) in southwestern Turkey, which is noted for its winding course. Meanders occur most frequently in channels that lie in fine-grained river sediments and have gentle gradients. The meandering pattern reflects the way the river minimizes resistance to flow and dissipates energy as uniformly as possible along its course.

If you try to wade or swim across a meandering river, you will soon realize that the velocity of the flowing water is not uniform in all parts of the river. Velocity is lowest along the bottom and sides of the channel because that is where the flow encounters the greatest frictional resistance. Along a straight segment of the channel, the highest velocity is usually found near the surface in midchannel. Wherever the water rounds a bend, however, the zone of highest velocity swings toward the outside of the channel. This type of variation makes it very difficult to obtain accurate measurements of the average velocity of flow in a large river like the Mississippi.

Over time, meanders migrate slowly down a valley. As water sweeps around a bend, strong turbulence causes undercutting and slumping of sediment where the fast-moving water meets the steep bank. Meanwhile, along the inner side of the bend, where the water is shallow and the flow is less rapid, coarse sediment accumulates to form a *point bar* (Fig. 8.6B). As a result of these processes, meanders slowly change shape and shift position as sediment is subtracted from and added to their banks. Sometimes the water in the channel manages to find a shorter route downstream, cutting off a meander. As sediment is deposited along the margin of the new channel, the bypassed meander is transformed into a curved *oxbow lake* (Fig. 8.7).

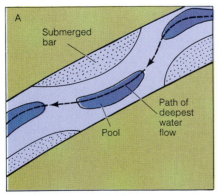

Straight channel

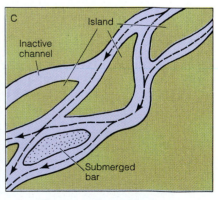

Meandering channel

Braided channel

▲ **F I G U R E 8.6**
Features associated with (A) straight, (B) meandering, and (C) braided channels. Pools are places along a channel where the water is deepest. Arrows indicate the direction of flow and trace the path of the deepest water.

▲ FIGURE 8.7
A meandering river near Phnom Penh, Cambodia, flows past agricultural fields that cover the river's floodplain. Light-colored sandy bars lie opposite cutbanks on the outsides of the meander bends. Two oxbow lakes, the products of past meander cutoffs, lie adjacent to the present channel.

Nearly 600 km of the Mississippi River's channel have been abandoned through cutoffs since 1776. However, contrary to the analysis offered by Mark Twain at the beginning of the chapter, the river has not been shortened appreciably, because the loss of channel resulting from cutoffs has been balanced by lengthening of the channel as other meanders have grown.

Braided Channels

The intricate geometry of a *braided stream* resembles the pattern of braided hair, for the water repeatedly divides and reunites as it flows through two or more interconnected channels separated by bars or islands (Fig. 8.6C). Braiding is related to a river's ability to transport sediment. If a river is unable to move all the available load, it tends to deposit the coarsest sediment in the form of a bar that divides the flow and concentrates it in the deeper segments of the channel on either side. As the bar builds up, it may emerge above the surface as an island and become stabilized by vegetation that anchors the sediment and inhibits erosion.

A braided pattern tends to form in rivers with a highly variable discharge and easily erodible banks that can supply abundant sediment to the channel system. It seems to represent an adjustment by which a river increases its efficiency in transporting sediment. Large braided rivers typically have numerous constantly shifting shallow channels (Fig. 8.8). Although at any given time the active channels may cover no more than 10 percent of the width of the entire channel system, within a single season all or most of the surface sediment may be reworked by the laterally shifting

▲ FIGURE 8.8
Satellite false-color image showing the intricate braided pattern of the Brahmaputra River where it flows out of the Himalaya en route to the Ganges Delta. Noted for its huge sediment load, the river may be 8 km wide during the rainy season.

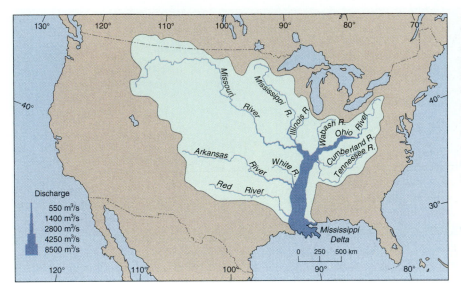

◀ **F I G U R E 8.9**
The drainage basin of the Mississippi River encompasses a major portion of the central United States and extends into southern Canada. In this diagram the width of the river and its major tributaries reflect discharge values, that is, the average amount of water carried by each portion of the river per unit of time (given in m³/s).

channels. The continuously shifting braided channels of the Brahmaputra River in the Ganges Delta represent a major flood hazard for residents of that densely populated region.

Drainage Basins and Divides

Every river is surrounded by its **drainage basin,** the total area that contributes water to the river. The line that separates adjacent drainage basins is a **divide.** Drainage basins range from very small (less than a square kilometer) to vast areas of subcontinental dimension. The drainage basin of the Mississippi River encompasses more than 40 percent of the area of the contiguous United States (Fig. 8.9). Not surprisingly, the area of a drainage basin is proportional to both the length and the mean annual discharge of the river into which its waters drain.

As a drainage system develops, details of its pattern change. Rivers acquire new *tributaries*—smaller streams that flow into it. Some old tributaries are lost as a result of *capture,* the interception and diversion of one stream by another. When capture occurs, some river segments are lengthened and others are shortened. Just as the river channel and its discharge and load are constantly changing, so too is the drainage system. Like a river channel, a drainage system is a dynamic system tending toward a condition of equilibrium.

Not surprisingly, there is a close relationship between drainage patterns and the nature of the underlying rocks. The ease with which a particular rock is eroded depends on its structure and composition. The course a river takes across the land is therefore strongly influenced by these factors. Structures such as faults and fractures in the underlying rocks can influence the direction of flow and the type of

drainage pattern that develops. Fig. 8.10 shows some of the most common drainage patterns and the geologic factors that control them.

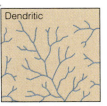

Irregular branching of channels ("treelike") in many directions. Common in massive rock and in flat-lying strata. In such situations, differences in rock resistance are so slight that their control of the directions in which valleys grow headward is negligible.

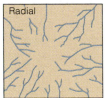

Channels radiate out, like the spokes of a wheel, from a topographically high area, such as a dome or a volcanic cone.

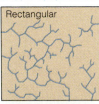

Channel system marked by right-angle bends. Generally results from the presence of joints and fractures in massive rocks of foliation in metamorphic rocks. Such structures, with their cross-cutting patterns, have guided the directions of valleys.

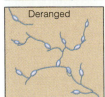

Streams show complete lack of adjustment to underlying structural or lithologic control. Characteristic of recently deglaciated terrain whose preglacial features have been remodeled by glacial processes.

▲ **F I G U R E 8.10**
Some common drainage patterns and their relationship to rock type and structure.

Dynamics of Streamflow

If you stand outside during a heavy rain, you can see that water initially tends to move down slopes in broad, thin sheets, a process called *overland flow*. After traveling a short distance, overland flow begins to concentrate into well-defined channels, thereby becoming **streamflow.** Streamflow and overland flow together constitute **runoff,** the portion of precipitation that flows over the surface of the land. The factors that control runoff determine how much of the water that falls as precipitation eventually reaches a given channel; the dynamics of streamflow determine how well that channel can handle the water.

Several basic factors control the way a river behaves. The most important are gradient, the channel's cross-sectional area (width × average depth), the average velocity of water flow, discharge, and load. Unlike the sediment that forms the bulk of a river's load, dissolved matter generally does not affect the behavior of a river. The relationship among discharge, velocity, and channel shape is expressed by the following equation:

$$
\begin{array}{ccccc}
Q & = & a & \times & v \\
\text{Discharge} & & \text{Cross-sectional area} & & \text{Average} \\
(\text{m}^3/\text{s}) & & (\text{m}^2) & & \text{velocity} \\
& & & & (\text{m/s})
\end{array}
$$

This equation is critical to understanding streamflow and, therefore, flooding. It tells us that when discharge changes, one or more of the factors on the right side of the equation must also change if equilibrium is to be maintained. For example, if discharge increases, both velocity and channel cross-sectional area are likely to increase so that the channel can accommodate the added flow. Conversely, a decrease in discharge can lead to a corresponding decrease in channel dimensions and velocity.

Flooding can produce dramatic changes in these factors. In one 6-month period the channel of the Colorado River at Lees Ferry, Arizona, underwent major changes in dimensions as discharge increased and then fell (Fig. 8.11). Before this flood, the channel averaged about 2 m in depth and 100 m in width. As discharge increased in late spring, the water rose in the channel and erosion scoured the bed until the channel was about 7 m deep and 125 m wide. With an associated increase in velocity, the enlarged channel was now able to accommodate the increased flood discharge and carry a greater load. As discharge fell, the river was unable to transport as much sediment and the excess load was dropped in the channel, causing its floor to rise. At the same time, decreasing discharge caused the water level to fall, thereby returning the cross-sectional area to its preflood dimensions.

RIVER FLOODING

Now that a basic understanding of river systems and streamflow has been established, let's look more closely at the processes involved in river flooding.

Flood Stage

Hydrologists use the term **stage** to refer to the height of a river (or any body of water) above a locally defined reference surface. When a river's discharge increases as a result of prolonged or intense precipitation or meltwater from a spring thaw, its channel may fill completely, so that any further increase in discharge results in water overflowing the banks of the river. In this condition the river is said to have reached **bankfull stage,** or **flood stage.**

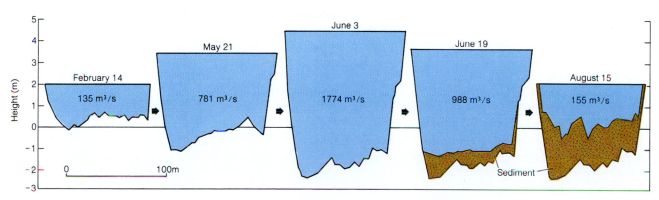

▲ **FIGURE 8.11**
Change in the cross-sectional area of the Colorado River at Lees Ferry, Arizona, during a 6-month period in 1956. As discharge increased from February to June, the channel floor was scoured and deepened and the water level rose higher against the banks. During the falling-water phase, from June to August, the river level fell and sediment was deposited in the channel, decreasing its depth.

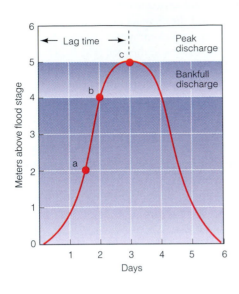

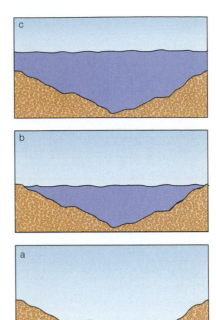

◀ **F I G U R E   8.12**

A hypothetical example of a flood. The discharge hydrograph (discharge plotted against time after the beginning of the storm) shows the discharges measured over several days at a single hydrologic gauging station. Point *a* shows preflood discharge, *b* shows the discharge at bankfull stage, and *c* shows peak discharge. The cross-sections and water levels in the channels at each of these times are shown for comparison. The peak or crest of the flood occurred about 3 days after the storm began (lag time), and the flood crested at 1 m above the local flood stage.

The discharge of a river is typically displayed on a **hydrograph,** a graph in which river discharge (*Q*) is plotted against time. Figure 8.12 shows a plot of discharges measured at one gauging station over a period of 6 days. The three channel cross-sections show what the river's cross-section and stage would have looked like on the 3 days corresponding to the points marked on the hydrograph. In this hypothetical example, the river rose from its normal level (at point *a*) to flood stage (*b*). It then exceeded flood stage (*c*), flowing over the banks and onto adjacent land and reaching its maximum discharge about 3 days after the beginning of the storm. The time that elapses between the

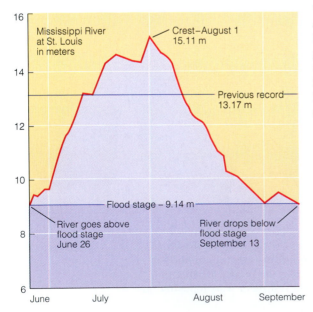

◀ **F I G U R E   8.13**

A discharge hydrograph for the Mississippi River flooding of 1993, taken from the *St. Louis Post-Dispatch.* The hydrograph shows the river cresting on August 1 at 49.58 feet (more than 15 m), 19.58 feet (about 6 m) above the local flood stage of 30 feet (just over 9 m). The river did not drop below flood stage until September 13, and local flooding continued even after that date.

onset of precipitation and the peak flood stage (3 days in Fig. 8.12) is called **lag time.** The lag time makes it possible to track the progress of the storm and prepare for the flood or evacuate the area if necessary.

The discharge measured when the river is at bankfull stage (point *b* on the hydrograph in Fig. 8.12) is referred to as **bankfull discharge.** Point *c*, at which the maximum discharge for this particular event was reached, is called the **peak discharge.** If you were listening to a report of this flood on the news, you might hear that "the river is expected to peak (or *crest*) tomorrow at 5 meters." This means that the peak discharge is projected to be 5 m vertically above the local reference surface, or 1 m above flood stage as shown on the hydrograph. On August 1, 1993, the Mississippi River crested at more than 15 m at St. Louis, or 6 m above the local flood stage of just over 9 m (Fig. 8.13).

Floodplains and Levees

During a major flood, the water overflowing a river's banks inundates the adjacent **floodplain** (Fig. 8.14). The fine silts and muds that settle out of floodwaters to form the broad, flat floodplain produce the deep, arable soils that make floodplain development so attractive. The size of a river's floodplain is a function of the size of the river itself. For example, the floodplain of the Mississippi River encompasses some 80,000 km²; a smaller stream develops a much smaller floodplain.

Often the boundary between a channel and its floodplain is the site of a *natural levee*—a broad, low ridge of fine sediment built along the side of the channel by debris-laden floodwater. Natural levees are created only during floods that are high enough to submerge the river's banks completely and flow over them onto the floodplain. As sediment-laden water flows over the banks, its depth, velocity, and turbulence decrease abruptly. This results in sudden, rapid deposition of the coarser part of the river's load along the margins of the channel, creating the natural levee. Farther away, finer silt and clay settle out of the quieter water covering the floodplain.

Precipitation and Infiltration

As mentioned earlier, most flooding involves a period of exceptionally long or intense precipitation. Of particular significance is the balance between precipitation and infiltration. When it rains, some of the water is intercepted either by impermeable surfaces or by vegetation. Much of the intercepted water is returned to the atmosphere directly through evaporation and transpiration. Some of the rainfall enters the ground through the process of infiltration. The remainder flows over the surface as runoff, that is, overland flow and streamflow. Runoff thus can be expressed as follows:

Runoff = Precipitation – Infiltration – Interception – Evaporation

During a storm, problems arise when the ground's *infiltration capacity*—that is, the maximum rate at which water can be absorbed—is exceeded. When that happens river channels and other surface depressions fill to overflowing and flooding occurs. The nature, magnitude, and likelihood of flooding therefore depend not only on the amount and distribution of rainfall but also on factors affecting the infiltration capacity of the ground.

Rainfall Distribution: Upstream and Downstream Flooding

The distribution of rainfall over the area affected by a storm is sometimes depicted in the form of an **isohyetal map** (an *isohyet* is a line connecting points of equal precipitation). The intensity of rainfall in a particular area is plotted using contours of equal rainfall depth, either for a specified time interval or for the entire duration of the storm. An isohyetal map of the upper part of the Mississippi River Basin is shown in Fig. 8.15. In this example, rainfall during the first half of 1993 is expressed in terms of the percentage by which it exceeded the 30-year average rainfall for the same period (January to July). The isohyetal map is superimposed on a map showing the general area where flooding occurred during that period.

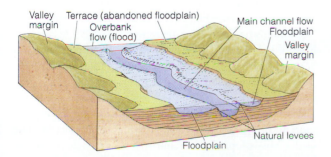

◀ F I G U R E 8.14
Main features of a river valley.

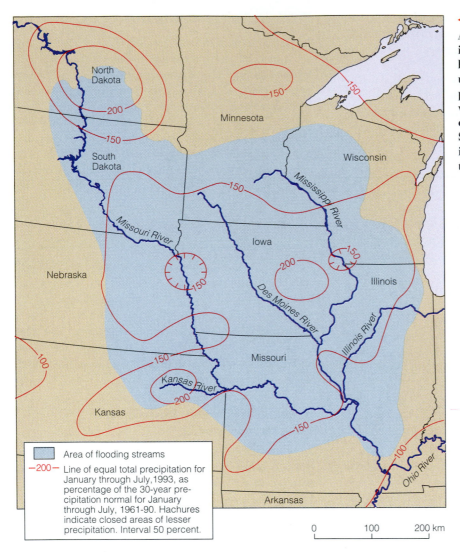

◀ **F I G U R E 8.15**

An isohyetal map showing rainfall in the upper Mississippi River basin during the period from January to July 1993. Rainfall is expressed in terms of the percent by which it exceeded the 30-year average rainfall for the same period. Superimposed is the general area in which flooding occurred during the summer of 1993.

Passing storms often generate brief intervals of intense rainfall. As the runoff from such a storm moves into the river channel, the river's discharge quickly rises. The crest of the resulting flood, the point at which peak discharge is reached, passes the gauging station about 2 hours after the storm has passed. It takes an additional 8 hours for the flood runoff to pass through the channel and for discharge to return to the normal level. This type of storm—intense, but infrequent and of short duration—often causes flooding that is severe but local in extent. This is called **upstream flooding,** because the effects of the storm runoff usually do not extend to the larger streams joining the system farther downstream (Fig. 8.16A).

Upstream flooding can be very hazardous. Such floods typically have a short lag time (the interval between maximum rainfall intensity and the peak of flooding). Floods in which the lag time is exceptionally short—hours or minutes—are called **flash floods.** A devastating flash flood occurred in the Black Hills of South Dakota in 1972, when

40 cm of rain, almost the total yearly average, fell in a period of under 6 hours. In this brief but intense event, 2932 people were injured and 237 killed; 750 homes were destroyed and as many as 2000 cars were damaged by floodwaters.

Storms that continue for a long time and extend over an entire region are responsible for **downstream flooding.** In this type of flooding, the total discharge increases downstream as a result of increased flows in the many tributaries joining the system (Fig. 8.16B). Downstream flooding is often intensified by spring meltwater. The 1993 Mississippi River flood was a case of downstream flooding. Persistent heavy rains began in the spring and continued on an almost daily basis throughout much of the summer. Precipitation in upstream areas, including Iowa, Nebraska, and the Dakotas, greatly exacerbated flooding downstream in Missouri and Illinois. The total area inundated by the flooding of the Mississippi and Missouri rivers and their tributaries has been estimated at more than 40,000 km².

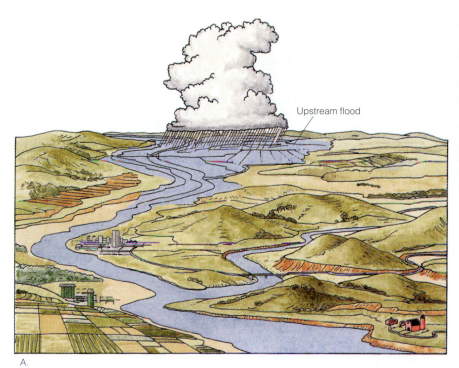

Upstream flood

A.

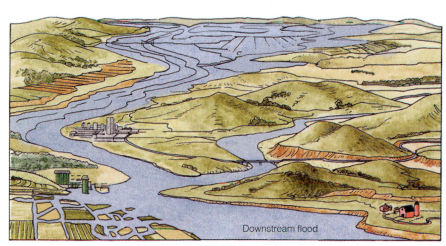

Downstream flood

B.

◄ F I G U R E 8.16
A. Upstream floods are generally local in extent, usually with short lag times. They typically result from intense storms of short duration. B. Downstream floods are regional in extent, with longer lag times and higher peak discharges. They result from regional storms of long duration or extended periods of above-normal precipitation.

Factors Affecting Infiltration

The capacity of the ground to absorb rainfall is influenced by a variety of factors. Of particular importance are the physical characteristics of underlying soil and rock formations. These include particle size and type (which in turn influence the porosity and permeability of the material), amount and type of vegetation, the amount of moisture and organic material in the soil, and the degree to which the material has been fragmented or disrupted by roots or burrowing animals. Climatic factors can also influence infiltration capacity. For example, soils in arid and semiarid regions sometimes develop a thick, highly impermeable crust composed primarily of calcium carbonate and variously referred to as *duricrust, hardpan,* or *caliche.* Permanently or seasonally frozen ground, even in the subsurface, also reduces the rate of infiltration.

Of all these factors, soil moisture is arguably the most important. Infiltration capacity may be high at the beginning of a storm, but in periods of intense or extended rainfall the ground can become completely saturated. A wet fall and big winter snows set the stage for the 1993 Mississippi River flood; by the time the heavy spring and summer rains began, large areas of ground were already well saturated, effectively preventing infiltration.

T A B L E 8.1 • **Fatalities from Some Disastrous Floods**[a]

River	Type of Flood	Date	Fatalities	Remarks
Huang He, China	River flooding	1887	ca. 900,000	Flood inundated 130,000 km² and swept many villages away
Johnstown, Pennsylvania	Dam failure	1889	2200	Dam failed. Wave 10–12 m high rushed down valley
Yangtze, China	River flooding	1911	ca. 100,000	Formed lake 130 km long and 50 km wide
Yangtze, China	River flooding	1931	ca. 200,000	Flood extended from Hankow to Shanghai (>800 km), leaving 10s of millions homeless
Vaiont, Italy	Dam failure	1963	2000	Landslide into lake caused wave that overtopped dam and inundated villages below

[a]Source: *Encyclopedia Americana* (1983).

Factors related to human use are critical in determining the absorptive characteristics of the ground. The primary effect of development is to increase runoff by increasing the amount of impervious cover, that is, roads, buildings, parking lots, and other structures that are impermeable to water.

HAZARDS ASSOCIATED WITH FLOODING

We can divide the hazards associated with flooding into primary, secondary, and tertiary effects. The primary impacts of flooding are those caused by actual contact with flowing water; they include death by drowning, structural damage to buildings, crop loss, and erosion of flooded areas. The secondary and tertiary effects are longer term impacts that are indirectly related to the flooding. As shown in Tables 8.1 and 8.2, river flooding can be disastrous to human interests, causing both loss of life and extensive property damage.

Primary Effects

Major floods can have a truly devastating impact on humans. The Huang He in China (Fig. 8.17), sometimes called the Yellow River because of the yellowish-brown color produced by its heavy load of silt, has a long history of disastrous floods. In 1887, the river inundated 130,000 km² and swept away many villages in the heavily populated floodplain. In August 1931, another flood killed a staggering 3,700,000 people.

The strength of flood currents can be surprising. People affected by flooding often express astonishment at the incredible force exerted by the flowing water, which may tear out trees by their roots or knock buildings off their foundations. Somehow one expects floodwaters to spread peacefully across the floodplain. Unfortunately, the twisted

T A B L E 8.2 • **Some Primary, Secondary, and Tertiary Impacts of the 1993 Flooding of the Mississippi**[a]

Primary impacts
Water damage to household items
Structural damage to buildings
Destruction of:
 Roads
 Rail lines
 Bridges
 Engineered structures (levees, etc.)
 Boats, barges, moorings
Historical sites destroyed
Crop loss
Cemeteries flooded, graves disrupted
Loss of life

Secondary and tertiary impacts
Destruction of farmlands
Destruction of parklands and wildlife habitat
Health impacts:
 Disease related to pollution
 Injuries (back, electric shock, etc.)
 Fatigue
 Stress, depression
Disruption of transportation services
Gas leaks
Disruption of electrical services
Lack of clean water
Impacts on crop prices; food shortages
Job loss and worker displacement
Economic impacts on industries:
 Construction (beneficial impact)
 Insurance (negative impact)
 Legal (beneficial impact)
 Farming (negative impact)
Misuse of government relief funds
Changes in river channels
Collapse of whole community structures

[a]This is not a comprehensive list but rather a sampling taken from newspaper reports during the event.

◀ F I G U R E 8.17
A large suspended load of sediment gives the Huang He a very muddy appearance and its English name (Yellow River). The Huang He has a long history of disastrous floods.

wreckage left behind in flooded towns tells a very different story.

As discharge increases during a flood, so does velocity. The increased velocity enables the river to carry not only a greater load but also larger particles. Damage caused by moving debris thus is one of the main hazards associated with flooding. One of the most frightening episodes in the 1993 Mississippi River flooding involved a group of large propane tanks that were torn from their moorings by the force of rising water. The tanks' steel cable moorings snapped, so they were linked to their supports only by connecting pipes. The tanks, which had been filled almost to capacity in order to make them heavier, bobbed and strained at the pipes and began to develop leaks. On August 2 the entire surrounding area was evacuated; residents were given only 3 minutes to leave their homes. The hazard was extremely serious: if a large cloud of propane vapor had

ignited, much of south St. Louis could have been obliterated in a fiery explosion. However, when the floodwaters began to recede, emergency crews managed to retether the tanks and repair the leaks.

The collapse of the St. Francis Dam in 1928 provides another example of the exceptional force of floodwaters. When the dam gave way, the water behind it rushed down the valley, moving blocks of concrete weighing as much as 9000 metric tons over distances of more than 750 m.

The massive load of finer particles carried by a river during flooding represents a different type of hazard. After the Mississippi and Missouri river floodwaters had receded, many people in Missouri, Illinois, and Iowa found their homes almost knee-deep in fine river mud. Volunteer crews from all over North America helped homeowners dig out their belongings (Fig. 8.18). Floodwaters can also concentrate garbage, debris, and pollutants, as described in this

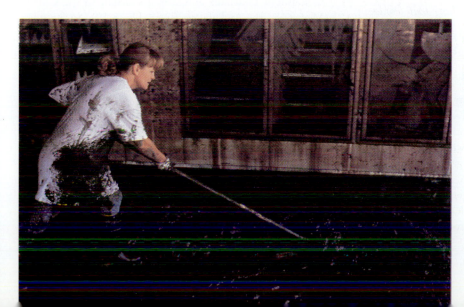

◀ F I G U R E 8.18
Cleaning up after the great Mississippi River flood of 1993. When flood waters finally receded they left behind a gooey blanket of muddy sediment.

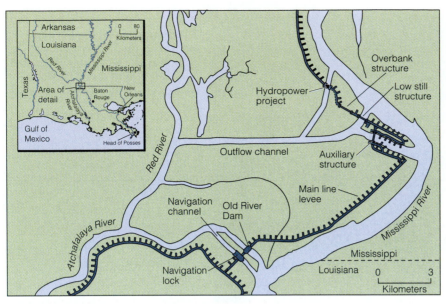

▲ **F I G U R E 8.19**

The Old River flood-control and channel-control structures at the convergence of the Mississippi and Atchafalaya rivers. The Atchafalaya takes water and suspended sediment from the Mississippi, its "parent," but leaves behind the river's bottom load of sediment. Meanwhile, the Atchafalaya is scouring its bottom, looking for sediment to match its flow. The scouring causes the Atchafalaya's channel to deepen, allowing it to draw even more water from the Mississippi. Eventually, when the flows in the two rivers are approximately equal, a great flood will cause the main channel to switch; the parent channel will quickly silt up and the new channel will enlarge rapidly and irretrievably. The engineering structures at this location are partly intended to avoid this type of occurrence.

eyewitness account of the aftermath of a flood caused by a dam failure in Buffalo Creek, West Virginia, in 1972:

> *Well, there was mud all over the house, no place to sleep. There was mud in the beds, about a foot and a half of it on the floors. And I had all that garbage. It was up over the windows. In the yard I had poles and trees and those big railroad ties from where they had washed out at Becco—furniture, garbage cans, anything that would float in the water.*

Secondary and Tertiary Effects

Disruption of services such as electricity, water, and gas is a common secondary hazard associated with flooding. One of the main challenges during any flood is maintaining the delivery of drinkable water to areas where water supplies have been contaminated. Road, bridge, and rail closures can contribute to the shortage of basic supplies in flooded communities. In regions where emergency response procedures are not well orchestrated, problems such as disease, hunger, and homelessness may result from flooding and the associated disruption of services.

Flooding can also have longer term effects, such as loss of wildlife habitat or permanent changes in local ecosystems. River channels can be permanently changed by flooding, sometimes quite dramatically, as illustrated by the Colorado River example discussed earlier (Fig. 8.11). During the 1993 Mississippi River flooding, a major concern was that the river might actually abandon its main channel in southern Louisiana and seek a shortcut to the Gulf of Mexico via the Atchafalaya River (Fig. 8.19). So far, engineering structures at the juncture of the two channels have prevented this from occurring.

Muddy deposits left by receding floodwaters can clog channels, interfering with transportation or recreational use. Some agricultural fields in the Midwest are now covered by thick deposits of white sand, the remnants of sand bars that built up behind artificial levees. When the levees were breached, the floodwaters scoured deep holes and carried immense loads of sand out onto the fields; some of the fields were irretrievably damaged.

In the long run, however, the silt deposited by floodwaters is extremely beneficial. It replenishes the rich, arable soil that is typical of floodplains. Before the construction of the Aswân High Dam and other dams on the Nile River, annual flooding deposited sediment over the river's floodplain and delta, adding to the soil at a rate of 6 to 15 cm per century. The construction of the dams eliminated seasonal flooding, but it also cut off the natural source of enrichment and replenishment of the floodplain soil; Egyptian farmers must now use artificial fertilizers and soil additives to keep their land productive.

PREDICTING RIVER FLOODING

There are three main approaches to flood prediction. They involve (1) using statistical techniques to predict the frequency of floods of a given magnitude; (2) using models and mapping to determine the areal extent of hazards associated with floods of a given magnitude; and (3) monitoring the progress of a storm in order to provide a forecast or early warning to those who may be affected by a flood.

Frequency of Flooding

Predictions of the frequency of flooding are based on statistical analyses of records of hydrologic events at a specific locality. For one particular gauging station, a range of discharge values will have been measured over the space of a year; among these, one value will be the *annual maximum discharge*—the highest value measured at that station in that year. A record of annual maxima over a period of years is an annual maximum discharge *series* or *array*.

Recurrence Interval

The first step in the analysis of a maximum discharge series is to rank the annual maxima. Rank is given the notation *m*; the highest annual maximum discharge in the series is given the rank $m = 1$, the next largest $m = 2$, and so on. The smallest annual event thus will have a rank that is equivalent to the number of years (*n*) represented in the data series, or $m = n$. These values are used to calculate the **recurrence interval (R),** which is the average interval between occurrences of two hydrologic events of equal magnitude. Recurrence interval is determined from the following relationship, called the *Weibull equation:*

$$R = \frac{n + 1}{m}$$

It is important to remember that *R* represents the *average* interval between two floods of equal magnitude, not the actual *recorded* interval. In other words, calculations based on the existing data series may indicate that a flood with a discharge of 2500 m³/s should recur approximately once every 50 years, but this is just a statistical average; in reality, two floods of that magnitude could occur in successive years, or 10 years apart, or 100 years apart, or even twice in the same year. For example, in 1986–1987 two floods with calculated recurrence intervals of 100 years occurred within one 10-month period in the Chicago area.

A different form of the Weibull equation can be used to

determine the probability that a discharge value of given magnitude will be exceeded in any given year:

$$P_e = \frac{m}{n + 1}$$

This is simply the reciprocal of *R* (i.e., 1/R). The term P_e is referred to as the *annual exceedance probability*. A discharge with a recurrence interval of, say, 5 years would be expected to occur approximately 10 times over a 50-year time span; in any single year, the probability that a discharge of equal or greater magnitude will occur is given by 1/R, or 0.20 (20 percent). A discharge with a recurrence interval of 100 years would have an annual exceedance probability of 0.1, so there is a 1 percent probability that a flood of that magnitude will occur or be exceeded in any given year.

Flood Frequency Curve

Flood magnitudes (i.e., discharges) are often plotted with respect to the recurrence interval calculated for a flood of that magnitude at the given locality. This type of graph produces a **flood frequency curve.** Figure 8.20, for example, shows the flood frequency curve determined from the annual maximum discharge series for the Skykomish River at

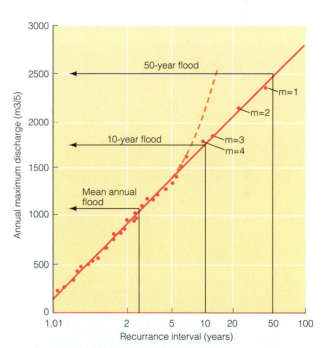

▲ **F I G U R E 8.20**
Curve of flood frequency for the Skykomish River at Gold Bar, Washington, plotted on a probability graph. A flood with discharge of 1750 m³/s has a 1 in 10 probability of occurring in any year (a 10-year flood), whereas a flood of 2500 m³/s has a 1 in 50 probability (a 50-year flood). The dashed line shows a possible extrapolation of the curve if the annual maximum discharge array from which the graph has been constructed did not include the four largest events. This extrapolation would lead to a misinterpretation of the expected size and recurrence interval for a flood of large magnitude.

Gold Bar, Washington. The curve shows that a flood with a discharge of at least 1750 m^3/s has a 1 in 10 (10 percent) probability of occurring in any given year, whereas a flood with a discharge of at least 2500 m^3/s has a 1 in 50 (2 percent) probability. We refer to a flood that has a recurrence interval of 10 years as a 10-year flood; if the interval is 50 years, it is a 50-year flood; and so on.

Flood frequency curves can also be used to predict the recurrence interval for events that are of greater magnitude than any historically recorded event. This is done by extending the flood frequency curve. For example, the curve shown in Fig. 8.20 suggests that a flood with a discharge of 2700 m^3/s would have a recurrence interval of about 100 years and, therefore, an annual exceedance probability of 1 percent. The longer the available data set, the more valid are the predictions based on the data. A long data set is critical in determining the recurrence interval for very large floods. For example, if the series represented in Fig. 8.20 did not include the four largest events (ranks m = 1, 2, 3, and 4), a projection of the flood frequency curve might suggest a very different recurrence interval and probability for a major flood.

Flood Hazard Mapping

Another important aspect of flood prediction is *flood hazard mapping*, the use of mapping to determine the impact and areal extent of a potential flood. Hazard mapping is concerned with the following types of questions: Given the occurrence of a flood of a specific magnitude in this locality, what will the event be like? How high will the water levels rise? What areas will be under water, and how deep will the water be in those areas? Flood hazard mapping can be based on known, recorded events or on prehistoric events.

Hazard mapping based on known events relies on recorded observations. Probably the most important sources of observational data for recent floods are aerial photographs and satellite images (Fig. 8.21). Remote sensing can provide information not only about the areal extent of floodwaters but also about the amounts of debris and silty deposits left behind after the floodwaters recede. On-site observations are also valuable sources of information concerning both recent floods and earlier ones, for which satellite images may not be available. This type of observation includes, for example, the elevation of high-water

A.

B.

▲ **F I G U R E 8.21**
A pair of satellite images shows the region where the Missouri River joins the Mississippi River at St. Louis, Missouri, (A) in a typical summer (July 1988) and (B) during the disastrous flood of July 1993, when weeks of torrential rains caused the rivers to overflow protective levees and inundate numerous towns and vast areas of farmland.

marks on the sides of buildings, or the location of a deposit left by floodwaters.

Experts also use physical models of river systems in their efforts to predict floods and the associated hazards. The U.S. Army Corps of Engineers formerly used a 13-km-long, 200-acre scale model of the Mississippi River and its tributaries to predict the system's flooding behavior, but the model became outmoded because of the extent of engineering interventions and modifications in the river system. Computer programs now enable experts to account for many variables in an integrated, coordinated manner; to constantly update their information; to predict the downstream impacts of events such as levee breaks; and to determine the potential effects of various flood management scenarios. Both the Army Corps of Engineers and the National Weather Service were criticized for failing to upgrade their computer modeling capabilities before the 1993 flooding.

Hazard mapping can be extended to encompass floods of greater magnitude than any in recorded history for the area under study. One way to accomplish this is by extrapolating from a known event. For example, if water depths and the area of inundation have been measured for a particular discharge value, the areal extent and depth of a flood of greater magnitude can be extrapolated by calculating the additional volume of water involved, taking into account the local topography. Another approach to flood hazard mapping is **paleoflood hydrology,** the study of ancient floods. Paleoflood hydrology relies on the study of landforms and sediment deposits to determine the areal extent, volume, depth, direction, and rate of flow of ancient floodwaters. The enormous prehistoric flooding in the Channeled Scablands of eastern Washington State was studied in this manner.

Monitoring the Progress of a Flood

So far we have looked at the procedures used to determine the frequency, probability, extent, and characteristics of floods of given magnitudes. Another important aspect of flood prediction is real-time monitoring of storms, which sometimes leads to the issuance of a *forecast* or *early warning* of flooding. Hydrologists use the term *forecast* to refer to a short-term prediction specifying the actual time when the peak or crest of a flood will pass a particular location, as well as the magnitude of the peak. An *early warning* is a statement issued by authorities to the general public. It is based on the forecast as well as on other types of flood hazard information, and it includes recommendations for actions such as evacuation of specific areas.

Forecasting depends on accurate estimates of the expected magnitude of peak discharge. These, in turn, depend on constant monitoring of the extent and distribution of rainfall. Information about the characteristics of the storm must be combined with knowledge of the area, including such features as topography, vegetation, and impermeable cover. This information can be used in predicting the amount of storm runoff, its velocity, and its probable course. The rate at which the peak is moving downstream can be determined by taking successive readings from a series of hydrologic gauging stations. The issuance of an early warning to downstream areas thus takes advantage of the lag time between the beginning of the storm and the arrival of the flood crest. Forecasting is particularly difficult in floods with short time lags, that is, flash floods.

HUMAN INTERVENTION

Flooding is highly susceptible to the effects of human intervention. Urban development and other human activities can worsen or even cause flooding in a variety of ways. People also intervene in natural systems for the purpose of controlling or mitigating flood hazards, and these interventions also have consequences. Therefore, approaches to the prediction and control of flooding must take human factors into account.

Channel Modifications

River channels are modified or "engineered" for many different reasons. Some common goals of channel modification include flood control, enhancing drainage in wetlands, increasing access to floodplain lands for development, facilitating transportation, and controlling erosion. Occasionally the goal is to improve the appearance of the river channel. The modifications usually consist of some combination of straightening, deepening, widening, clearing, or lining of the natural channel. These processes are collectively referred to as **channelization** (or sometimes *channel enhancement*).

In the context of flood control, channelization is generally undertaken with the aim of increasing the channel's cross-sectional area. Given the relationships among discharge, velocity of flow, and cross-sectional area discussed earlier, an increase in the width and depth of the channel should enable it to handle a greater discharge at a higher velocity. Straightening of the channel can also lead to higher flow velocities, thereby also facilitating the passage of more water per unit of time.

The Channelization Controversy

Channelization is a controversial process. Many opponents believe that such modifications interfere with natural habi-

VALMEYER: TRADING FLOODS FOR SINKHOLES

*T*he people of Valmeyer, Illinois, had had enough. After Mississippi River floodwaters inundated their town not once but twice last summer, destroying 90 percent of the homes, offices, and public buildings, the 900 citizens decided to move to higher ground. In October 1993 they asked the Illinois State Geological Survey (ISGS) to examine a new site for geologic hazards.

The new site is a 500-acre tract about 3 km inland from the present town on a limestone bluff 90 m above the Mississippi floodplain. After studying well logs, borehole data, and maps, eight ISGS scientists walked the area, drilled for subsurface samples, ran tests to determine the depth to bedrock and nature of the bedrock surface, and determined the likelihood of collapse at a nearby limestone mine. Four potential types of geologic hazards were identified:

- About 9 m of highly erodible loess on the surface; therefore, the town will have to be careful about water disposal, including storm water and runoff from lawns.
- Karst features, such as sinkholes and other collapsible structures that must be avoided during construction, including 14 possible sinkholes that were mapped, as well as others that probably exist.

- Some radon (radioactive soil gas), but not enough to be a health hazard.
- The remote possibility that the limestone mine will someday extend under the town.

At a town meeting in December 1993 the citizens voted to apply for relocation. Approval came in May 1994; 95 percent of Valmeyer's buildings and homes will be rebuilt at the new site (a few families chose to remain on the floodplain). The site on the floodplain will be graded and seeded with prairie grasses and other native vegetation. The Federal Emergency Management Agency (FEMA) is moving or reconstructing about 6600 structures damaged in the floods, but only a few towns have asked for relocation.

Valmeyer's relocation costs are estimated at $9.6 million; FEMA will provide $7.2 million. As a condition of the grant, the town must adopt drainage, sediment, and erosion controls; a storm-water management plan; construction guidelines; and other mitigation ordinances to guard against damage to the loess, construction around sinkholes, and other actions that might result in activation of known geologic hazards.

---

tats and ecosystems (Fig. 8.22). The aesthetic value of the river can be degraded, the groundwater regime disrupted, and water pollution aggravated by channelization. Paradoxically, it has also been demonstrated that, although channelization may control flooding in the immediate area, it may actually contribute to more intense flooding further downstream.

An interesting example of channelization gone awry is the Kissimmee River in Florida. In 1962 a channelization project was undertaken by the Army Corps of Engineers, primarily for the purpose of flood control. Over a period of 9 years and at a cost of $24 million, the river was straightened and parts of it were lined with concrete. Thirty years later, the river had been degraded aesthetically; drainage problems and water pollution had worsened, with ensuing public health problems; 40,000 acres of wetland habitat had been lost; and the original goal of flood control had not been fully achieved. Now the South Florida Water

Management District has received approval from the Florida government to restore parts of the Kissimmee River to its original condition. Work has already begun on a test segment of the restoration. If the restoration is completed, the final cost will far exceed that of the original channelization project.

Controversy aside, perhaps the most important thing to understand regarding channelization is that *any* modification of a channel's course or cross-section renders invalid any series of hydrologic data collected there in the past. During the Mississippi River floods of 1973 and 1993, the actual water levels were much higher than had been predicted for the discharges recorded during those events. For example, the 1973 flood was referred to as a "200-year flood" on the basis of its extent and water levels, even though it occurred during a period when measured discharges were only at the 30-year level. Most analysts agree that extensive modifications in the river's channel over

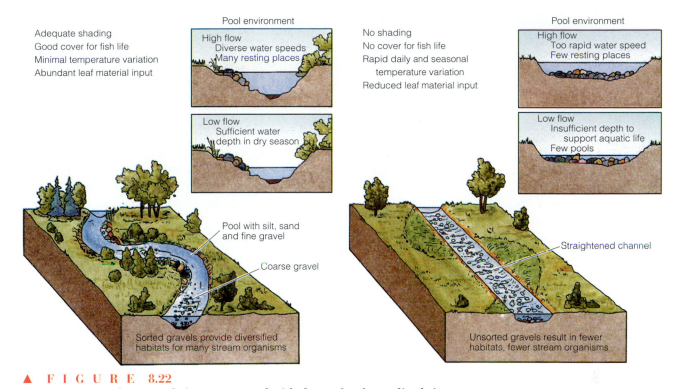

Adequate shading
Good cover for fish life
Minimal temperature variation
Abundant leaf material input

Pool environment
High flow
Diverse water speeds
Many resting places

Low flow
Sufficient water
depth in dry season

Pool with silt, sand
and fine gravel

Coarse gravel

Sorted gravels provide diversified
habitats for many stream organisms

No shading
No cover for fish life
Rapid daily and seasonal
temperature variation
Reduced leaf material input

Pool environment
High flow
Too rapid water speed
Few resting places

Low flow
Insufficient depth to
support aquatic life
Few pools

Straightened channel

Unsorted gravels result in fewer
habitats, fewer stream organisms

▲ F I G U R E 8.22
Characteristics of a natural river, compared with those of a channelized river.

many years contributed significantly to the failure to predict the extent of damage associated with these floods.

Effects of Development on Flood Hazards

Development can contribute to flooding in a variety of ways. As mentioned earlier, urban construction on unconsolidated, compressible sediments, often accompanied by withdrawal of groundwater, can lead to subsidence and urban flooding. Floodplain development that decreases the cross-sectional area of the river channel can also increase

flood hazards because it reduces the amount of discharge the river can handle (Fig. 8.23).

Although the effects of development, particularly urbanization, can be complex, flood hazards are associated mainly with the problems created by impermeable cover and storm sewer drainage. An increase in the amount of impermeable (or impervious) cover reduces infiltration capacity. In urban areas, impermeable ground cover—buildings, roads, parking lots, and the like—is ubiquitous, and runoff from paved areas can add substantially to total runoff. Storm sewers also can be important contributors to

◀ F I G U R E 8.23
Development on a floodplain can alter the cross-section of the channel, which, in turn, may increase water levels in subsequent floods.

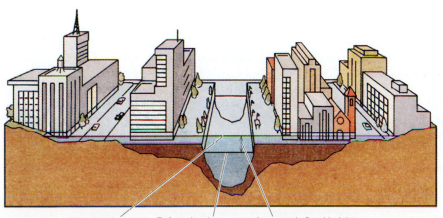

After development   Before development   Increase in flood height

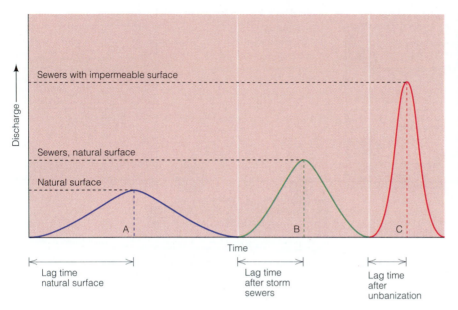

◀ **F I G U R E 8.24**
**Discharge hydrographs for a given
hydrologic event on (A) a natural
surface, (B) a surface with storm
sewers installed, and (C) an impermeable surface with sewers.
The increase in impermeable surface and enhanced runoff through
storm sewers is associated with a
decrease in lag time and an increase in peak discharge and total
discharge for the event.**

flooding because they allow the runoff from paved areas (sometimes called *urban runoff*) to reach the river channel more quickly.

Because of the effects of development, floods in urban areas differ from floods in natural river valleys in several important respects. Urbanization typically leads to a decrease in lag time, an increase in the peak discharge, and an increase in the total discharge for a particular flood. These relationships are illustrated in Fig. 8.24.

Organized Response to Flood Hazards

Organized response to flood hazards takes two main forms: (1) structural and engineering approaches aimed at controlling flooding and (2) regulations, emergency response programs, and compensation packages aimed at decreasing vulnerability and adjusting to the hazards.

Structural Approaches

A wide range of structural approaches can be used to control flooding or protect human interests from its impacts. In addition to the channel modifications discussed earlier, structural approaches include dams and barrages; retention ponds and reservoirs designed to release drainage from paved areas in a controlled manner; and levees, dikes, and flood walls designed to keep out floodwaters. Sometimes these approaches backfire. In the 1973 Mississippi River flooding, for example, levees preserved the downtown area of St. Louis from damage, while surrounding areas were inundated by floodwaters. The flood walls managed to protect St. Louis again in the summer of 1993, but some experts feel that the flood control system may have worsened flooding both upstream and downstream, essentially squeezing the river into higher crests. This raises several obvious questions: To what extent can or should structural approaches to flood control be carried out? Do such approaches create a false sense of security? Can they worsen problems elsewhere along the river's path? And when does the cost—economic or environmental—of engineering the river channel become prohibitive?

Adjustment and Reduction of Vulnerability

Since the passage in the United States of the National Flood Insurance Act in the late 1960s, with resulting legislation at the state level, significant efforts have been made to implement more nonstructural means of reducing damage from flooding. Nonstructural approaches have gained renewed support since the damaging floods in the Midwest in 1993 and the southeastern United States in 1994, which led many people—experts as well as the general public—to question the effectiveness of structural interventions for flood control. Among the nonstructural approaches now in use are floodplain zoning, specialized building codes, open-space planning in flood zones, floodplain buyout programs, mortgage limitations, and development rights transfers. Federal Emergency Management Agency (FEMA) regulations stipulate that if a home or business owner is compensated for more than 50 percent of the value of a building, the structure must be demolished and the site rezoned to prohibit future development. The goal of all these approaches is to limit floodplain development to applications that are appropriate or adaptable to recurrent inundations, such as recreational uses or certain types of farming. Most experts agree that engineering interventions are still needed, at least for the protection of urban areas already existing in the floodplains. However, calls for the buyout and

HOW FEMA WORKS

*T*he U.S. Federal Emergency Management Agency (FEMA) has two broad mandates: to ensure preparedness for a nuclear attack and to respond to natural disasters such as hurricanes and floods. Until the end of the Cold War in the late 1980s, FEMA's primary focus was on preparedness for nuclear war. Since 1989, however, the emphasis has shifted to responding to natural disasters, largely because of several closely timed catastrophes: the Loma Prieta earthquake, the Midwest floods, and Hurricanes Hugo, Andrew, and Iniki.

A military mission has some things in common with a natural disaster mission, such as dealing with the details of evacuation, medical attention, and shelter for victims. However, the National Performance Review (1993) concluded that FEMA's original dual mission policy had diluted the effectiveness of the agency staff and created a barrier to agency communications during recent emergencies. The review recommended that FEMA shift the efforts of its employees and resources from the national security component of their mission to preparation for and response to the consequences of *all* disasters, natural and human generated.

The primary responsibility for disaster response rests with state and local governments. The federal government, through FEMA, offers only supplemental support during an emergency. When a disaster overwhelms the resources of a local or state jurisdiction, FEMA steps in to coordinate the efforts of 26 federal agencies and the Red Cross. As it now exists, however, the system encourages state and local elected officials to apply for maximum federal disaster assistance.

Annual budget appropriations for FEMA total nearly $1 billion ($827 million in 1993), of which one-third is designated for disaster relief and two-thirds for military preparedness. The U.S. Congress has the authority to supplement these funds with special appropriations. For example, Congress appropriated nearly $3 billion in additional funding for FEMA operations in response to Hurricanes Andrew and Iniki in 1992 and $1.7 billion for assistance to areas affected by the Midwest floods in 1993. Requests to FEMA have increased by approximately 50 percent in the past 10 years, and even minor emergencies have been awarded full compensation. For example, 17 states and the District of Columbia were awarded a total of $126 million to cover the costs of snow removal following the storms of March 1993.

The disasters that struck the United States during the early 1990s highlighted the need for communities to increase their general level of preparedness. According to FEMA reports, fewer than 15 percent of the people affected by the Midwest floods had flood insurance, even though most lived on a recognized high-risk floodplain. Better building construction and zoning regulations can also reduce losses during disasters; noncompliance to building codes accounted for $5 billion of the $20 billion insurance costs resulting from Hurricane Andrew.

In response to criticism of its handling of recent national emergencies, FEMA has adopted a broad plan for reorganization. The new organizational design will consolidate the military and natural disaster divisions into an all-hazards approach to emergency management. In addition, to limit federal assistance to situations of real need, FEMA is developing objective criteria for declaring emergencies and "major" disasters. The resulting "disaster hierarchy" will assist in the allocation of federal aid. FEMA also intends to implement a competitive program of state-level grants for emergency preparedness and to enforce existing building codes and insurance requirements for high-risk areas.

restoration of bottom lands to their original wetland state are becoming increasingly common.

Following a particularly devastating flood in Brisbane in 1974, the Australian Minister for Environment and Conservation stated that

Floodplains are for floods, although this is not to say that there is nothing we can do to minimize the damage which flooding can cause. [We] should remember that floods and droughts are in many areas the norm; it is good years which are the exception. . . . There are often perfectly sound reasons for using land subject to flooding for agricultural and other activities. However, it is essential in planning such enterprises that full provision is made for foreseeable losses.

His statement summarizes the philosophy behind the regulation of floodplain development, that is, to obtain full and reasonable use of floodplain resources while maintaining a sensible attitude concerning the hazards associated with flooding. An integrated approach is probably the only realistic way to reduce vulnerability to the risks of living on a floodplain. Such an approach combines floodplain zoning regulations, scientific information and public education about flood hazards, engineered flood-control methods, efficient early-warning and emergency response mechanisms, and compensation for loss caused by flooding.

SUMMARY

1. The advantages of living near water include fertile soil; access to water, food, energy, and other resources; waste absorption and dispersal; and transportation routes. However, flooding can take a high toll in loss of life and damage to structures.

2. The majority of floods are caused or exacerbated by intense precipitation, usually in combination with other geologic factors. Processes that lead to flooding in coastal zones include tsunamis, hurricanes, cyclones, and unusually high tides. Urban development on compressible sediments can lead to subsidence, with associated flooding. Failure of dams and protective structures can also cause flooding.

3. A river's channel is an efficient conduit for the movement of water and sediment. Channels continuously change shape and orientation in response to variations in discharge and load. The dimensions and characteristics of a river's channel determine its ability to handle the flow of water, influencing the nature and extent of flooding.

4. No two rivers are alike. The development of different types of channels—straight, meandering, or braided—depends on variations in gradient, sediment load, and discharge.

5. The area of land drained by a given river—its drainage basin—depends on the river's size and discharge. The type of drainage pattern that develops depends on the composition and structure of underlying rocks.

6. The main factors that control streamflow are gradient, channel cross-sectional area, average velocity of water flow, discharge, and load. Discharge, velocity, and cross-sectional area are interrelated such that when discharge changes, the product of the other two factors also changes.

7. During floods, streams overflow their banks and construct natural levees, which grade laterally into the silt and clay that make up the adjacent floodplain.

8. Typically there is a lag between the onset of precipitation and the peak of flooding downstream. This makes it possible to track the progress of the storm and make preparations for flooding. The lag time is typically shorter in upstream flooding than in downstream flooding.

9. When the rate or amount of precipitation exceeds the ground's infiltration capacity, runoff is increased and flooding may occur. The nature, magnitude, and likelihood of flooding depend on the amount and distribution of rainfall and factors that affect infiltration capacity.

10. The primary impacts of flooding are those caused by actual contact with flowing water, including death by drowning, structural damage to buildings, crop loss, and erosion of flooded areas. The secondary and tertiary effects are longer term impacts indirectly caused by the flooding, such as interruption of services, permanent changes in the river channel, or loss of wildlife habitat.

11. Prediction of flood frequency is based on statistical analysis of hydrologic records. Such records permit the calculation of recurrence intervals and annual exceedance probabilities for a discharge of given magnitude. These are plotted on flood frequency curves, which allow for extrapolation to events of greater magnitude than those represented in the historical records.

12. Flood hazard mapping, which is based on paleoflood hydrology and observations of known events, is used to predict the extent of impacts for discharges of a given magnitude. Real-time monitoring of floods allows ex-

perts to forecast the arrival time and magnitude of the flood peak and, if necessary, issue early warnings.

13. Human activities influence flooding in many ways. Development can change runoff patterns and affect infiltration capacity and other characteristics of the ground. Channel modifications can alter the relationship between a channel's cross-section and its discharge, thereby changing flood patterns and invalidating historical records.

14. Structural approaches to flood control include channelization; dams and barrages; retention ponds and reservoirs; and levees, dikes, and flood walls designed to keep out floodwaters. Recently efforts have been made to implement more nonstructural means of flood damage reduction, including floodplain zoning, specialized codes, floodplain buyout programs, mortgage limitations, and development rights transfers.

TERMS TO REMEMBER

bankfull discharge (p. 201)
bankfull stage (p. 199)
channel (p. 195)
channelization (p. 209)
discharge (p. 195)
divide (p. 198)
downstream flooding (p. 202)
drainage basin (p. 198)
flash flood (p. 202)

flood (p. 192)
flood frequency curve (p. 207)
floodplain (p. 201)
flood stage (p. 199)
gradient (p. 195)
hydrograph (p. 200)
isohyetal map (p. 201)
lag time (p. 201)
load (p. 195)

paleoflood hydrology (p. 209)
peak discharge (p. 201)
recurrence interval (p. 207)
river (p. 195)
runoff (p. 199)
stage (p. 199)
stream (p. 195)
streamflow (p. 199)
upstream flooding (p. 202)

QUESTIONS AND ACTIVITIES

1. Has your family ever been affected by flooding? Was it coastal or river flooding? Upstream or downstream? Did human activities have anything to do with causing the flood (as in a dam failure) or worsening it?

2. If you have lived through a devastating flood, write an account of the event. Even if your own family was not affected, it is worthwhile to record your impressions of events that affected your neighbors or nearby communities.

3. Become aware of your own environment. To what extent are the rivers, streams, or coastlines in your area artificially controlled? You don't have to live near a major river or seacoast to answer this question; even small streams are often heavily engineered. Try to discern the motivation for the intervention: Is it primarily for flood control? ease of transportation? erosion control? How successful do you think the intervention has been?

4. What do you think should happen to a family whose home or business is destroyed in a major flood? What if they have no flood insurance? Who should be responsible for the cost of rebuilding or relocating? Should they be allowed to rebuild in the same location?

5. Imagine the discussion that went on in Valmeyer, Illinois, concerning whether or not the town should relocate. What do you think were the main arguments for and against the plan? How might you have voted?

6. Investigate the Federal Emergency Management Agency. What kinds of emergencies besides floods does FEMA respond to? How good is its track record? Do you think past criticisms of FEMA have been valid? Do you think the proposed changes in FEMA's structure are appropriate?

7. Table 8.2 was constructed from newspaper clippings during and after the 1993 Mississippi River flood. Try the same thing yourself the next time a major natural disaster occurs (it doesn't necessarily have to be a flood). Keep track of newspaper, television, and radio reports and make a list of the reported impacts of the event. Then divide your list into primary, secondary, and tertiary hazards.

8. What do you think of the plan to restore parts of the Kissimmee River to their natural state? Will the restoration justify its high cost? The restoration will be an integrated program involving the buyout of privately owned lands along the channel. In what ways might the program's coordinators convince private landowners to cooperate with the restoration project? These would be good topics for a group discussion or report. Start with some individual research into the original goals of the channelization, the negative impacts of the original project, the reasons for undertaking the restoration, and the geologic setting of the south Florida region.

HAZARDS OF OCEAN AND WEATHER

For Hot, Cold, Moist, and Dry, four champions fierce,
Strive here for mastery . . .
• John Milton (1608–1674)

*A*round Christmastime along the Pacific coast of South America, the temperature of the ocean and air rises, heavy rainfall begins, and changes in marine life are observed. This weather pattern and its ecological effects have been recognized by fishers since ancient times. They call it *El Niño*—"the child"—after the Christ child, because of the season when the changes begin.

El Niño begins when warmer, less saline tropical water starts moving toward the south along the coast of South America, deflecting the cooler, more nutrient-rich waters of the Humboldt Current that normally flow northward along the coast. In most years, the trend reverses itself by March or April, but once every 3 to 7 years El Niño is more intense, extensive, and prolonged. Ocean waters in the eastern equatorial Pacific become anomalously warm, the trade winds weaken, and the normally westward-flowing South Pacific Equatorial Current weakens or even reverses its flow. These exceptional El Niños can have devastating effects on marine life and on coastal communities that depend on the sea. They also bring torrential rains, drought, and other disruptions of normal weather patterns, whose effects are by no means restricted to the South American coastal region.

The most severe El Niño of this century occurred in 1982–1983. It caused heavy rains in Colombia, Ecuador, and northern Peru, drought in southern Peru and Bolivia, and storms that lashed the western coast of the continent. Its effects were also felt around the Pacific. There were droughts in Australia, New Guinea, and Indonesia and increased storm activity over the Pacific. Unusual weather patterns in other parts of the world—intensified drought in northeast Brazil and the Sahel region of Africa, warm winters in North America and Europe—may have been caused by El Niño as well.

There are two important lessons to be learned from El Niño. First, complex interactions and feedback loops between the oceans and the atmosphere play a critical role in determining climate and weather; in effect, the oceans and the atmosphere are interdependent elements of the global climate system. Second, El Niño serves as a reminder of human vulnerability to environmental change. When weather patterns fluctuate as dramatically as they do during an exceptional El Niño, the ecologic and economic impacts can be devastating.

THE OCEAN–ATMOSPHERE SYSTEM

In many respects, the oceans and the atmosphere are inextricably linked; they are complementary parts of one huge, complex, dynamic system. They are the two great water reservoirs fundamental to the functioning of the global hydrologic cycle as well as that of other global biogeochemical cycles. The oceans play a critical role in climate and weather systems, particularly in regulating the temperature and humidity of the lower part of the atmosphere. Atmospheric circulation, in turn, drives ocean waves and currents.

In this chapter we describe the atmosphere and the oceans, consider their interrelationships, and show how oceanic and atmospheric processes function together in the formation of weather and climatic patterns. The goal of the chapter is to examine some of the many ways in which weather, waves, currents, and tides affect the lives of people

Fishers on the Pacific coast of South America have poor catches during an El Niño event. These fishers in northern Chile are sorting the day's meager catch.

around the world. We focus in particular on geologic processes in coastal zones, which are especially vulnerable to the hazards of ocean and weather.

In this chapter we are concerned with natural oceanic and atmospheric processes that are sometimes damaging to human interests. As with other types of natural hazards, the impacts of some ocean and weather processes—notably desertification and coastal erosion—can be worsened by human activities. Air pollution, ozone depletion, and global warming, which are primarily technological or anthropogenic hazards, are addressed in Chapter 18. Similarly, offshore dumping of wastes and other types of pollution in the marine environment—which are wholly anthropogenic—are discussed in Chapter 17.

Weather and Climate

Weather is an extremely complicated phenomenon. It is much studied but not very well understood and even less successfully predicted. The term **weather** refers to conditions in the atmosphere at any given time, particularly temperature, barometric pressure, humidity, clouds, precipitation, and wind velocity. The Sun's rays do not strike the surface of the Earth with equal force everywhere. Weather fluctuations and the restlessness of the atmosphere simply represent the natural, but never completed, process by which the system attempts to smooth out these differences in surface temperature.

Weather phenomena are local and transient. When the characteristic weather patterns for a given region are averaged over a significant period the term **climate** is used. Thus, the patterns and processes that we think of as weather—wind, rain, snow, sunshine, storms, and even floods and droughts—are temporary, local aberrations when viewed against the more stable, longer term background of global climate.

The "Butterfly Effect"

The interactions among air, land, and water that create weather are not only complex but also highly sensitive to changing conditions. Meteorologists sometimes characterize this sensitivity by saying that a change as small as the draft created by a butterfly's wing can become magnified through a network of feedback systems, ultimately developing into a wind or even a cyclone. They even have a name for this phenomenon: the *"butterfly effect."* The variability, complexity, and extreme sensitivity of atmospheric processes make weather prediction a very tricky business.

THE ATMOSPHERE

The Earth's atmosphere consists of a mixture of gases dominated by nitrogen (N_2) and oxygen (O_2), with small amounts of other gases, water vapor, and clouds of water droplets. Scientists describe the atmosphere as composed of

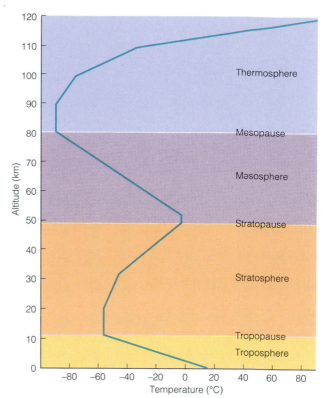

▲ **F I G U R E   9.1**

The atmosphere is divided into four zones on the basis of variations of temperature with altitude. The outermost zone, the thermosphere, continues to an altitude of about 700 km.

several layers, each with its own physical, chemical, and temperature characteristics (Fig. 9.1). From the bottom up, these layers are: (1) the *troposphere,* from the ground surface to approximately 15 km; (2) the *stratosphere,* up to 50 km; (3) the *mesosphere,* up to 90 km; and (4) the *thermosphere,* up to 700 km. Most weather-related phenomena originate in the lowest layer, the **troposphere**. The troposphere also contains about 90 percent of the actual mass of the atmosphere, including virtually all the water vapor and clouds. Although little mixing occurs between the troposphere and the layers of the atmosphere above it, the troposphere itself is extremely dynamic and thoroughly mixed.

Circulation in the Atmosphere

The reason the troposphere is so dynamic is that more of the Sun's heat is received per unit of land surface near the equator than near the poles. This unequal heating gives rise to *convection currents.* The heated air near the equator expands, becomes lighter, and rises. Near the top of the troposphere it spreads outward in the direction of both poles. As the upper air travels northward and southward, it gradually cools, becomes heavier, and sinks. On reaching the Earth's surface, this cool, descending air flows back toward the equator, warms up, and rises, thereby completing a convective cycle (Fig. 9.2A). In **meteorology** (the study of the Earth's atmosphere and weather processes), the term

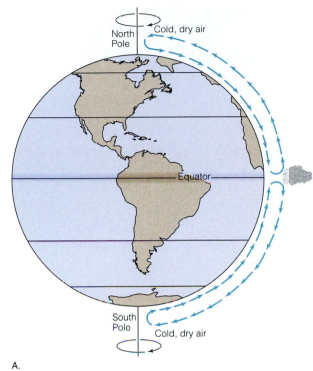

A.

◀ **F I G U R E 9.2**
A. Global circulation as it would happen on a nonrotating Earth. Huge convection cells would transfer heat from equatorial regions, where the input of solar energy is greatest, to the poles, where the solar input is least.
B. The Earth's global wind systems, shown schematically. Because the Earth is not stationary, the flow of air toward the poles and the return flow toward the equator are influenced by the Coriolis effect. Convection does operate, but the flow is not as simple as the case shown for a nonrotating Earth. Moist air, heated in the warm equatorial zone, rises convectively and forms clouds that produce abundant rain. Cool, dry air descending at latitudes 20°–30° produces a belt of subtropical high pressure in which many of the world's great deserts lie.

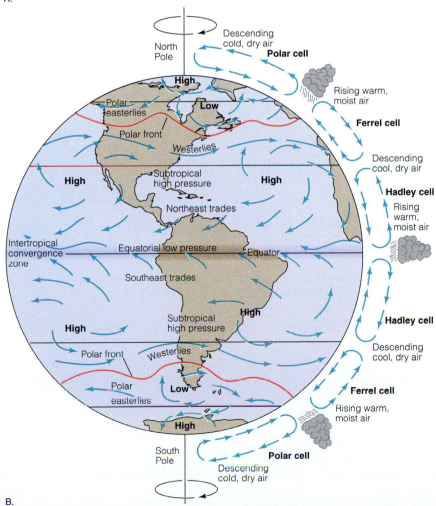

B.

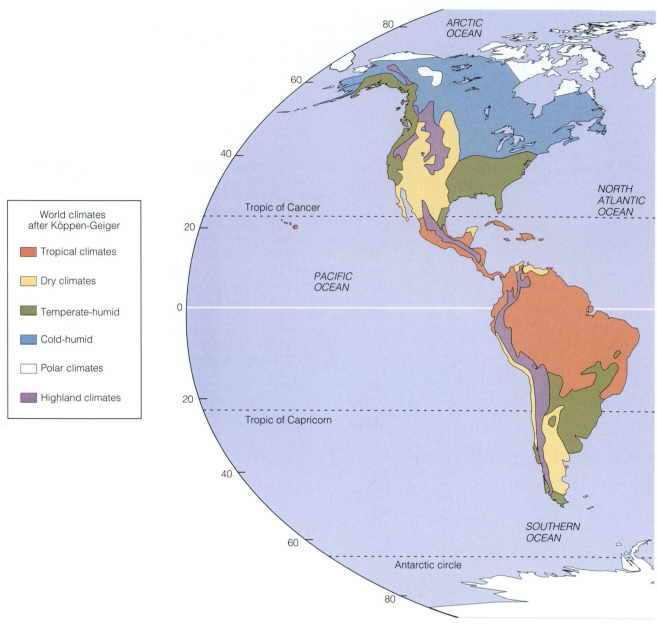

World climates after Köppen-Geiger

- Tropical climates
- Dry climates
- Temperate-humid
- Cold-humid
- Polar climates
- Highland climates

▲ **F I G U R E 9.3**
The Köppen climatic classification system. There are five basic types of climates: (A) tropical climates, (B) dry climates, (C) temperate humid climates, (D) cold humid climates, and (E) polar climates.

convection is used specifically to refer to the vertical component of an atmospheric convective cycle. The horizontal motion of air in the cycle is referred to as *advection*.

The Coriolis Effect

The Earth's rotation modifies what would otherwise be a simple convective cycle (Fig. 9.2). The **Coriolis effect**, named after the nineteenth-century French mathematician who first analyzed it, causes any body that moves freely with respect to the rotating Earth to veer to the right in the Northern Hemisphere and to the left in the Southern

Hemisphere. This is true regardless of the direction in which the body may be moving. Both flowing water and flowing air respond to the Earth's rotation; the global patterns of ocean currents and wind systems therefore are influenced by the Coriolis effect.

Wind Systems

In the atmosphere, the Coriolis effect breaks up the flow of air between the equator and the poles into belts (Fig. 9.2B). The result is a large belt or cell of circulating air lying between the equator and about 30° latitude in both the

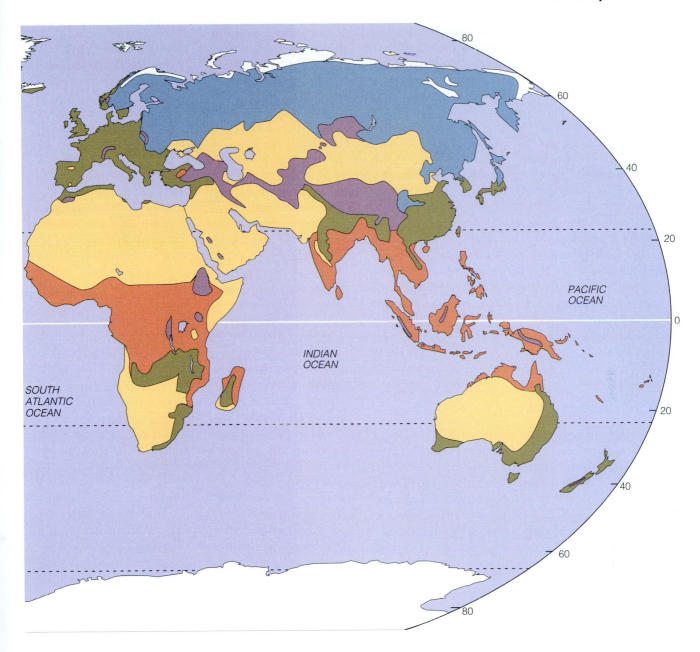

Northern and Southern hemispheres. In these low-latitude cells (called *Hadley cells*), the prevailing winds are northeasterly in the Northern Hemisphere (i.e., they flow *from* the northeast toward the southwest), whereas in the Southern Hemisphere they are southeasterly. These winds are called **trade winds** because their consistent direction and flow carried trade ships across the tropical oceans at a time when winds were the chief source of power.

In each hemisphere, a second cell of circulating air lies poleward of the low-latitude cell. In these middle-latitude cells (called *Ferrel cells*), westerly winds prevail (i.e., blowing *from* the west). Cold upper air flowing toward the equator descends near 20°–30° latitude in both hemispheres, and northward-moving surface air rises in higher latitudes at

the point where it meets dense, cold air flowing from the polar regions.

A third set of circulating air cells (called *polar cells*) lies over the polar regions. In each polar cell, cold, dry, upper air descends near the pole and moves toward the equator in a wind system called the polar easterlies. As this air slowly warms and encounters the belt of westerlies, it rises along the *polar front* (the zone where the polar cells and the Ferrel cells meet) and returns toward the pole.

Effects of Air Circulation on Climate

The global patterns of air flow ultimately control the variety and distribution of the Earth's climatic zones. Those patterns, in turn, are determined by the nonuniform heating of the Earth's surface, the Coriolis effect, the distribution of land and sea, and the topography of the land. For example, the cold upper air converging where the low- and mid-latitude cells meet cannot hold as much moisture as warm air, so at the point where this cold air descends, dry conditions are created at the land surface. As a result, much of the arid land in both the Northern and Southern hemispheres is centered between latitudes 20° and 30°. By contrast, abundant moisture in the warm air surrounding the equator condenses as the air rises and becomes cooler and denser, creating clouds that release their moisture as tropical rains.

If the Earth had no mountains and no oceans to affect the moving atmosphere, the major climatic zones would all lie parallel to the equator. However, the pattern of climatic zones is distorted by the distribution of oceans, continents, high mountains, and plateaus (Fig. 9.3). As a result, average temperature, precipitation, cloudiness, and windiness vary greatly from one place to another and give rise to an array of distinct climatic regions.

OCEANS AND COASTAL ZONES

By now it should be clear that the processes that produce weather and climate are extremely complex. There is a delicately balanced, dynamic interplay between the oceans and the atmosphere that involves a constant exchange of heat, moisture, and gases. Oceanic processes and their influence on precipitation and wind systems have a major influence on local weather fluctuations and regional climatic patterns. In this section we examine the ocean ecosystem and oceanic processes in detail.

The Ocean Ecosystem

Close to 97 percent of the world's surface water is salt water, which circulates throughout the oceans in currents, wave systems, and tides. The Southern Hemisphere is dominated by oceans, and the Northern Hemisphere by land masses. Because of the role of the oceans in regulating atmospheric temperature and moisture, this difference between the hemispheres is very important in determining the distribution of weather and climatic patterns around the world.

The oceans are all interconnected, and water is able to flow from one into the other; in a sense, they can be considered a single huge ecosystem. However, the oceans are far from a uniform environment; it is more precise to view them as a set of different but interconnected ecosystems.

The study of the physical, chemical, biologic, and geologic aspects of the Earth's oceans is called **oceanography.** Although there are many bodies of salt water on the Earth, oceanographers recognize four main ocean basins: the Atlantic, the Pacific, the Arctic, and the Indian oceans. The Pacific is the largest and deepest; it is greater in area than all of the land masses on Earth put together and contains more than half of the world's surface water.

Oceanic Environments

The ocean ecosystem comprises a number of separate but interrelated environments, each with its own physical and biologic characteristics. Offshore from the continental coasts are the shallow, barely sloping *continental shelves* (Fig. 9.4). These typically are fertile environments that support abundant fisheries because of the presence of nutrient-rich sediments that have been washed off the land into coastal waters. The steeper *continental slopes* lead downward to the ocean floor.

The ocean floor itself is remarkably variable in topography. Ocean floor terrains range from the extremely flat *abyssal plains* to rugged underwater volcanic mountains and long, narrow *oceanic trenches* (Fig. 9.4). The deepest spots on the Earth's surface, almost 12 kilometers below sea level, occur within these trenches. The ocean floor is a virtually unexplored source of mineral resources. The depths of the oceans also support unique and exotic life forms, some of which have only recently been observed for the first time. The variability of oceanic environments reflects the fact that the ocean floors are among the most tectonically active regions in the planet's geologic cycle.

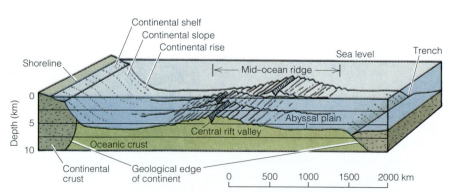

◄ F I G U R E 9.4

Schematic drawing of the ocean floor showing major topographic features.

The Nature of the Coastal Zone

A coastal zone represents the interface between land and sea and typically includes communities that span the biotic spectrum from terrestrial to marine. Physically, the coast is a constantly changing environment. Coastal zones encompass a wide variety of ecosystems and landforms, such as mangrove swamps, estuaries, salt marshes, beaches, barrier islands, rocky cliffs, and coral reefs. Each type of coastal ecosystem exhibits its own physical and biologic dynamics. Each contributes its own unique resources, and each performs environmental services, such as buffering the shoreline against the erosive force of storm-generated waves.

A majority of the world's population lives within 100 km of the ocean. This reflects our dependence on the oceans and the economic benefits afforded by easy access to ocean resources and services, as well as the particular richness of resources in coastal zones. However, the concentration of such large numbers of people in coastal areas means that the coastal environment must absorb the impacts of a wide range of human activities. It also means that human vulnerability to hazards can be particularly high in coastal zones; infrequent events such as large storms can cause major loss of life and damage to property. For these reasons, it is important to understand the special types of geologic processes that characterize coastal zones as well as the hazards of ocean and atmosphere to which the inhabitants of these areas are particularly susceptible.

TIDES, CURRENTS, AND WAVES

If you visit almost any coastal zone on two occasions a year apart, you will see changes. Sometimes the changes are small, but often they are substantial. Large dunes may have shifted; sand may have built up behind barriers, or been eroded away; steep sections of coastline may have collapsed; channels may have broken through from the sea to lagoons on the landward side, where no channels were before. The energy driving these changes comes from ocean currents and waves, which in turn derive their energy from winds and tides.

Tides and Water Levels

Tides are the cycles of regular rise and fall of the level of water in oceans and other large bodies of water. They result from the gravitational attraction of the Moon, and to a lesser degree the Sun, acting on the Earth. Gravitational attraction causes ocean water to bulge upward on the side of the Earth nearest to the Moon. On the opposite side of the Earth, inertia (the force that tends to maintain a body in uniform linear motion) created by the Earth's rotation also raises a bulge in the ocean, but in the opposite direction. The result is two *tidal bulges,* one on either side of the Earth.

To visualize how tides work, consider the tidal bulges oriented with their maximum amplitude lying along a line extending through the center of the Earth and the center of the Moon, as shown in Fig. 9.5. While the Earth rotates, the tidal bulges remain essentially stationary beneath the Moon. Thus, during a full rotation of the Earth, any given point of land will move westward through both tidal bulges each day. Every time a land mass encounters a tidal bulge, the water level along the coast rises. As the Earth continues to rotate, the coast passes through the highest point of the tidal bulge (high tide) and then the water level begins to fall.

At most places on the margins of the oceans, two high tides and two low tides are observed each day. There are two tidal cycles because any point on a coast passes across both tidal bulges during every complete rotation of the Earth. The Sun's gravitational force also affects the tides, sometimes opposing the Moon by pulling at a right angle and sometimes aiding it by pulling in the same direction. Because the distance from the Earth to the Sun is so much

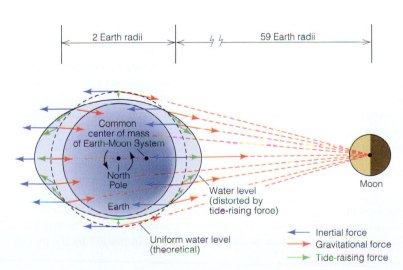

▲ **F I G U R E   9.5**
Tidal forces. Tides are raised by the Moon's gravitational attraction and by inertial force. On the side toward the Moon, both forces combine to distort the water level, creating a tidal bulge. On the opposite side of the Earth, where inertial forces are greater than the Moon's gravitational force, a tidal bulge forms in the opposite direction. As the Earth rotates, the tidal bulges remain essentially stationary and land masses essentially "run into" them twice each day, creating two high tides and two low tides for one full rotation of the Earth.

A.   B.

▲ **F I G U R E   9.6**
The tidal range in the Bay of Fundy in eastern Canada is one of the largest in the world.
A. Coastal harbor of Alma, New Brunswick, at high tide. B. The same view at low tide.

greater than that from the Earth to the Moon, the Sun is only about half as effective as the Moon in producing tides. Therefore, the opposing tidal effects never entirely cancel each other out.

Tidal Bores

In the open sea the effect of tides is small. However, the configuration of a coastline can greatly influence tidal *run-up* height, the highest elevation reached by the incoming water. Narrow openings into bays, rivers, estuaries, and straits can amplify normal tidal fluctuations. At the Bay of Fundy in Nova Scotia, tidal ranges (i.e., the difference between high and low tide) of up to 16 m are reported (Fig. 9.6). The bay has a very long, narrow configuration that causes the incoming tide to rush in, forming a steep-fronted, rapidly moving wall of water called a **tidal bore.** The extreme tidal range at the Bay of Fundy makes it one of the few localities in the world that may be suited for the exploitation of tidal energy. However, a wall of water moving as fast as 25 km/h can easily move large quantities of sediment. Minimizing the impacts of water-borne sand is a major engineering challenge in the development of tidal energy technologies.

Fluctuations in Water Level

Tides are not the only cause of fluctuations in water level in seas and large lakes. Global sea levels change in response to a wide variety of factors, including long-term climatic fluc-

tuations, wind-driven atmospheric forces, seismic and submarine landslide activity, and even changes in the size and shape of the ocean basins. Of these, long-term climatic variations—especially glaciation—are the most important. At the height of the most recent glacial period, about 18,000 years ago, sea levels were about 120 m lower than they are today, reflecting the fact that a great quantity of water was tied up in ice sheets at the time.

Changes in water level that are global in extent are referred to as *eustatic* changes. As we will see in Chapter 18, eustatic changes in sea level may prove to be one of the most problematic effects of accelerated global warming caused by human activities.

Water levels in other large bodies of water, such as the Great Lakes, also change in response to factors such as barometric pressure, seasonal changes, amount of precipitation, and storm waves. Wind-driven processes, such as the piling up of water in the downwind part of a lake, can also be important in causing water-level fluctuations. In Lake Ontario, for example, this effect can raise water levels by as much as 60 cm. When combined with other seasonal or meteorological effects that cause high water levels, the result can be significant flooding in coastal areas.

Ocean Currents

Surface ocean **currents** are broad, slow drifts of water in a particular direction. They are set in motion by the prevailing surface winds. Air that flows across a water surface drags the water slowly forward, creating a current of water as broad as the current of air, but rarely more than 50 to 100 m deep.

In low latitudes surface seawater moves westward with the trade winds (Fig. 9.7). The general westerly direction of the North and South equatorial currents, which are driven by these winds, is reinforced by the Earth's rotation. The moving currents are influenced by the Coriolis effect and are deflected wherever they encounter a coast. On reaching middle latitudes they travel eastward, moved by the prevailing westerly winds. The result is a circular motion of water in each major ocean basin, both north and south of the equator.

Along most continental margins the predominant flow of water roughly parallels the coast. Warm surface water originating in the equatorial region moves north or south along the eastern margins of continents to about latitude 45°, whereas cold, east-flowing water at higher latitudes encounters the western margins of continents and is deflected south (see Fig. 9.7). At still higher latitudes, cold currents generally prevail.

Ocean Waves

Like surface currents, ocean waves receive their energy from winds. The size of a wave depends on how fast, how far, and how long the wind blows across the water surface. A gentle breeze blowing across a bay may ripple the water or form low waves less than a meter high. By contrast, storm waves produced by intense winds blowing for days across hundreds or thousands of kilometers of open water may become so high that they tower over ships unfortunate enough to be caught in them.

The Dynamics of Wave Motion

In deep water, each small parcel of water in a wave moves in a loop, returning very nearly to its former position as the wave passes (Fig. 9.8). The distance between successive wave crests or troughs is the *wavelength,* usually denoted λ. Downward from the surface of the water a progressive loss of energy occurs; this is expressed as a decrease in the diameter of the looplike motion of the water parcels. Eventually, at a depth of about $\lambda/2$, the motion of the water becomes negligible. This effective lower limit of wave movement

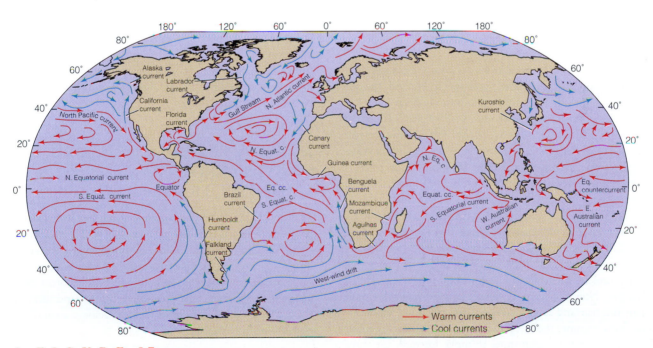

▲ **F I G U R E 9.7**
Surface ocean currents form a distinctive pattern, curving to the right (clockwise) in the Northern Hemisphere and to the left (counterclockwise) in the Southern Hemisphere because of the Coriolis effect. The westward flow of tropical Atlantic and Pacific waters is interrupted by continents, which deflect the water toward the poles. The flow then turns away from the poles and returns as eastward-moving currents in the middle latitudes.

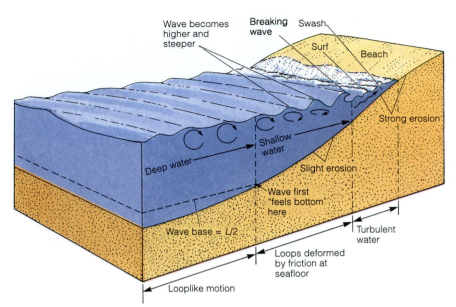

◄ F I G U R E 9.8
Waves change as they travel from deep water through shallow water to the shore. In the process the circular motion of water parcels, found in deep water, changes to elliptical motion as the water becomes shallower and the wave encounters frictional resistance to forward movement. In this figure, vertical scale is exaggerated, as is the size of the loops relative to the scale of the waves.

(which, by extension, is also the lower limit of erosion by the bottoms of waves) is called the **wave base**. Wavelengths as long as 600 m have been measured in the Pacific Ocean. For such large waves the lower limit of wave action ($\lambda/2$) is about 300 m, one and a half times as deep as the average depth of the outer edges of continental shelves (200 m). Thus, it is possible for large waves to affect even the outer parts of the continental shelves.

Wave Action Along Coastlines

Waves change as they travel from deep water through shallower water toward the shore. As shown in Fig. 9.8, the circular loops that characterize wave motion in deep water become flatter as the water becomes shallower and frictional resistance against the bottom restricts forward movement. Landward of depth $\lambda/2$, the circular wave motion is influenced by the increasingly shallow seafloor, which restricts vertical movement. As depth decreases, the orbits of the water parcels become progressively more elliptical until the movement of water is limited to a back-and-forth motion.

As a wave approaches the shore it undergoes a rapid transformation. The frictional resistance of the shallow seafloor interferes with wave motion and distorts the wave's shape, causing the height to increase and the wavelength to decrease. Now the front of the wave is in shallower water than the rear and is also steeper than the rear. Eventually the front becomes too steep to support the advancing wave, and as the rear part continues to move forward, the wave collapses, or *breaks* (see Fig. 9.8).

When a wave breaks, the motion of the water instantly becomes turbulent, like that of a swift river. Such "broken water," called **surf**, is found between the line of breakers and the shore, an area known as the *surf zone*. Each wave finally dashes against rock or rushes up a sloping beach until its en-

ergy is expended; then the water flows back toward the open sea. Water that has piled up against the shore returns seaward in an irregular and complex way, partly as a broad sheet along the bottom and partly in localized narrow channels as *rip currents*, which are responsible for dangerous undertows that can sweep unwary swimmers out to sea.

The geologic work of waves is accomplished primarily by the direct action of surf. Surf is a powerful erosional agent because it possesses most of the original energy of each wave that created it. This energy is eventually consumed in turbulence, in friction at the bottom, and in movement of sediment that is thrown violently into suspension from the bottom.

COASTAL EROSION AND SEDIMENT TRANSPORT

The world's coastlines are dynamic zones of conflict where land and water meet. Through erosion and the creation, transport, and deposition of sediment, the form of a coastline changes, often slowly but sometimes very rapidly. Erosional forces tear away at the land while other forces move and deposit sediment, thereby adding to the land. Because the conflict is unending, few coastlines achieve a condition of complete equilibrium.

Erosion by Waves

Most erosion along a seacoast is accomplished by waves moving onshore. Wave erosion takes place not only at sea level, in the surf zone, but also below and—especially during storms—above sea level. Most coastal erosion is con-

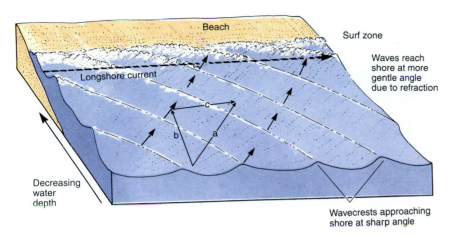

◄ F I G U R E 9.9
A longshore current develops parallel to the shore as waves approach a beach at an oblique angle. A line drawn perpendicular to the front of each approaching wave (a) can be resolved into two components: the component oriented perpendicular to the shore (b) produces surf, whereas that oriented parallel to the shore (c) is responsible for the longshore current. Such a current can transport considerable amounts of sediment along a coast.

fined to a zone that lies within 10 m above and 10 m below mean sea level. Ocean waves typically break at depths that range between wave height and 1.5 times wave height. Because waves are seldom more than 6 m high, the depth of vigorous erosion by surf should be limited to 6 m times 1.5, or 9 m below sea level. This theoretical limit is confirmed by observation of coastal structures such as seawalls, which are only rarely affected by surf to depths of more than 7 m.

An important kind of erosion in the surf zone is the wearing down of rock by particles transported by waves. Through continuous rubbing and grinding with these particles, the surf wears down and deepens the bottom and eats into the land. At the same time, the particles themselves become smoother, rounder, and smaller. Because surf is the active agent, this activity is limited to a depth of only a few meters below sea level. In effect, the surf is like an erosional saw cutting horizontally into the land.

Transport of Sediment by Waves and Currents

Sediment, either produced by waves pounding against a coast or brought to the sea by rivers, is redistributed by currents that build distinctive shoreline deposits or move sediment offshore onto the continental shelves. The transport of sediment is accomplished by several processes, including longshore transport and beach drift.

Longshore Transport

Most waves reach the shore at an oblique angle (Fig. 9.9). The path of an incoming wave can be resolved into two directional components, one oriented perpendicular to the shore and the other parallel to the shore. The perpendicular component produces the crashing surf. The parallel component sets up a **longshore current** within the surf zone that flows parallel to the shore. While surf erodes sediment at the shore, the longshore current moves the sediment along the beach. The direction of longshore currents may change sea-

sonally if the prevailing wind directions change, thereby causing changes in the direction of the arriving waves.

Beach Drift

Meanwhile, on the exposed beach, incoming waves produce an irregular pattern of water movement along the shore. Because waves generally strike the beach at an angle, the *swash* (uprushing water) of each wave travels obliquely up the beach before gravity pulls the water down the slope of the beach (Fig. 9.10). This zigzag movement of water carries sand and pebbles first up, then down the beach

▼ F I G U R E 9.10
Surf swashes obliquely onto a Brazilian beach and forms a series of arc-shaped cusps as the water loses momentum and flows back down the sandy slope.

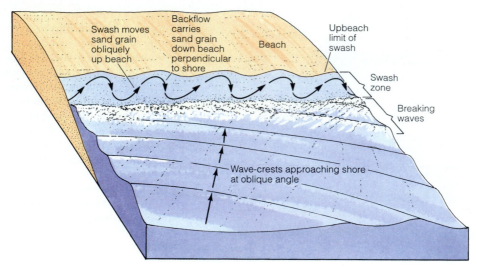

▲ F I G U R E 9.11

As surf rushes up a beach with each incoming wave, sand grains are picked up and carried toward the shore. Surf approaching the shore at an angle travels obliquely up the beach. The return flow, pulled by gravity, flows back nearly perpendicular to the shoreline. A grain of sand therefore moves along a zigzag path as successive waves reach the shore. Net motion down the beach is called beach drift.

slope. The net effect of successive movements of this type is the progressive transport of sediment along the shore, a process known as **beach drift** (Fig. 9.11). The greater the angle of waves to shore, the greater the rate of drift. Marked pebbles have been observed to drift along a beach at a rate of more than 800 m/day. When the volume of sand moved by beach drift is added to that moved by longshore currents, the total can be very large.

Coastal Dynamics During Storms

The approximate equilibrium among the erosional and depositional forces operating along coasts is occasionally interrupted by exceptional storms that erode cliffs and beaches at rates far greater than the long-term average. After a single storm, cliffs of compact sediment on Cape Cod were observed to have retreated up to 5 m—more than 50 times the normal annual rate of retreat. Infrequent bursts of rapid erosion of this type can have a significant impact on the inhabitants of coastal areas (Fig. 9.12).

The west coasts of Ireland and Britain are exposed to the full force of Atlantic storm waves. During one great storm in Scotland a solid mass of stone, iron, and concrete weighing 1200 metric tons was ripped from the end of a breakwater and moved inshore. The damage was repaired with a block weighing more than 2300 metric tons, but 5 years

later storm waves broke off that block and moved it too. These incidents involved pressures of about 27 metric tons/m². Even waves with much less force can break bedrock from sea cliffs. Waves pounding against a cliff compress the air trapped in fissures, and the force of the compressed air can be great enough to dislodge blocks of rock.

In calm weather an exposed beach is likely to receive more sediment than it loses and therefore to become wider. But during storms the increased energy in the surf erodes the exposed part of a beach and makes it narrower. Because storminess is seasonal, changes in beach profiles are more likely to occur at some times of the year than at others. Along parts of the Pacific coast of North America, winter storm surf tends to carry away fine sediment, and the remaining coarse material assumes a steep profile. In calm summer weather, fine sediment drifts in and the beach assumes a gentler profile.

A Variety of Coastlines

The end result of the constant interplay between erosive and depositional forces operating along coastlines is a wide variety of shorelines and coastal landforms. The world's coasts do not fall easily into identifiable classes. Their configurations depend on the active geologic processes at work; on the structure and erodability of coastal rocks; and on the

◀ **F I G U R E 9.12**
Storm waves damage the pier, Seal Beach, California.

length of time these processes have operated. Eustatic sea level changes can also influence the development of coastal features; many coastal and offshore landforms are relics of times when sea level was either higher or lower than it is now (Fig. 9.13). Repeated emergence and submergence of coastlines over many glacial–interglacial cycles, each accompanied by erosion and redeposition of shoreline deposits, has resulted in complex assemblages of coastal landforms. The types of embayments that occur along a coastline also contribute to the variety of coastal types.

Despite the variability of coasts and shorelines, three basic types are most common. They are the rocky (cliffed) coast, the lowland beach and barrier island coast, and the coral reef. Each of these is characterized by a particular set of erosional and depositional landforms.

Rocky (Cliffed) Coasts

The most common type of coast, comprising about 80 percent of ocean coasts worldwide, is a rocky or cliffed coast (Fig. 9.14). Seen in profile, the usual elements of a cliffed coast are a wave-cut cliff and wave-cut bench or terrace, both the work of erosion. A **wave-cut cliff** is a coastal cliff cut by wave action at the base of a rocky coast (Fig. 9.15). As the upper part of the cliff is undermined, it collapses and the resulting debris is redistributed by waves. An undercut cliff that has not yet collapsed may have a well-de-

◀ **F I G U R E 9.13**
An uplifted wave-cut bench at Tongue Point, New Zealand. Crustal uplift along this coast has raised the former seafloor to expose a broad bench. Light-colored beach sediment overlies the darker rocks of the wave-cut cliff at the seaward edge of the bench. Below the uplifted bench, a younger one is forming.

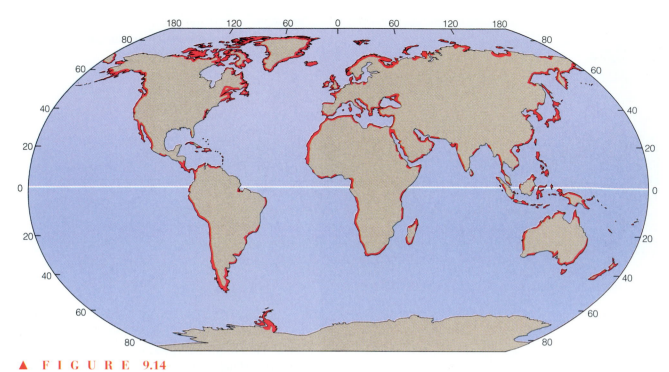

▲ F I G U R E 9.14
World map showing location of sea-cliff coasts (red lines).

veloped notch at its base. Below a wave-cut cliff you can often find a *wave-cut bench* or *terrace*, a platform that has been cut across bedrock by the surf. The shoreward parts of some benches are exposed at low tide. If the coast has been uplifted, a wave-cut bench and its sediment cover can be completely exposed (see Fig. 9.13).

The rocky character of cliffed coasts may be misleading to those in search of a stable, unchanging platform on which to build a home. Cliffed shorelines are susceptible to frequent landslides and rock falls as erosion eats away at the base of the cliff. Roads, buildings, and other structures built too close to such cliffs can become casualties when sliding occurs.

Beaches and Barrier Islands

Beaches are characteristic features of many coasts, even those with steep, rocky cliffs. Along sandy coasts that lack cliffs, the beach constitutes the primary shore environment. A beach is regarded by most people as the sandy surface above the water along a shore. Actually, a **beach** is defined as wave-washed sediment along a coast. A beach thus includes sediment in the surf zone, which is continually in motion.

On low, open shores an exposed beach typically has several distinct elements (Fig. 9.16). The first is a rather gently sloping *foreshore*, a zone extending from the level of lowest tide to the average high-tide level. Here is found a *berm*, a nearly horizontal or landward-sloping bench formed from sediment deposited by waves. Beyond the berm lies the *backshore*, a zone extending inland from the berm to the farthest point reached by surf. On some coasts, beach sand is blown inland by onshore winds to form belts of coastal dunes.

Sediment that has been transported to the beach and offshore areas by rivers or by longshore currents is reworked

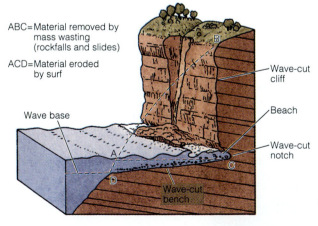

▲ F I G U R E 9.15
Principal features of a shore profile along a cliffed coast. Notching of the cliff by surf action undermines the rock, which collapses and is reworked by surf. Note the large proportion of material removed by mass-wasting (ABC) relative to that eroded by surf (ACD).

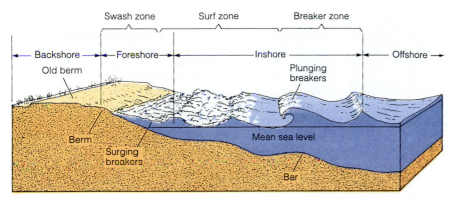

Swash zone   Surf zone   Breaker zone

Backshore → ← Foreshore → | Inshore | Offshore →

Old berm

Plunging breakers

Berm

Mean sea level

Surging breakers

Bar

▲ **F I G U R E   9.16**

Section across a beach showing the principal elements of the shore profile. The section is about 75 m long. Vertical scale is exaggerated in this figure.

by wave action, forming a variety of depositional landforms. A landform commonly associated with beaches is the **barrier island**, a long, narrow, sandy island lying offshore and parallel to the coast (Fig. 9.17). A barrier island generally consists of one or more ridges of sand dunes associated with successive shorelines. Barrier islands are found along most of the world's lowland coasts. For example, the Atlantic coast of the United States consists mainly of a series of barrier beaches ranging from 15 to 30 km in length and 1.5 to 5 km in width, located 3 to 30 km offshore. Sand dunes are typically the highest topographic features along this coastline, which includes Coney Island, New York City's coastal playground, and the long chain of islands cen-

tered at Cape Hatteras on the North Carolina coast. As seen in Figure 9.18, land along this coastline is eroding in some places and *accreting* (building up) in others.

During major storms, surf washes across low places on barrier islands and erodes them, cutting inlets that may remain open permanently (Fig. 9.19; also see Box 9.1). At such times fine sediment is washed into the lagoon between the barrier island and the mainland; this is called *overwash*. In this way the length and shape of barrier islands is always changing. Studies of the response of barrier islands to changing environmental conditions suggest that island development is closely related to the amount of sediment in the system, the direction and intensity of waves and

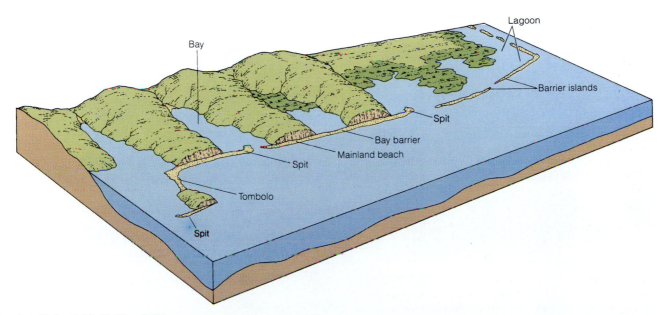

Lagoon

Bay

Barrier islands

Spit

Bay barrier

Mainland beach

Spit

Spit

Tombolo

Spit

▲ **F I G U R E   9.17**

Some depositional features along a stretch of coast. The local direction of beach drift is toward the free end of the spits.

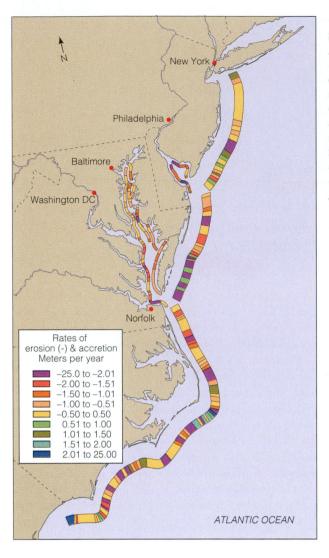

◄ FIGURE 9.18
Shoreline erosion and accretion along the middle Atlantic coast of the United States. The warmer colors (purple, red, orange) show segments of the coast where erosion is occurring, while the cooler colors (green, blue) show segments where the land is accreting through deposition of sediment. Rates of erosion and deposition are given in the legend. Overall, 44 percent of the shoreline on this map is eroding at rates greater than 0.6 m/year and 17 percent is accreting. "Shoreline Erosion and Accretion of the Middle-Atlantic Coast," U.S.G.S. open-file report 92-377, 1972, by Robert Dolan and Judith Peatross.

◄ FIGURE 9.19
Storm-driven erosion on Fire Island, a barrier beach on the south side of Long Island, New York.

PEA ISLAND AND THE BEACH STABILIZATION CONTROVERSY

*I*f engineers, politicians, scientists, environmentalists, and beachfront property owners were to meet on the beach at the Pea Island National Wildlife Area, they would soon be embroiled in intense discussions. This tiny barrier island south of Oregon Inlet, North Carolina, provides a setting for the most difficult issues related to beach stabilization: private versus public rights, dune grasses versus dune buggies, ecology versus economy, and beach stabilization versus nonintervention.

Oregon Inlet (Fig. B1.1) was cut during a storm in 1846 and has migrated south at a rate of almost 30 m/year to its current position, which is stabilized through the ongoing efforts of the Army Corps of Engineers. These efforts include constant dredging to maintain a navigation channel to the mainland, construction of a rock groin on the north shore of the island, and plans to construct two jetties—each 2.4 km long—on either side of the inlet. The projected cost of the latter project is more than $100 million.

Barrier islands like Pea Island are dynamic places. The islands are washed over, dunes shift, inlets migrate, and structures are perpetually pounded by ocean waves. Rising sea level, even at gradual rates, is pushing many barrier islands toward the mainland. Without new sources of sediment, anchored barrier islands will be submerged. Beachfront property owners see this as a problem; however, the movement of the islands is a natural process and becomes a problem only when static structures are placed on unstable surfaces.

The north end of Pea Island has been designated as a national wildlife refuge and hence cannot be used for commercial or residential development. But many questions remain about the stabilization of other parts of the island: What environmental impacts will it have? How long should it be continued? Will it be successful? Are the costs, both economic and environmental, justified? And, most importantly, will the long-term costs of stabilization eventually exceed the value of the property that is being protected?

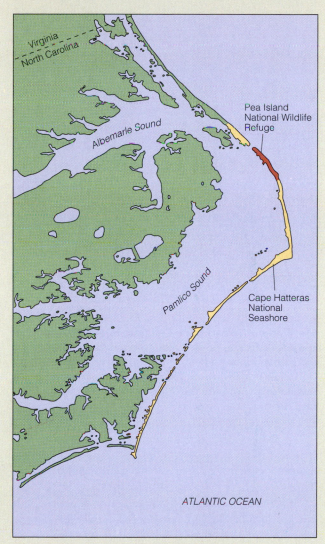

▲ F I G U R E B1.1
The distribution of federally owned and managed lands along the Outer Banks, showing locations mentioned in the text.

▲ FIGURE 9.20

The complex spit of Cape Cod. Waves and currents rework sediment eroded from the peninsula, forming the south side of Cape Cod Bay, and transport the sediment northward and southward. An eddy carries sediment around the north point of the spit and into the bay.

nearshore currents, the shape of the seabed, and the stability of sea level.

Other common depositional landforms associated with beaches include the *spit* (an elongated ridge of sand or gravel that projects from land and ends in open water) and the *tombolo* (a spitlike ridge of sand or gravel that connects an island to the mainland); these are also illustrated in Figure 9.17. A well-known example of a large, complex spit is Cape Cod, Massachusetts (Fig. 9.20).

The elongate bay lying inshore from a barrier island or other low, enclosing strip of land (such as a coral reef) is called a *lagoon.* Lagoons are commonly fed by *estuaries,* the wide, funnel-shaped mouths of rivers in the tidal zone where fresh and salt water meet. Lagoons and estuaries are important habitats for a wide variety of plants, birds, and animals. They also play an important role in the protection of mainland shorelines because they serve as buffers against the erosive impact of storm waves. Unfortunately, these sensitive environments are particularly susceptible to the impacts of human activities.

Coral Reefs

Many of the world's tropical coastlines consist of limestone **reefs** built by vast colonies of tiny organisms, principally corals, that secrete calcium carbonate. Reefs are built up very slowly over thousands of years. Each of the tiny coral animals, called *polyps,* deposits a protective layer of calcareous material; the layers eventually build to form a complex reef structure. *Fringing reefs* form coastlines that closely border the adjacent land, while *barrier reefs* are separated from the land by a lagoon, as in the case of the Great Barrier Reef off Queensland, Australia (Fig. 9.21).

Reefs are highly productive ecosystems inhabited by a diversity of marine life forms. They also perform an important role in the recycling of nutrients in shallow coastal environments. They provide physical barriers that dissipate the force of high-energy waves, protecting the ports, lagoons, and beaches that lie behind them, and they are an important aesthetic and economic resource.

Corals require shallow, clear water in which the temperature remains above 18°C. Reefs therefore are built only at or close to sea level and are characteristic of low latitudes. Because of their very specific water temperature and light level requirements, coral reefs are particularly susceptible to damage from human activities as well as from natural causes such as tropical storms.

Adapting to Coastal Erosion

Shoreline homes command premium prices in the real estate market. If they are built on solid rock, they can be a lasting investment; however, coastlines are among the most dynamic places on the Earth's surface, and a house built on a coast composed of erodible sediment can prove to be a poor bargain. There are several ways of stabilizing the shoreline and protecting coastal property; all of them have some drawbacks.

Protection of the Shoreline

Measures for protecting the shoreline fall into two general categories: *hard stabilization,* which relies on engineered structures; and *soft stabilization,* or nonstructural approaches to shoreline stabilization.

Hard Stabilization The structures involved in hard stabilization are of two main types: those that interrupt the force of the waves and those that interrupt the flow of sand along the shore. Examples of the first type are *seawalls* and *breakwaters* (Fig. 9.22A). Since these are built parallel to the shoreline, they deflect wave energy away from the protected property (Fig. 9.22B). However, these structures can actually accelerate loss of beach sand on the ocean side, making the beach steeper and narrower until it is finally destroyed.

The second type of hard stabilization structure includes *groins* and *jetties* (Fig. 9.23A). These structures are built

◀ F I G U R E 9.21
The Great Barrier Reef, a barrier coral reef on the continental shelf of northeastern Australia, is one of the world's most diverse marine ecosystems.

A.

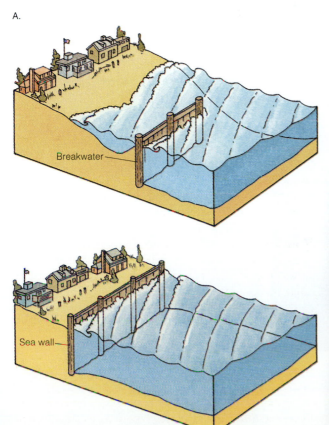

Breakwater

Sea wall

▼ F I G U R E 9.22
Hard stabilization of a shoreline: seawalls and break-waters. A. Structures such as seawalls and breakwaters are built parallel to the shoreline to interrupt the force of the waves. B. Breakwaters constructed offshore from Tel Aviv, Israel, protect the beach zone from incoming waves. Sediment builds up behind each breakwater, creating a scalloped coastline. Meanwhile wave action at the base of such structures can hasten erosion and steepen the beach on the ocean side.

B.

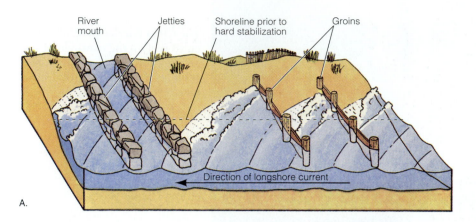

River mouth • Jetties • Shoreline prior to hard stabilization • Groins

Direction of longshore current

A.

B.

▲ F I G U R E 9.23

Hard stabilization of a shoreline: groins and jetties. A. Structures such as groins and jetties are built perpendicular to the shoreline to interrupt the flow of sediment with the longshore current. B. Groins have been built perpendicular to the shoreline of Miami Beach, Florida, to prevent excessive loss of sand by longshore drift at this popular resort area.

perpendicular to the shoreline and widen protected beaches by trapping sand updrift of the groin or jetty. At the same time, however, they degrade the adjacent beaches downdrift (Fig. 9.23B). Once a property owner has installed a groin, neighboring downdrift property owners are cut off from their sand supply and may be forced to add a groin field to their beach fronts.

Soft Stabilization Soft stabilization refers primarily to the process of beach "nourishment," in which sand is dredged offshore and deposited onto the beach as a slurry via pipelines. The newest techniques use offshore hopper dredges instead of onshore pipeline systems to transfer sand to a point close to the beach from which natural wave dynamics distribute the sand onto the beach itself.

Beach nourishment and other soft stabilization approaches, such as the stabilization of dunes by planting certain types of grasses, have generally been considered more desirable than hard, structural approaches. However, beach nourishment raises several concerns, not the least of which is high cost; once it is begun, beach nourishment must be repeated periodically or it will not succeed. The dredging of millions of cubic meters of sand can also cause extensive disruption to shoreline ecosystems. Some species, such as sea turtles, are very susceptible to changes in the characteristics of the sand: if the sand is too coarse, they may have trouble nesting; if it is too dark, their eggs may overheat and die in the nest. Also, it is unclear how resistant beach nourishment is to the effects of storms. For example, the Army Corps of Engineers was in the process of replenishing Folly Beach, North Carolina, when a major hurricane hit. In a single day most of the piped-in sand was washed into shallow offshore areas.

The Controversy over Coastal Erosion

The long-term effectiveness of hard stabilization techniques for stabilizing shorelines is a subject of intense controversy. Many experts argue that, in the long run, extensive interference in the natural processes affecting shorelines can lead to expensive and possibly irreversible damage. Because of their high recreational value, beaches in densely populated regions may justify the costs of stabilization and protection against erosion. A beach, however, presents a special sort of problem. As a result of longshore currents and beach drift, what happens on one part of a beach affects all the parts lying in the downdrift direction. The net result of intervention therefore may be to protect one part of the beach at the expense of another part; small beaches have been completely destroyed in this way in only a few years.

Beaches around the world are deteriorating because of human interference. Along North Carolina's Outer Banks, for example, there is a noticeable difference between the barrier islands that have been built on and those that have not: the undeveloped islands tend to have beaches 100–200 m wide, whereas developed beaches, such as those on Hat-

teras Island, have been reduced to widths of 30 m or less by accelerated erosion. In southern California, most of the sand on beaches is not supplied directly by the erosion of wave-cut cliffs but is carried to the sea by streamflow during floods. However, because buildings and other structures in stream valleys are vulnerable, dams have been built across the streams to control flooding. The dams trap sand and gravel carried by the streams, preventing it from reaching the sea. As a consequence, the natural balance among the factors involved in the longshore transport of sediment has been upset, resulting in significant erosion of some beaches (Fig. 9.24).

A further dramatic example of human interference can be seen along the coast of the Black Sea. Of the sand and pebbles that form the natural beaches there, 90 percent used to be supplied by rivers as they entered the sea. During the 1940s and 1950s three things occurred: large resort developments were built at the beaches; large breakwaters were constructed so that two major harbors could be extended into the sea; and dams were built across some rivers inland from the coast. All this construction upset the equilibrium among the supply of sediment to the coast, longshore currents and beach drift, and deposition of sediment on beaches. By 1960 it was estimated that the combined area of all the beaches along the coast had decreased by 50 percent. Then beachfront buildings began to sag or collapse as the surf ate away at their foundations. An ironic twist to this chain of events lies in the fact that large volumes of sand and gravel were removed from beaches and used as concrete aggregate, not only to construct the resort buildings but also to build the dams that eventually cut off the supply of sediment to the coast.

▲ **F I G U R E 9.24**
Damage to homes on Fire Island, New York caused by coastal erosion. The storm that caused the damage occurred in December 1992.

There is, of course, another way to adjust to coastal erosion and protect shorelines: leave them alone and move away. This may seem like an extreme measure, but more and more communities are adopting this stance. Shoreline protection measures sometimes succeed in protecting homes and other structures along the beach, but they are rarely able to protect the beach system itself in the long term. Zoning laws and policies for some coastal areas specify that undeveloped beachfront sections on barrier islands must be left to natural processes, rather than built upon and "stabilized." For beachfront areas where construction is permitted, building codes can dictate appropriate site-specific structural designs. If the buildings predate the codes, insurance programs will sometimes pay a property owner to move the structure or to not rebuild a house in a location threatened by beach erosion.

EXCEPTIONAL WEATHER

Many of the exceptional weather-related occurrences that can be disastrous for humans are actually normal, natural aspects of weather formation—explicable and, in fact, inevitable in the broader context of global oceanic, atmospheric, and climatic processes. In this last section of the chapter we shall look at the natural causes and human impacts of weather-related occurrences of exceptional intensity. These occurrences include a variety of intense storms; drought and desertification; and El Niño, a complex weather pattern with far-reaching consequences.

Cyclonic Storms

A **cyclone** is an atmospheric low-pressure system that gives rise to roughly circular, inward-spiraling wind motion, called *vorticity*. Although we tend to think of cyclones as being destructive tropical storms, the scientific connotation of the term is more general and includes a variety of broader, weaker low-pressure systems. Because of the Coriolis effect, such systems rotate in a counterclockwise direction in the Northern Hemisphere and clockwise in the Southern Hemisphere (Fig. 9.25).

Tornadoes

A **tornado** is a cyclonic storm with a very intense low-pressure center. Tornadoes are short-lived and local in extent, but they can be extremely violent (Fig. 9.26). They typically follow a very narrow, sharply defined path, usually in the range of 300–400 m wide. U.S. National Weather Service records show that tornadoes have the strength to drive 2 x 4 wooden boards through brick walls, lift an 83-ton railroad car, and carry a home freezer over a distance of 2 km. Their rate of forward motion is highly variable; some

◄ F I G U R E 9.25
A low-pressure center (cyclone) centered over Ireland and moving eastward over Europe. The counterclockwise winds of a Northern Hemisphere low are clearly shown by the spiral cloud pattern.

◄ F I G U R E 9.26
**A tornado crossing the
plains of North Dakota.**

are almost stationary, whereas others race along at over 100 km/h. The strength of a tornado can be estimated from the damage it causes by referring to the *F-scale,* which was devised by Professor T. Theodore Fujita of the University of Chicago (Table 9.1).

Although they can occur anywhere in the world, tornadoes are particularly common in the central and southeastern United States. Tornadoes typically form along the *cold front* of a fast-moving mid-latitude cyclonic storm system. When a moving mass of cold air overtakes and traps an underlying layer of warm, moist air, the warm air is drawn up into the core of the storm in a spiraling upward motion. At the same time the cold air spirals downward, creating a *vortex,* or twisting funnel cloud.

The destructive capabilities of tornadoes come partially from their extremely high wind velocities, which have been clocked at 450 km/h, and partially from the near-vacuum that exists within the vortex. The air pressure inside the vortex may be as low as 60 percent of normal atmospheric pressure; buildings can actually explode and disintegrate as a result of the pressure differential between the inside and outside air. The partial vacuum also causes the tornado to suck up soil and debris, which give the funnel cloud its typical dark, ominous appearance.

T A B L E 9.1 • **The F-Scale for Tornado Intensity**

F-Scale	Category	Estimated Wind Speed, km/h	Damage
0	weak	65–118	Minor. Twigs broken.
1		119–181	Trees down, mobile homes moved off foundation.
2	strong	182–253	Demolish mobile home; roof off frame houses.
3		254–332	Lift motor vehicles. Destroy well-constructed building.
4	violent	333–419	Level buildings, toss automobiles around.
5		420–513	Lift and toss around houses.

▲ **F I G U R E   9.27**
Hurricane Andrew, one of the largest and strongest hurricanes of modern times, formed over the Atlantic Ocean and slammed into Florida in August 1992. The hurricane, here photographed from above, packed winds in excess of 200 km/h.

Typhoons and Hurricanes

Tropical cyclones that form over the ocean are also characterized by low-pressure centers and cyclonic wind circulation. In contrast to tornadoes, however, tropical cyclones are longer lived and much more regional in extent. Their low-pressure centers are less intense than those of tornadoes, but the total power of such a storm can be awesome—the energy flow in a tropical cyclone in one day can be equivalent to the energy released by 400 20-megaton hydrogen bombs.

Tropical cyclones are called **hurricanes** in the Caribbean and North America, **typhoons** in the western Pacific, and tropical cyclones in the Indian Ocean. When fully developed, a tropical cyclone is a circular storm resembling a huge whirlpool up to 600 km in diameter (Fig. 9.27). Such storms travel erratically and can last several weeks. The intense low-pressure center creates winds with velocities up to 300 km/h. At the center itself, called the *eye* of the storm, the inward-rushing winds are drawn upward; they never actually reach the center of the storm. This phenomenon is responsible for the eerie calm found in the eye of a storm.

Damage from hurricanes and typhoons comes from their intense winds and torrential rains as well as from the battering and erosional effects of ocean waves (Fig. 9.28). Flooding—both river flooding caused by intense rainfall in a short period and coastal flooding caused by *storm surges* (to be discussed shortly)—is another hazard commonly associated with hurricanes.

▲ **F I G U R E   9.28**
All that remained of a Florida town after Hurricane Andrew passed through. All the damage seen here was caused by high-speed winds.

In countries with adequate early-warning systems, the main hazard associated with hurricanes and typhoons is property damage. When Hurricane Andrew roared through Florida, Louisiana, and the Bahamas in 1992, it caused an estimated $20 billion in property damage, but only 25 people were killed. In contrast, a hurricane that hit Galveston, Texas, in August 1900 reportedly killed 8000 people, mainly because there was no early warning. In 1991 cyclones devastated the coast of Bangladesh and killed 200,000 people; perhaps 500,000 others died there in a cyclone in 1970. Bangladesh is particularly vulnerable to cyclone hazards because of its high population density in low-lying coastal areas, combined with the lack of an effective early warning system or emergency response mechanism.

Nor'easters

Another devastating type of cyclonic storm is the **nor'easter.** Nor'easters are *extratropical*—that is, they form outside of tropical regions. They originate in regions of atmospheric instability in middle latitudes over land or coastal regions and are named for their winds, which blow from the northeast. The low-pressure system of a nor'easter is typically weaker and more diffuse than that of a tropical cyclone, and the associated winds are less intense. However, they can cover very large areas, mostly along the eastern coast of North America, and are associated with very high waves that often cause significant damage. Nor'easters often generate storm surges as high as 5 m above normal tidal levels. These can be particularly damaging because the storms themselves often last several days (i.e., through several tidal cycles).

A particularly damaging nor'easter, known as the Ash Wednesday storm, hit the Atlantic coast of the United States on March 7, 1962. The storm's 10-m-high waves, which battered parts of the coast for days, caused damage estimated at over $300 million. Like tornadoes and other types of storms, nor'easters are classified by their intensity, from I (weak) through V (extreme). The Ash Wednesday nor'easter was a class V storm; it created many new tidal in-

lets and caused extensive erosion of dunes and beaches. Table 9.2 shows the Dolan/Davis scale, which relates storm classes to the types of coastal damage observed along the Outer Banks of North Carolina.

Storm Surges

A phenomenon often associated with intense wind systems is the **storm surge**, an abnormal, temporary rise in water levels in oceans or lakes. As mentioned earlier, friction between wind and the surface of the water can cause water to "pile up" in the downwind portion of an enclosed body like a lake or bay. A sharp drop in atmospheric pressure, such as that associated with an intense storm, can also create a storm surge. The very low pressures in the eye of a cyclone cause the surface of the water to bulge upward as much as 1 m.

Storm surges can cause extensive damage and coastal flooding, particularly if they coincide with high tides or standing waves *(seiches)* in an enclosed body of water. The configuration of the coastline and the angle at which the storm is approaching can also influence the height of the surge. Much of the loss of life associated with storms is actually caused by coastal flooding resulting from storm surges. The devastating 1970 cyclone in Bangladesh created a storm surge with water levels officially reported at 4.5 m above normal; survivors reported water levels up to 9 m above normal. The extensive flooding of lowland coastal areas in the Netherlands in 1953 (discussed in Chapter 8) was associated with a storm surge in the North Sea, which is essentially a closed body of water. The surge was generated by a mid-latitude cyclone off the southern coast of Iceland.

Droughts, Dust Storms, and Desertification

In contrast to the exceptional weather occurrences we have considered thus far, a **drought** is a period of weather *inactivity*—specifically, an extended period of exceptionally low precipitation. The effects of drought can be devastating to local agriculture and livestock and the people who depend on them. Severe droughts that occurred in China from

T A B L E 9.2 • The Dolan/Davis Scale: The Relationship Between Storm Class and Coastal Damage, Inferred from Observations Along the Outer Banks of North Carolina[a]

Storm Class	Beach Erosion	Dune Erosion	Overwash	Property Damage
I (weak)	Minor	None	None	None
II (moderate)	Moderate	Minor	None	None
III (significant)	Extending across beach	Significant	None	Moderate
IV (severe)	Severe with recession	Severe or localized destruction	On low-profile beaches	Loss of structures at community scale
V (extreme)	Extreme	Dunes destroyed over extensive areas	Massive, in sheets and channels	Extensive regional-scale losses in millions of dollars

[a] From Davis and Dolan, Nor'easters, *American Scientist,* September–October 1993, pp. 428–439.

1876 to 1879 caused an estimated 10 million deaths from famine and malnutrition.

The Nature of Drought

Three categories of drought are recognized by the U.S. National Weather Service: (1) a *dry spell,* a minimum of 15 consecutive days during which less than 0.8 mm of rain falls; (2) a *partial drought,* or 29 consecutive days in which the mean daily rainfall does not exceed 0.2 mm; and (3) an *absolute drought,* a period of at least 15 days without any measurable rainfall. Obviously, classifications of this type depend on the ability to define the "normal" amount of precipitation for a particular locality. Often this is easier said than done; some regions, particularly semiarid regions adjacent to large deserts, are characterized by extreme variability in precipitation.

Parts of the world where there are distinct wet and dry seasons may experience *seasonal droughts.* Seasonal droughts are more predictable than other droughts but may be equally stressful for residents of the area if the dry season is unusually long or if precipitation has been below average during the preceding wet season.

Although lack of moisture is associated with both droughts and deserts, it is important to distinguish between the two. The word "desert" literally means a deserted (relatively uninhabited) region that is nearly devoid of vegetation. However, the development of artificial water supplies has changed the meaning of the word by making many desert regions suitable for agriculture. As a result, the term **desert** is now generally used as a synonym for land where annual rainfall is less than 250 mm or in which the rate of evaporation exceeds the rate of precipitation, regardless of whether the land is "deserted" (Fig. 9.29). Thus the term "drought," which refers to an exceptional weather occurrence, may not be strictly applicable to true desert areas where low precipitation and moisture deficiency are everyday phenomena. Drought conditions, during which precipitation is *abnormally* low for a period ranging anywhere from weeks to decades, may occur at any time in any part of the world.

Dust Storms

One of the most striking features of deserts is dust storms, in which visibility at eye level is reduced to 1000 m or less by dust raised from the ground by blowing winds. Such storms are most frequent in the vast arid and semiarid regions of central Australia, western China, Central Asia, the Middle East, and North Africa (see Fig. 9.29). In North America, blowing dust is especially common in the Great Plains and in the desert regions of the southwestern United States.

In the "dust-bowl" years of the mid-1930s, many farm families in the southern Great Plains region of the United States abandoned their homes and lands and trekked west,

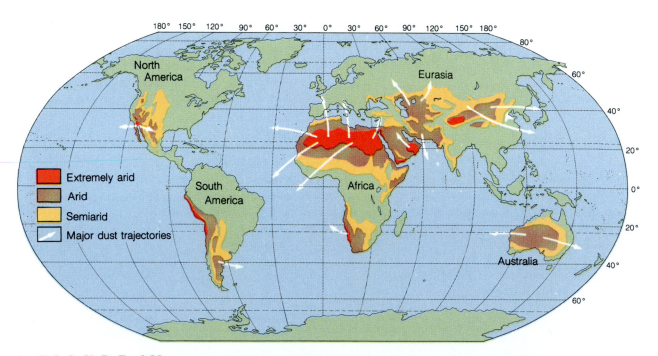

▲ **F I G U R E   9.29**

Major dust storms are most frequent in arid and semiarid regions that are concentrated in the areas of subtropical high-pressure belts north and south of the equatorial zone. Arrows show the most common trajectories of dust transported during major storms.

part of a great migration described by John Steinbeck in his award-winning novel *The Grapes of Wrath*. A primary factor behind the exodus was a severe drought, which led to major dust storms that destroyed crops and buried formerly productive fields with drifting sand and dust. In one storm on March 20, 1935, a cloud of suspended sediment extended 3.6 km above Wichita, Kansas. The dust load in the lowermost 1.6 km of this cloud was estimated at 35 million kg per cubic kilometer. Samples of sediment collected from flat roofs showed that on the day of the storm about 280,000 kg of rock particles, or about 5 percent of the load suspended in the lowermost layer of air, was deposited on each square kilometer of land. Enough sediment was carried eastward to cause a temporary twilight at midday over New York and New England.

The frequency of dust storms is related to cycles of drought, with a marked rise in atmospheric dust concentrations coinciding with severe droughts. The frequency has also risen with increasing agricultural activity, especially in semiarid lands. The effectiveness of the wind in creating the 1930s dust bowl was aided by decades of poor land-use practices. Grasses growing on the prairies when the original settlers arrived protected the rich topsoil from wind erosion. However, over time the grasses were replaced by plowed fields and seasonal grain crops that left the ground bare and vulnerable part of the year. Although today the land is still potentially vulnerable in drought years, improved farming practices should reduce the likelihood of similar catastrophes in the future.

Desertification and Land Degradation

Desertification, the spread of desert into nondesert areas, can result from natural environmental and climatic changes as well as from human activities. The major symptoms are declining water tables, increasing saltiness of water and topsoil, decreasing surface water supplies, unnaturally high rates of erosion, and destruction of vegetation. Although we can find abundant evidence of natural desertification in the geologic record, there is increasing concern that human activities can promote desertification regardless of natural climatic trends.

In the region south of the Sahara lies a belt of dry grassland known as the Sahel (Arabic for *border* or *shore*). There the annual rainfall is normally only 100 to 300 mm, most of it falling during a single brief rainy season. In the early 1970s the drought-prone Sahel experienced the worst drought of this century (Fig. 9.30). For several years the annual rains failed to appear, causing the adjacent desert to spread southward as much as 150 km. The drought zone extended from the Atlantic to the Indian Ocean, a distance of 6000 km, and affected at least 20 million people, many of them seminomadic herders. The remarkable persistence of the Sahelian drought was among the most outstanding climate anomalies of the second half of the twentieth century.

The results of the drought and associated desertification were intensified by the fact that between about 1935 and

▲ F I G U R E 9.30

Overgrazing during years of drought killed most of the vegetation around wells in the Azaouak Valley of Mali. Without vegetation, soil blows away and the desert advances.

1970 the human population of the Sahel region had doubled, and the number of livestock had also increased dramatically. This increase led to severe overgrazing, so that with the coming of the drought the grass cover almost completely failed. About 40 percent of the cattle died. Millions of people suffered from thirst and starvation, and many died as vast numbers migrated southward in search of food and water. In the mid-1970s the rains returned briefly, but in the 1980s drought conditions resumed. Ethiopia and the Sudan were especially hard hit, and mass starvation was alleviated only by worldwide relief efforts.

Desertification is a natural process, but like other geologic hazards it can be worsened or even initiated by human actions. At a certain point it becomes difficult to distinguish the natural parts of the process from the human impacts. It may be more appropriate to use the term *land degradation* to refer to human activities, such as inappropriate or overly intensive agricultural or forestry practices, that lead to the deterioration and eventual desertification of formerly productive lands.

El Niño, La Niña, and the Southern Oscillation

The intricate interrelationships between oceanic and atmospheric processes are dramatically illustrated by an exceptional weather phenomenon known as **El Niño**. In simple terms, El Niño is an anomalous warming of surface waters in the eastern equatorial Pacific. However, as indicated at the beginning of the chapter, both the causes and effects of El Niño are complicated. They center on the be-

BOX 9.2
•
THE HUMAN PERSPECTIVE

PREDICTING EL NIÑO

A particularly intense El Niño can be a devastating event, upsetting weather patterns, as well as the ecosystems and human industries that depend on them, worldwide. For this reason, scientists who study ocean–atmosphere interactions are devoting a great deal of effort to the prediction of El Niño. A major problem is that the recurrence interval for El Niños is irregular, which makes them particularly difficult to predict using statistical methods. The interval generally ranges from 3 to 7 years, and a single event will typically last 18 to 24 months. But even the length of an event is difficult to quantify and depends on the exact definition chosen for the "beginning" and "end" of the phenomenon.

In general, the first noticeable anomaly associated with the onset of an El Niño is the weakening of the pressure differential between the region centered over Indonesia and the one centered over the southeastern Pacific. This difference is called the *Southern Oscillation Index.* An exceptionally low index is probably the most reliable early indicator of the onset of El Niño, followed by the weakening of trade winds and changes in sea surface temperatures (Fig. B2.1). As of early 1995, one El Niño had recently ended (in 1993) and another had just gotten under way.

The series of atmospheric and oceanic phenomena associated with El Niño is fairly well known, and the sequence of events is usually more or less predictable. However, the processes that produce the anomalies are not understood. Clearly the Southern Oscillation—the seesawing of atmospheric circulation between the two pressure systems—is fundamental to the process. Closely coupled with the changes in atmospheric circulation are changes in sea surface temperature. What is not clear is exactly what initiates the entire process. Research indicates that changes in sea-surface temperature influence the location of atmospheric convection zones because large-scale convection is most vigorous over the warmest waters. This suggests that a thermal disturbance in the ocean may initiate the process—but what, in turn, controls ocean water temperatures?

One intriguing suggestion is that magmatic heat—that is, heat from submarine lava flows or hydrothermal activity at sea floor spreading centers—may affect ocean temperatures enough to begin the El Niño cycle. The input of

magmatic heat to ocean water is comparable to other contributors to the equatorial ocean heat budget. The intermittent time scale and variable size of submarine volcanic eruptions may prove to be consistent with the apparently erratic pattern of variations in the El Niño–Southern Oscillation cycle. It has recently been confirmed that El Niño events correspond to periods of exceptionally intense seismic activity, which may, in turn, be indicative of periods of active submarine volcanism. Although there is much that is not yet understood concerning the contribution of magmatic heat to ocean thermal patterns, the idea is intriguing in that it may represent a coupled process involving the dynamics of the atmosphere, the oceans, *and* the lithosphere of the Earth.

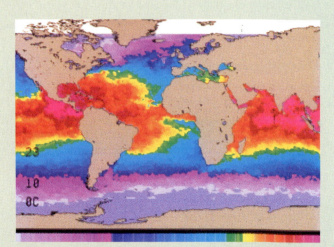

▲ F I G U R E B2.1

False color, satellite image of the El Niño event of 1983. El Niño events occur when the cold, South Equatorial current fails to reach the equator on the Pacific coast of South America. The image, made July 1, 1983, shows sea surface temperatures, color coded according to the scale on the bottom of the image. A tongue of cold water (blue and green) reaches the equator on the Atlantic coast of Africa, but there is no equivalent on the Pacific coast of South America, as there would be in a normal year.

havior of trade winds and associated currents in the equatorial zone of the Pacific Ocean. The causes of normal annual fluctuations in these processes are not well understood; exceptional El Niños are even more difficult to analyze and predict.

The Normal Scenario

Under ordinary conditions (Fig. 9.31A), trade winds blow from the southeast along the coast of South America, then west along the equator toward Indonesia and Australia. The trade winds create currents that push warm tropical waters

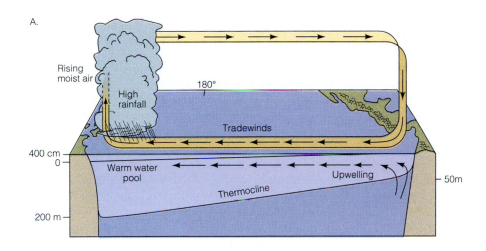

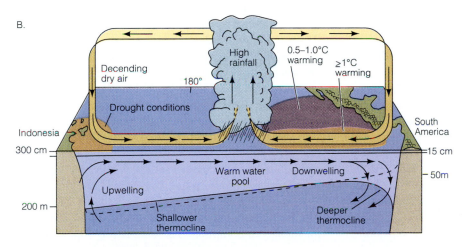

▲ **F I G U R E   9.31**
El Niño is a manifestation of the Southern Oscillation, an atmospheric pressure seesaw between a high-pressure center in the southeastern Pacific and a low-pressure center over Indonesia. A. Under ordinary conditions the pressure differential between these centers drives easterly trade winds along the equator. The winds pile up warm water, raising the sea level in the western Pacific and depressing the thermocline there. Off South America, where the trade winds drive surface water offshore, the thermocline is shallow and cool water wells up to the surface. Near Indonesia the trade winds converge with westerly winds, producing rising air and heavy rains. The air flows eastward at high altitudes and sinks in the central and eastern Pacific, where the weather is dry. B. During an El Niño the east–west pressure difference becomes so low that the trade winds collapse in the western Pacific. The warm water piled up there flows back toward the east. At the same time, the thermocline is depressed off South America, where the upwelling water becomes warm. Both effects warm the surface of the sea. During the 1982–1983 El Niño, the severest in a century, the wind directions, and hence the weather pattern, were completely reversed.

north along the coast of Peru and then west along the equator. These currents cause warm water to "pile up" in the western Pacific, raising sea levels there by as much as 40 cm. The associated increase in surface water temperature in the western Pacific also causes a depression in the *thermocline,* the boundary between warm surface water and the underlying cooler water. Moisture-laden air in the western Pacific causes seasonal heavy rains to fall; the moisture-depleted air travels back at high altitude to the eastern Pacific, where the resulting weather is normally sunny and dry.

As shown in Fig. 9.31A, the offshore movement of warm surface waters from the coast of South America allows cool water to well up from the depths of the ocean to the surface. These upwelling waters are rich in the nutrients that nourish plankton and ultimately support the Peruvian fishing industry. Every year around Christmastime, warm currents flow southward along the coast of South America and interfere with the upwelling of cold water. This signals an end to the fishing season—but only temporarily. By March or April the situation has usually reversed itself, and the trade winds and water temperatures are back to normal.

The normal pattern is complicated by the *Southern Oscillation,* an atmospheric pressure "seesaw" between a high-pressure system (where air masses are descending) centered in the southeastern Pacific and a low-pressure system (where air masses are rising) centered over Indonesia and northern Australia. Superimposed on the normal annual variations is a back-and-forth, seesaw-like fluctuation of air pressure between these two systems.

The exact causes and timing of the seesaw effect are not well understood, but clearly the two systems are linked. Sometimes the pressure difference between them is very great. In this case the trade winds will be very intense, and the temperature of equatorial coastal waters in the eastern Pacific will be unusually low. An exceptionally intense event of this type is referred to as *La Niña* (the girl). Like El Niño, La Niña can have far-reaching weather effects.

In other years, however, the pressure differential between the two atmospheric systems drops to anomalously low levels. This generally signals the beginning of an El Niño (Fig. 9.31B). The decrease in pressure differential causes the southeasterly trade winds to weaken; sometimes they become so weak that they reverse, blowing from west to east along the equator. The warm water piled up in the western Pacific flows back toward the east. Along the coast of South America the thermocline is depressed; temperatures of both surface water and upwelling water are unusually warm. The rising, moisture-laden part of the convective pattern in the atmosphere shifts from the western Pacific to the east.

The Impacts of an Exceptionally Intense El Niño

The weakening or reversal of wind and current patterns during El Niño wreaks havoc on normal regional weather systems. In the El Niño of 1982–1983—the worst of this century—the normal weather pattern was completely reversed. The region of rising air (normally situated near Indonesia and Australia) shifted all the way across the Pacific to the coast of South America, causing heavy rains to fall in that normally dry region. Islands in the central Pacific received month after month of record rainfalls, and Peru reported its heaviest rainfall in 450 years. Meanwhile, Indonesia, Melanesia, and eastern Australia suffered exceptionally long droughts. The normal tracking patterns of tropical storms were also disrupted. For example, Hawaii was hit by an unusual northward-moving hurricane in November 1982; the last major storm with similar characteristics had occurred 25 years before, also during an El Niño. Record-breaking weather anomalies developed in other regions as well; the eastern Pacific jet stream current became much more intense than usual, causing extensive wet spells and storminess all the way from California to Cuba.

When temperature zones in the ocean shift, normal zones of marine life also shift; early accounts of El Niños in Peru contained reports of exotic life forms carried south by the warm currents, together with the disappearance of normally abundant bird and marine species. In some El Niños the anomalously warm water temperatures can cause widespread coral mortality and a deadly algal bloom called a *red tide,* which contaminates shellfish and renders them toxic for human consumption. The long-lasting suppression of nutrient-rich cool water upwelling along the coast of South America meant that the 1982–1983 El Niño was also disastrous for the Peruvian anchovy fishing industry which, prior to that time, had accounted for one fifth of the world's marine catch.

SUMMARY

1. The oceans and the atmosphere are complementary parts of a huge, complex, dynamic system. They are fundamental to the functioning of the global hydrologic cycle. The oceans play a critical role in climate and weather systems. Atmospheric circulation, in turn, drives ocean waves and currents.

2. *Weather* refers to the local condition of the atmosphere at any given time. Most weather-related phenomena originate in the troposphere, the lowest layer of the atmosphere. When the characteristic weather patterns for a given region are averaged over a significant time interval, they are referred to as *climate*.

3. Because of uneven heating of the Earth's surface, air is always in motion. The Coriolis effect breaks up the simple convective flow of air between the equator and the poles into a set of beltlike wind systems. The global pattern of air circulation, interacting with oceans and land masses, determines the distribution of climatic zones.

4. Close to 97 percent of the world's surface water is salt water, most of which resides in four great ocean basins: the Atlantic, Pacific, Indian, and Arctic Oceans. The oceans are interconnected to form a single global oceanic system, but there are dramatic local variations in topography and other characteristics.

5. Coastal zones encompass a wide range of ecosystems and landforms. People tend to inhabit coastal zones in order to have access to ocean resources, but this can increase their vulnerability to hazards caused by oceanic and atmospheric processes.

6. Tides are created by the rhythmic rise and fall of water in oceans and other large bodies of water. They result from the gravitational attraction of the Moon and (to a lesser degree) the Sun acting on the Earth. Two tidal bulges, one on either side of the Earth, are raised by gravitational forces. As the Earth rotates, these bulges remain essentially stationary. Each major land mass therefore "runs into" a tidal bulge twice in each rotation. Water-level fluctuations in major bodies of water can also be caused by long-term climatic changes, especially those related to glacial–interglacial cycles.

7. Surface ocean currents are broad, slow drifts of water set in motion by the prevailing surface winds. Both north and south of the equator, currents are driven by the trade winds but are influenced by the Coriolis effect and deflected wherever they encounter a coast. The result is a circular motion of water in each major ocean basin.

8. Like currents, waves receive their energy from the wind. In the open ocean, water moves in a circular motion, the diameter of the circle becoming smaller with depth. When a wave approaches the shore, it begins to "feel" the bottom; its wavelength shortens and wave height steepens until eventually it breaks in the surf zone.

9. Erosional forces tear away at the land, whereas other forces move and deposit sediment, thereby adding to the land. Most erosion is accomplished by wave action in the surf zone. Longshore currents and beach drift move the eroded sediments downdrift along the shoreline. The approximate equilibrium among the erosional and depositional forces that operate along coasts is interrupted by exceptional storms that erode cliffs and beaches at rates far greater than the long-term average.

10. The dynamic interplay between erosion and deposition produces a wide range of coasts. The most common are rocky or cliffed coasts, beaches, and coral reefs. Rocky coasts are characterized by wave-cut cliffs and benches. Depositional landforms associated with beaches include dunes, barrier islands, spits, and tombolos. On the mainland side of barrier islands are found coastal environments such as lagoons and estuaries. Coral reefs, which are particularly susceptible to changes in water temperature and levels of sedimentation, are of two main types: fringing reefs and barrier reefs.

11. Techniques for protecting shorelines from erosion include hard stabilization (engineered structures such as seawalls, breakwaters, groins, and jetties) and soft stabilization (beach nourishment and dune stabilization). Beach stabilization is expensive and may lead to the degradation or destruction of the beach in the long run. Another approach to shoreline protection is nonintervention, that is, protection of coastal areas through zoning laws and other regulations.

12. A cyclone is an atmospheric low-pressure system that gives rise to roughly circular, inward-spiraling wind motion. This broad definition encompasses a variety of intense cyclonic storms, including tornadoes, tropical cyclones, hurricanes, typhoons, and nor'easters. Such storms cause damage through high winds and wave action. A hazard commonly associated with cyclonic storms is storm surges, which often cause coastal flooding.

13. Drought is a condition in which there is little or no rainfall for an extended period. Low rainfall is a characteristic of arid and semiarid regions and the great

deserts associated with them. Dust storms are common in deserts and drought-stricken regions. Desertification occurs when deserts advance onto formerly productive lands. It is a natural process, but it can be worsened or even initiated by human actions.

14. El Niño is an anomalously warm ocean current in the Pacific, with associated weakening of trade winds and widespread meteorologic effects. It is an example of an exceptional weather occurrence driven by complex interactions among the atmosphere, oceans, and even the lithosphere.

IMPORTANT TERMS TO REMEMBER

barrier island (p. 231)
beach (p. 230)
beach drift (p. 228)
climate (p. 218)
Coriolis effect (p. 220)
currents (p. 224)
cyclone (p. 238)
desert (p. 242)
desertification (p. 243)
drought (p. 241)

hurricane (p. 240)
El Niño (p. 243)
longshore current (p. 227)
meteorology (p. 218)
nor'easter (p. 241)
oceanography (p. 222)
reef (p. 234)
storm surge (p. 241)
surf (p. 226)
tides (p. 223)

tidal bore (p. 224)
tornado (p. 238)
trade winds (p. 221)
troposphere (p. 218)
typhoon (p. 240)
wave base (p. 226)
wave-cut cliff (p. 229)
weather (p. 218)

QUESTIONS AND ACTIVITIES

1. Do you live in a coastal zone? If so, what type of coast is it (reef, beach, rocky cliff)? Visit the coast. Can you recognize some of the landforms commonly associated with this type of coast? Which of the landforms appear to be erosional? Which ones are depositional?

2. The next time you are near the shore of an ocean or a large lake, take a close look at the processes of wave action and sediment transport. At what angle are the waves hitting the shore? What is the direction of the longshore currents? Do you observe any signs of coastal erosion (relative to beachfront buildings, for example)? Have attempts been made to stabilize the shore using structures such as groins, jetties, or breakwaters? What effects have these structures had? If the structures are newly installed, start your own program of periodic monitoring to assess their impact on the movement of sediment and the stability of the shoreline.

3. If you can, visit a shoreline before and after a storm. What changes do you notice? What erosional effects did the storm have? Were those effects caused primarily by wave action or did high winds also play a role? What kind of storm was it, what was its intensity, and where did it originate?

4. Investigate the linkages among human activities, land degradation, and desertification (we will also look more closely at this problem in Chapter 14). You may want to carry out your investigation in the context of a case study: the dust bowl years in the American Great Plains or the ongoing drought and desertification in the Sahel are good examples.

5. Do you live near a large lake? Is the lake susceptible to storm surges? What is the orientation of the lake (i.e., what is the map direction of a line drawn through the longest dimension of the lake)? In general, large lakes oriented parallel to the direction of prevailing winds are most susceptible to storm surges. Is this true of your lake?

6. Conduct an experiment to demonstrate the Coriolis effect. Use a large, flat rotating plate (you can put a large, circular piece of cardboard on top of a rotating

cake server, for example). Set the plate rotating, then roll a small rubber ball across it. Observe the ball's motion from above. Does it move directly across the plate in a straight line, or is its path deflected? In what direction is it deflected? Compare your results with what you know about the impacts of the Coriolis effect on air and water circulation.

7. Synthesize what you have learned in this and previous chapters in Part II by creating a table in which you compare and contrast the hazards occurring on the east and west coasts of North America. Which natural hazards are most common on each coast? Are they the same or different? If you wish, you can add the southern Gulf and northern Arctic coasts to your table.

CHAPTER 10

METEORITE IMPACTS

It is easier to believe that Yankee professors would lie,
than that stones would fall from heaven.
• attributed to Thomas Jefferson
(upon hearing two Yale professors report a meteorite fall)

Approximately 210 million years ago a large meteorite, probably several kilometers in diameter, tore through the atmosphere and landed in the northern wilderness of Québec with a massive explosion. The impact kicked dust and debris high into the atmosphere and left a scar some 100 km across in the hard rock of the Canadian shield. The ancient, ringlike scar is now filled by the waters of the Lac Manicouagan reservoir. At the center of the lake is a plateau topped by melted rock and a series of uplifted peaks, also remnants of the impact. The entire event probably lasted little more than a minute. At the time, of course, there were no human witnesses, but the event coincides approximately with the mass extinction of marine species at the end of the Triassic Period, one of the greatest extinctions of species in geologic history. Could there be a connection between the meteorite impact and the mass extinction?

On June 30, 1908, a mysterious explosion occurred in the Tunguska region of Siberia. An eyewitness more than 100 km away from the blast site gave the following account of the event:

Suddenly the whole sky was split in two and above the forest the whole northern part of the sky appeared to be covered with fire. I felt a great heat as if my shirt had caught fire. There was a mighty crash. I was thrown onto the ground about [7 m] from the porch. A hot wind, as from a cannon, blew past the huts from the north. Many panes in the windows [were] blown out, and the iron hasp in the door of the barn [was] broken.

The witnesses closest to the blast were reindeer herders asleep in their tents about 80 km from the site of the explosion. They and their tents were blown into the air, and some of the men were knocked unconscious by the force of the explosion. They reported that "everything around was shrouded in smoke and fog from the burning fallen trees." Even at a distance of 500 km, the sound was described as "deafening," and fiery clouds were observed. Scientists as far away as 3600 km reported anomalous air pressure waves, and seismic vibrations were detected as far as 1000 km from the site. The tremendous blast—approximately equivalent to a 15-megaton bomb—knocked down trees like matchsticks in a circular pattern radiating outward from its center, yet it left no discernible crater. After years of controversy, the consensus among investigators is that a fragment of a comet, probably about 20 to 60 m in diameter (10^7 to 10^8 kg in mass), exploded and shattered in the atmosphere just above the Earth's surface.

History is full of eyewitness accounts of fiery falls of extraterrestrial material from the sky. Fragments of such material have been recovered from all over the globe, and there are documented accounts of people, houses, and cars being struck by falling stones. How common are such events? What is their cause? What are the risks associated with a major event like the meteorite impact that caused Manicouagan Crater? These are some of the questions we will address in this chapter as we examine a phenomenon that has its origin not on Earth but in outer space.

WHAT IS A METEORITE?

A **meteorite** is a fragment of extraterrestrial material that strikes the surface of the Earth. Before the fragment hits the ground, it is referred to as a *meteoroid*—a small (i.e., smaller than a planet) solid body floating in space. The scientific study of meteorites and meteoroids is called *mete-*

Manicouagan Crater in Québec was created 210 million years ago by a large meteorite impact. The original crater, now a ring lake, was 75 to 100 km in diameter.

▲ **F I G U R E   10.1**
Two meteor tracks (top right and center right) belong to the Perseid meteor shower, 1993. The shower reaches a peak about August 12 of each year. Such regular showers occur when the Earth crosses the orbit of a comet and its debris enters into the atmosphere. The Perseid shower is associated with the Comet Swift–Tuttle.

▲ **F I G U R E   10.2**
Woodcut showing a meteorite fall near the town of Ensisheim, France, in 1492. The stone was recovered and preserved in a local church.

oritics. These terms (as well as *meteorology,* the study of weather phenomena) are all derived from the Greek word *metēoron,* meaning "phenomenon in the sky."

When meteoroids enter the atmosphere they are heated by friction. The surface material of the fragment may become ionized, causing it to glow. Glowing fragments of extraterrestrial material passing through the atmosphere are **meteors,** commonly known as *shooting stars;* exceptionally bright meteors are called *fireballs.* At certain times of the year, large swarms of meteors, all coming from roughly the same direction, can be observed; these are called *showers.* An example is the Perseid shower, which occurs every year in mid-August (Fig. 10.1).

▼ **F I G U R E   10.3**
Worshipers circling the Kaaba (rectangular building), the holy site at the center of the mosque in Mecca, Saudi Arabia. The stone inside the Kaaba is believed to be a meteorite.

▲ FIGURE 10.4
Dr. Ursula Marvin, a scientist at the Smithsonian Astrophysical Observatory, examines a meteorite discovered on the icy surface of Antarctica. The meteorite was discovered in 1981.

Throughout history there have been reports of stones falling from the sky (Fig. 10.2). Such falls were observed and recorded in early Chinese, Greek, and Roman writings. Recovered meteorites were venerated and worshipped in a number of ancient civilizations, including that of the Pueblo Indians in Arizona. The black stone enshrined in the Kaaba, the sacred shrine of Islam in Mecca, is reputed to be a meteorite that was recovered before A.D. 600 (Fig. 10.3). However, the scientific community did not acknowledge the extraterrestrial origin of meteorites until the late 1700s.

Several close encounters with meteorites have been reported in recent times as well. In 1938 a small meteorite smashed through the roof of a garage in Illinois. In 1954 a 5-kg meteorite fell through the roof of a house in Alabama, bounced off the radio, and hit the owner of the house on the head. Another small meteorite demolished a car in a New York suburb in 1992.

Although meteorite fragments have been found all over the world, even in sediments dredged up from the sea floor, Antarctica outstrips all other collection sites in importance. In 1969, Japanese geophysicists discovered that meteorite fragments are scattered over the Antarctic ice sheet—some on the surface, others frozen into the subsurface ice (Fig. 10.4). For the most part, the fragments have been preserved from weathering and have suffered little contamination because of the subzero temperatures of their surroundings. Even the smallest samples are relatively easy to detect. Overall, several times more meteorites have been found in Antarctica than the total amount recovered prior to 1969.

Meteorite impacts have played a major role in the formation and geologic history of the Earth. It now appears that they may also have had significant consequences for the evolution, and possibly even the origin, of life. In this chapter we examine where meteorites come from, what they are made of, and their role in the formation of planetary surfaces. We also consider what happens when a large meteorite strikes the Earth, and the probability of such an event occurring during a human lifetime.

Composition and Classification of Meteorites

Meteorites are commonly named after the location in which they fell. For example, the famous Allende meteorite, the most primitive piece of solar system material known, was recovered near Pueblito de Allende, Mexico. In a very general classification scheme, meteorites can be grouped into three categories, as shown in Table 10.1: (1) *stones*; (2) *irons*; and (3) *stony-irons*.

T A B L E 10.1 • Simplified Classification of Meteorites

Meteorite Type		Description	Probable Origin
Stones	Primitive: chondrites	Unaltered by geologic processes such as metamorphism; contain chondrules and mixtures of high- and low-temperature minerals; some chondrites contain amino acids	Formed from primitive materials in the solar nebula early in solar system history
	Differentiated: achondrites	Silicate minerals similar to those found in terrestrial rocks; alteration and metamorphism evident	Broken off from outer layer of a differentiated parent body such as a large asteroid
Stony-irons		Mixtures of silicate minerals and iron–nickel metal	Contact between outer, rocky layer and metallic core of a differentiated parent body such as a large asteroid
Irons		Primarily iron–nickel metal	Metallic core of a differentiated parent body such as a large asteroid

Stones

The *stony meteorites,* or **stones**, resemble terrestrial igneous rocks. They are made primarily of silicate minerals, including many of the same minerals that are abundant in the Earth's crust. Although they are by far the most common type of meteorite (about 94 percent), because they look so much like terrestrial rocks they often are not recognized as meteorites unless the fall is actually witnessed.

Among the stones, the most abundant are the *chondritic meteorites* or **chondrites**, so named because they contain small, round, glassy-looking spheres called *chondrules,* from a Greek word meaning "seedgrain." Chondrites, especially the *carbonaceous chondrites,* tend to be very old, and most have remained unaltered since the time of their formation (see Box 10.1). Scientists believe that the chondrules in these meteorites are "frozen" droplets of the actual material that condensed from the gaseous solar nebula early in the formation of the solar system.

The other major group of stony meteorites is the *achondrites,* which lack chondrules. These meteorites have been heated, melted, subjected to pressure, and/or fractured to the extent that any chondrules that were present have been destroyed. Many achondrites closely resemble ordinary volcanic rocks of the Earth or Moon.

Irons

The *iron meteorites,* also called **irons**, are alloys of nearly pure metallic nickel and iron, sometimes with enclosed fragments of stony (i.e., silicate) material. They are easily recognizable because they are much denser than the rocks found in the Earth's crust. If you pick up an iron meteorite, it will feel *much* heavier than a terrestrial rock. Iron meteorites also tend to develop a distinctive rusty weathered surface that helps distinguish them from terrestrial rocks. When iron meteorites are cut and polished, they often reveal an intricate interlocking crystal texture called a *Widmanstätten pattern* (Fig. 10.5). These crystal growth pat-

▲ **F I G U R E   10.5**
Widmanstätten texture in a cut and polished iron meteorite. The crystal-growth texture indicates to scientists that iron meteorites cooled slowly, deep in the core of a large parent body. This photograph is about 10 cm across.

terns indicate that the iron meteorites cooled slowly, probably while buried in the core of a larger body.

Stony-Irons

As the name implies, the **stony-iron** meteorites are composed of intimate mixtures of stony (silicate) and metallic material. Their classification is based primarily on the minerals they contain and the proportions in which those minerals are present. Many of the stony meteorites consist of a type of material called *breccia,* a rock that has been fragmented and then recemented through heat and/or chemical cementation processes. The stony-irons present us with a mystery: How did the stony and metallic materials come to be intimately associated in a single fragment? We will attempt to provide some clues to this question as we consider the origin of meteorites in more detail.

THE ORIGIN OF METEORITES

Most meteorites are fragments of larger bodies, called *parent bodies.* It is believed that all known meteorites come from parent bodies within our own solar system, illustrating how isolated our solar system is from the rest of the vast Milky Way Galaxy.

The Asteroid Belt

Nestled between the orbits of Mars and Jupiter, at the boundary between the inner (terrestrial) and outer (Jovian) planets, is the **asteroid belt** (Fig. 10.6 and Table 10.2). The asteroid belt is a swarm of at least 100,000 **asteroids**—small, irregularly shaped rocky bodies orbiting the Sun (Fig. 10.7). About 4000 of these objects, those with known characteristics and reasonably predictable orbits, have been officially classified. Once the orbit of an asteroid has been determined with some accuracy, the asteroid is given a number and a name. The number indicates the order of discovery, and the name is honorary. Thus, "1 Ceres" was the first asteroid discovered (by an Italian astronomer named Piazzi, on New Year's Day, 1801); it is named after the patron goddess of Sicily. Asteroid "1000 Piazzi" was the 1000th asteroid to be discovered; its name honors the discoverer of Ceres.

The asteroids are probably either remnants of a planet that was broken up or rocky fragments that failed to gather together into a planetary mass. The latter possibility is more likely; in spite of the very large number of objects in the asteroid belt, their total mass is not even equal to that of the Moon. More important for our present discussion are these three characteristics of asteroids: (1) some of them are large enough to have undergone internal *differentiation;* (2) since the formation of the asteroid belt, they have continued to undergo collisions and mutual *fragmentation;* and (3) some of them have *Earth-crossing orbits.* Let's consider each of these in turn.

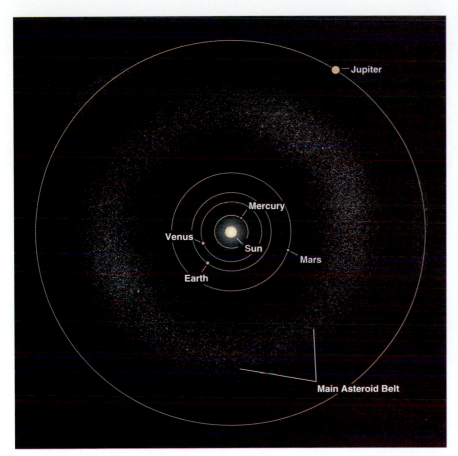

The location of the main asteroid belt, between the orbits of Mars and Jupiter.

T A B L E 10.2 • Orbits of the Planets and the Main Asteroid Belt, Stated in Terms of the Semi-major Axis of the Object's Orbit (Its Distance from the Sun)

Planets		Semi-major Axis (AU[a])
Mercury		0.387
Venus	Terrestrial planets	0.723
Earth		1.000
Mars		1.52
Asteroid belt		≈2.8
Jupiter		5.2
Saturn		9.58
Uranus	Jovian planets	19.1
Neptune		30.2
Pluto		39.44

[a]AU = Astronomical unit; 1 AU = 1.496×10^{11} m = mean distance of the Earth from the Sun.

▲ F I G U R E 10.7
Only recently have scientists been able to take high-resolution photographs of asteroids in orbit. This photograph of asteroid 951 Gaspra was taken in 1991 by the Galileo spacecraft. Gaspra is believed to be a fragment of a larger asteroid.

BOX 10.1

•

FOCUS ON . . .

CARBONACEOUS CHONDRITES: THE MOST PRIMITIVE MATERIAL KNOWN

*C*arbonaceous chondrites are a special group of stony meteorites with some interesting characteristics. As the name implies, they consist primarily of dark, fine-grained carbon-rich material. They are low in density and very easily broken. They contain large amounts of volatile materials and even some complex organic compounds. The ages of carbonaceous chondrites cluster around 4.6 billion years, the age of the solar system. These observations indicate that the carbonaceous chondrites formed in a cold part of the solar system and that they have not undergone significant geologic modification since their origins in the early stages of solar system formation. If they had been modified by geologic processes such as metamorphism, their volatile materials and organic compounds would have been driven off by heat.

The carbonaceous chondrites are scientifically important because they are among the least altered, or most *primitive,* of all known solar system materials. Allende, shown in Figure I.5, is a meteorite of this type. The discovery of protein-related amino acids of confirmed extraterrestrial origin in some carbonaceous chondrites suggests the possibility that incoming meteorites may have contributed some of the first building blocks of life on Earth. Even if

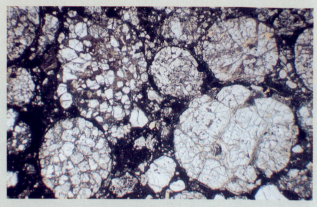

▲ **FIGURE B1.1**
Round chondrules seen in Tieschitz meteorite that fell on July 15, 1878, in what is now the Czech Republic. The field of view is 2.5 mm across.

the basic organic complexes needed for life were already present in the oceans, it is within the realm of scientific possibility that a meteorite impact supplied the energy needed to catalyze the chemical reactions leading to the formation of amino acids and proteins.

Differentiation

Differentiation refers to processes through which planetary bodies develop concentric layers that differ in their composition. The Earth, for example, differentiated early in its history into a dense, iron-rich metallic core, a silicate mantle, and a thin crust of much less dense silicate material. Some asteroids appear to be homogeneous, but others are large enough to have differentiated into a core, mantle, and crust, like the Earth did. It is probable that the different layers in asteroids have provided the material for the different classes of meteorites.

Fragmentation

With so many solid bodies orbiting in a close swarm, collisions and mutual fragmentation are inevitable. In fact, the size distribution of asteroids in the asteroid belt closely resembles that of fragments in a shattered rock. Asteroids in the main part of the belt encounter one another at velocities sufficient to cause mutual fragmentation upon impact. Figure 10.8 is a painting by astronomer and artist William K. Hartmann that shows the probable appearance of an asteroid collision spraying out meteoritic fragments.

Apollo Objects

Earth-approaching bodies are referred to as **Apollo objects**. Asteroids whose orbits pass within the orbit of the Earth are referred to as **Earth-crossing asteroids;** they are also known as the *Apollo group* of asteroids. About 150 Apollo objects with diameters of 1 km or more are known, but this is only a small fraction of the total number of Earth-crossing objects. The largest known Apollo asteroid is approximately 8 km in diameter. A subset of the Apollo group is the *Aten group* of asteroids, which have orbits that don't just

▲ **F I G U R E 10.8**
Painting by astronomer and artist William K. Hartmann showing the probable result of a collision between asteroids spraying meteorite fragments in all directions.

cross over but fall completely within the Earth's orbit. Another important orbital grouping of asteroids is the *Amor group* of Mars-crossing asteroids, which approach (but do not quite cross) the Earth's orbit.

Many Apollo asteroids will eventually strike the Earth. In general, these objects have rather unstable orbits because they are under the gravitational influences of Earth, Mars,

and Venus. If the object doesn't hit one of these three planets within 100 million years, it is likely that it will be ejected from its orbit as a result of a near miss with one of them. This means that any asteroid that started out as an Earth-crossing object early in the history of the solar system is probably long gone, so there must be a source of replenishment. The most obvious source of new Earth approachers is fragments that have been ejected from the asteroid belt, where collisions are common.

Asteroids as Parent Bodies of Meteorites

All of the characteristics of asteroids discussed so far coincide rather nicely with the characteristics we would look for in the parent bodies of meteorite samples.

1. We know from the Widmanstätten patterns in iron meteorites that the parent bodies of iron meteorites must have been large enough for minerals to cool very slowly in the core. We also know that the parent bodies of at least some of the meteorites must have been differentiated because of the modifications some of the meteorites have undergone and because of the chemical and mineralogic contrasts among the different types of meteorites.

2. The ongoing collisions and mutual fragmentation of asteroids in the asteroid belt provide a mechanism for breaking apart a differentiated parent body, with the metallic core of the body presumably yielding the iron meteorites, the mantle and crust yielding the stones, and the core–mantle boundary or the recementation of broken fragments yielding the odd mixtures of material found in the stony-irons (Fig. 10.9).

3. Finally, ongoing collisions serve as a mechanism by which some asteroid fragments may be kicked out of their normal orbits within the asteroid belt, yielding the Earth-approaching group of Apollo asteroids. Of

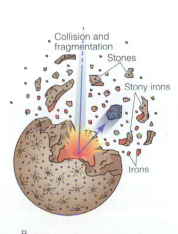

◀ **F I G U R E 10.9**
The fragmentation of a differentiated asteroid parent body leads to the formation of three different types of meteorites. A. The parent body differentiates, separating into a metallic (iron–nickel) core. B. The parent body collides with another asteroid in the asteroid belt, breaking into smaller pieces. Iron meteorites are remnants of the parent body's core; stones are remnants of the outer layer; and stony-irons are remnants of the boundary between the layers.

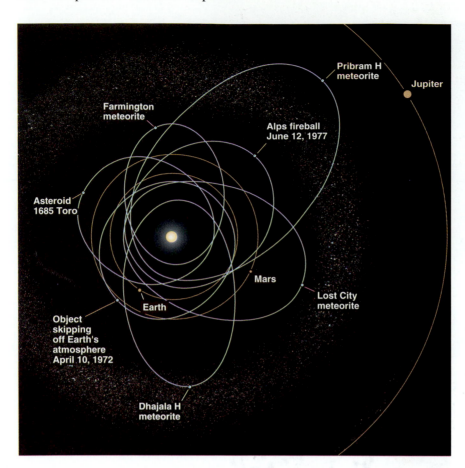

◄ F I G U R E 10.10
Orbits of meteorites whose incoming trajectories have been observed. Most cross through or originate in the asteroid belt.

the few recovered meteorites for which trajectories have been determined, all appear to have originated in or near the asteroid belt (Fig. 10.10).

The natural conclusion is that asteroids in the asteroid belt are the main source of most of the meteorites that strike the Earth. Yet there are some classes of meteorites that do not fit well into this scenario because of their unusual chemical or mineralogic characteristics. Some of these meteorites are virtually identical to samples brought back from the Moon. Other rare meteorites are thought to be derived from Mars. Meteorites of lunar and martian origin must have been broken off from the surfaces of their parent bodies by the impact of another body, which sent them flying toward the Earth. The impact must have been very large for the ejected fragments to exceed the velocities necessary to escape the gravitational attraction of the parent bodies. Another, larger group of unusual meteorites is thought to have been derived from comets.

Comets as Parent Bodies of Meteorites

A **comet** is a bright object with a long, wispy tail that always points away from the Sun (Fig. 10.11). The study of comets has an interesting and colorful history. In European folklore, comets were considered signs of evil and omens of

bad luck and catastrophe. Most of the classical Greek and Roman philosophers believed that comets originated within the atmosphere because the celestial sphere was unchanging and unchangeable. The extraterrestrial origin of comets was not recognized until the 16th century, when the Danish astronomer Tycho Brahe demonstrated that they originate far beyond the Earth's atmosphere.

Comets generally are not visible until they come at least as close as Mars. The closer they come to the Sun, the more they begin to take on their familiar appearance (Fig. 10.12), with a central core called the *nucleus* surrounded by a bright diffuse halo called a *coma*. The nucleus and the coma together make up the comet's *head*. As the comet approaches the Sun, icy material in the nucleus is vaporized; driven by the solar wind, it streams away from the Sun, carrying dust particles along with it. Thus, the comet develops two types of visible tails when it is near the Sun: a *gas* or *plasma tail* made of volatilized icy material from the nucleus, and a *dust tail* made of particles of dust carried along by the solar wind. The two types of tails can be seen in Figures 10.11 and 10.12.

Observations by an international flotilla of research probes during the 1986 appearance of Comet Halley (its thirtieth historically reported appearance and fourth predicted reappearance) revealed it to have an irregularly shaped nucleus about 16 km x 8 km in dimension (Fig. 10.13). Sur-

◄ F I G U R E 10.11
Comet West, showing its dust tail (white) and its gas tail (blue). The photograph was taken in 1976.

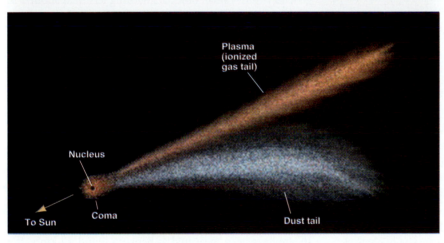

Plasma (ionized gas tail)

Nucleus

Coma

To Sun

Dust tail

◄ F I G U R E 10.12
The parts of a comet.

prisingly, it was found to be one of the darkest known objects in the solar system, its dark color resulting from a layer of black carbon-rich material on the outside of the nucleus. The nucleus was also found to have an extremely low density (es-timated at 0.1 to 0.4 gm/cm^3). These observations con-firmed scientists' notions of comets as fluffy masses of icy material loosely held together by a matrix of silicate dust—the so-called "dirty snowball" theory.

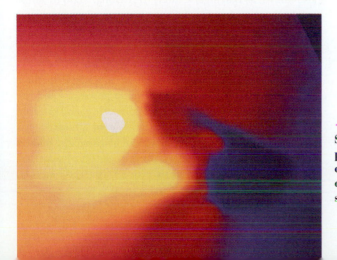

◄ F I G U R E 10.13
Spacecraft observing Halley's Comet during its 1986 trip past the Earth found its nucleus to be an irregular, dark object about 16 km long × 8 km wide. Jets of gas (red and orange), volatilized by the heat of the Sun, can be seen streaming out of the nucleus.

▲ **F I G U R E   10.14**
Destruction caused by the Tunguska event, in Siberia, June 30, 1908, was still very evident when this photograph was taken in 1927. More than 3000 square kilometers of forest were flattened.

The orbits of comets are quite different from those of most other bodies in the solar system. For one thing, they are highly *eccentric,* which means that they are elongated and elliptical, unlike the more nearly circular orbits of the planets. For another thing, comets typically do not orbit within the Sun's equatorial plane, as do most other bodies in the solar system. These observations and the very icy compositions of the comets suggest that they come from the far outer reaches of the solar system.

Comets are almost certainly the parent bodies of some of the objects that strike the Earth's surface. The nucleus of Halley's Comet was found to contain material whose composition is very similar to that of some chondrites. Some comets have been observed to split apart, fragment, and even disintegrate completely upon approaching the Sun. This may be due in part to the effect of jets of gas and volatilized material streaming out of the nucleus. Most periodic meteor showers are thought to consist of swarms of small, glowing fragments of comets that have entered the atmosphere.

The fact that comets are so loosely held together means that they tend to break apart when they enter the atmosphere. Nevertheless, the impact of a large cometary fragment can be devastating, as illustrated by the Tunguska explosion described at the beginning of the chapter (Fig. 10.14). In the Tunguska event, the absence of a discernible crater and the lack of significant amounts of meteoritic material in the vicinity of the blast suggest that the comet exploded above the surface.

IMPACT EVENTS

What happens when an extraterrestrial object strikes the Earth? In the process of *impact cratering,* the planetary surface is deformed as a result of the transfer of kinetic energy from the object to the surface. The result is an impact **crater**, a bowl-shaped depression that is approximately circular. The crater's width and depth and other effects associated with the impact depend primarily on the size and velocity of the incoming body.

Size and Velocity of Incoming Objects

The total *meteorite flux*—that is, the mass of meteorites that strikes the Earth—is about 10^7 to 10^9 kg/year. This amounts to hundreds of tons of meteoritic material a day! Most of this material consists of dust-sized particles, or *micrometeorites*. Because of the Earth's gravitational field, the meteorite flux on Earth is about 50 percent greater than that on the Moon. But friction from the Earth's atmosphere generally causes small objects, anything less than about 1 m in diameter, to disintegrate and burn up before reaching the ground. This, combined with the active modification of the surface by erosional processes, accounts for the fact that the Earth's surface is much less noticeably cratered than that of the Moon.

Objects large enough to make it through the atmosphere and cause significant cratering—that is, objects greater than 1 m in diameter—strike the Earth at a rate of about one a year. Great impact events, which cause craters of several kilometers or more, are much rarer. Meteorites ranging from 100 m to several kilometers in diameter, for example, strike the Earth less than once in every million years (1 in 10^6 years).

The velocities of small meteorites entering the atmosphere have been measured at between 4 and 40 km/s. If a large meteorite struck the Earth at such a velocity, the amount of energy released would be enormous. It has been calculated that the impact of a meteorite 30 m in diameter and traveling at a speed of 15 km/s would release as much energy as the explosion of 4 million tons of TNT. The resulting impact crater would be the size of Meteor Crater (also known as Barringer Crater) in Arizona, 1200 m across and 200 m deep (see Fig. I.6).

Cratered Surfaces

To a large extent, the Earth's surface is shaped by three processes: tectonic processes, magmatic processes, and the surface processes of weathering, mass-wasting, and erosion. To varying degrees, each of these sets of processes has played a role in shaping the surfaces of all the rocky planets and moons in the solar system. In the context of solar system history, however, a fourth process, impact cratering,

JUPITER'S ENCOUNTER WITH COMET(?) SHOEMAKER–LEVY 9

*T*wenty-one pieces of a fragmented object called Shoe-maker–Levy 9 smashed into the turbulent atmosphere of Jupiter over a 1-week period from July 16 to 22, 1994 (Fig. B2.1). The explosion created superheated atmospheric pools of magnesium, sulfur, and silicate on the side of Jupiter that was not directly visible from the Earth. At the point of collision the fragments were traveling at 60 km/s and released an amount of energy equal to 6 million megatons of TNT, more than the amount that could be released by all the world's nuclear weapons combined.

Scientists had a long time to prepare for this space spectacular. An intact Shoemaker–Levy 9 with an estimated radius of 5 km had approached Jupiter the previous year. At that time the object broke into smaller pieces because of tidal stresses caused by its proximity to the giant planet. More fragmentation resulted from collisions among particles. The comet that returned included 21 fragments ranging from 1 km to more than 2 km in radius as well as innumerable smaller fragments.

With the time and place of the comet's impact predicted a year in advance, the scientists were able to simulate the event, test theories, and prepare the global scientific and public communities to observe an impact similar to the one that could have accounted for mass extinctions on Earth millions of years ago. Amateur and professional astronomers pointed their telescopes at Jupiter. The best view, however, was from NASA spacecraft, which were positioned to have an unobstructed view of the far side of the planet. Signals were sent back from Galileo, an unmanned spacecraft on its way to a rendezvous with Jupiter in 1995, and from the Hubble Space Telescope, which produced the highest resolution images.

Even as Shoemaker–Levy crashed into Jupiter, scientists were scrambling to explain the data they were collecting. The first observations described upwardly expanding dark plumes at each impact site. To the observers' surprise, no water molecules were detected in these plumes. This led some scientists to speculate that Shoemaker–Levy might have been an asteroid instead of a comet. Comets are composed of chunks of water ice and other types of ice mixed with silicate dust and dark carbonaceous material; if Shoemaker–Levy was a comet, why didn't it leave a trail of water molecules in its impact plume? It may have been a more rocklike object, or possibly a burned-out comet.

Visible-light telescopes picked up dark impact marks in the Jovian clouds. These marks were distorted and merged over time by Jupiter's atmospheric winds, forming a black band that began to appear several weeks after the collision. Ultraviolet spectra taken from equipment on the Hubble Space Telescope revealed sulfur and hydrogen sulfide, two emissions never before detected on Jupiter.

When all the data have been analyzed and reviewed, astronomers expect to have a much better understanding of the stratification of Jupiter's lower atmosphere, the planet's magnetosphere, and storms and eddies occurring in its atmosphere. This information may eventually help scientists comprehend similar features of the Earth's atmosphere.

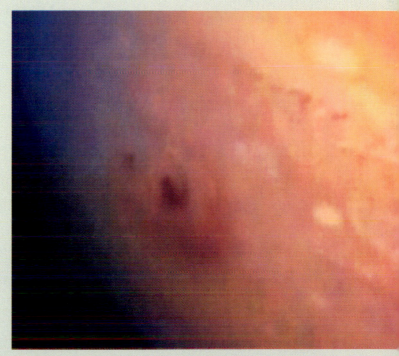

▲ F I G U R E   B2.1

The impact area of Shoemaker–Levy 9, fragment G, on Jupiter, photographed through the Hubble Space Telescope. The image, which was made about 20 hours after the impact, shows a center of boiling gas the size of the Earth.

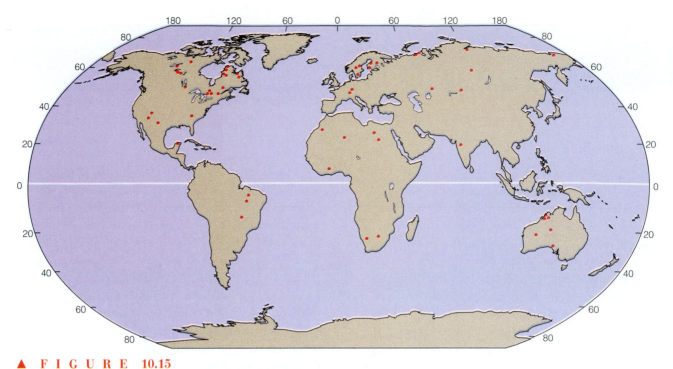

▲ F I G U R E 10.15
World map of large impact craters. More than 200 impact craters have been identified on the Earth.

has been equally important in the modification of planetary surfaces, including that of the Earth.

Terrestrial Impact Craters

More than 200 impact craters have been identified on the Earth (Fig. 10.15). However, there must have been many more impacts than this small number implies. We know, for example, that the Moon is heavily cratered as a result of a period of particularly intense meteorite bombardment early in the history of the solar system. The Earth experienced the same period of intense bombardment, but few craters remain because weathering and the rock cycle continually erase the evidence. Most recognizable terrestrial impact craters have been preserved because they are very large, very recent, or located in very stable geologic environments, or else because they were rapidly buried by protective sediment that has since been removed by erosion.

There are two basic types of craters (Fig. 10.16): simple

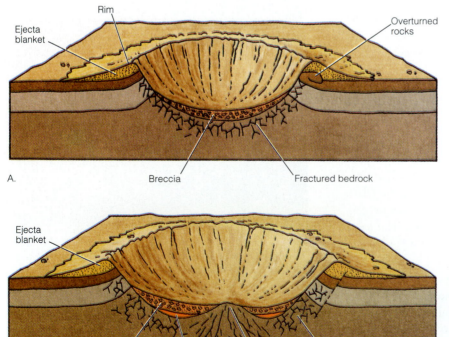

◄ F I G U R E 10.16
Two basic types of craters. A. Simple craters are relatively small, with simple, bowl-like shapes. B. Complex impact structures and basins are large (3 km or more in diameter), with central peaks and ringed structures.

craters, which are generally less than 5 km in diameter with a simple, bowl-shaped excavation; and complex impact structures and basins, which are generally 3 km or more in diameter, with a distinct central peak and/or ringed structure. Notable examples of simple craters are Meteor (Barringer) Crater in Arizona; New Québec Crater in Québec; and Roter Kamm in Namibia. Some examples of complex impact structures include Upheaval Dome in Utah; Sudbury in Ontario; and Gosses Bluff in Australia.

The Mechanics of Impact Cratering

Although many ancient impact craters have been discovered, no large, natural impact crater has ever been observed as it was being formed. What is known about the process of impact cratering comes largely from laboratory experiments (see Fig. 10.17). As a high-speed meteorite strikes and penetrates the surface of a planet, it causes a jet of rock and dust, called **ejecta**, to be strewn at high velocity away from the point of impact in all directions. This produces a blanket of ejecta that surrounds the crater and thins out at greater distances from the rim.

At the same time, the impact compresses the underlying rocks and sends intense shock waves outward. The pressures produced by the shock waves from a large meteorite are so great that the strength of the rock is exceeded; the re-

A.

B.

▲ **F I G U R E   10.17**
The sequence of events in the formation of an impact crater. A. The initial impact ejects a high-velocity jet of material from just below the surface. B. The passage of shock waves through the bedrock produces high pressures and compresses surrounding rock and sediment layers. In places the pressure exceeds the strength of the rock, causing fracturing. C. Rock layers along the rim of the crater are folded back and overturned by decompression. The ejected debris forms a circular blanket around the crater.

▲ **F I G U R E   10.18**
Meteorite impact craters. A. Meteor Crater, Arizona, is a simple crater. The raised surface surrounding the 1200 m wide crater is a blanket of rock ejected during the impact event. B. Upheaval Dome, Canyonlands National Park, Utah, is a complex crater consisting of more than one ring and a central, uplifted dome.

sult is a large volume of crushed and brecciated material. In very large impact events, vaporization may occur. Once the compressive shock waves have passed, rapid decompression occurs. In some cases the decompression causes melting. Molten rock may then rise along fractures produced by the impact and flood the floor of the crater.

Rock layers adjacent to the rim of the crater are overturned. In the largest impact craters, the central crater is surrounded by one or more raised rings of deformed rock. The outer rings and central peaks of large, complex impact craters (Fig. 10.18) are presumed to have formed as a result of the initial compression.

Cratering is a very rapid process; the event that produced Meteor Crater, for example, is estimated to have lasted only about 1 minute. After the initial impact, however, the crater is modified by several kinds of postimpact events. The walls may slump; rebound may produce changes in the floor and rim; and erosion may fill the crater with debris.

METEORITE IMPACTS AND MASS EXTINCTIONS

Whereas a major meteorite impact would surely ravage local ecosystems, the regional and global effects of such an event could be equally devastating. Many scientists believe that there is a direct connection between major impact events and the mass extinctions of species that have occurred periodically throughout the history of the Earth. Other scientists disagree vigorously with this theory. The controversy surrounding the impact theory of extinction has generated one of the liveliest multidisciplinary dialogues in recent scientific history.

Regional and Global Effects

First, let's consider whether a major meteorite impact could be devastating to life, in principle. It has been estimated that the impact of an object 10 km in diameter traveling at 20 km/s would instantaneously release about 100 megatons of kinetic energy—a nonnuclear explosion equivalent to about 1000 times the total yield of all existing nuclear weapons. Among the widespread and catastrophic effects of such an impact would be global darkness; extreme cold, followed by enhanced warming of the atmosphere; acidic rain; tsunamis; and global wildfires (Fig. 10.19).

Darkness Fine dust particles created by the impact would be ejected into the upper atmosphere, probably encircling the globe in a matter of weeks. On the basis of modeling of impacts and knowledge about dust injected into the atmosphere by volcanic eruptions, it has been suggested that the

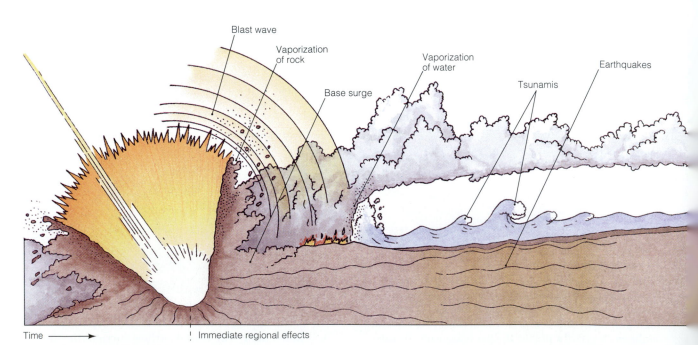

Time ⟶ | Immediate regional effects

▲ **F I G U R E   10.19**
Global and regional effects of a major impact.

entire world could be plunged into darkness for several months following a major impact event. The results would include the suppression of photosynthesis, collapse of food chains, and crop failure. The dust would eventually fall to the ground, depositing a thin blanket of sediment worldwide. This sediment band, preserved in the geologic record, would be the global "signature" of the impact.

Cold   The period of global darkness would probably be accompanied by intense cold because the dust would block out incoming solar radiation. (Researchers modeling these effects have noted similarities with the possible aftermath of a nuclear war, the so-called *nuclear winter* scenario.) If the impact occurred in the ocean, water vapor would also be distributed throughout the atmosphere, remaining there long after the dust particles had settled out. Water vapor is an important agent of warming in the Earth's atmosphere, known as the greenhouse effect. The presence of a large quantity of additional water vapor in the atmosphere could enhance this effect, so that the cold period would be followed by a period of extreme global warming.

Other Effects   The energy and shock heating resulting from a major explosion could cause nitrogen and oxygen in the atmosphere to combine, creating compounds such as nitric acid that would result in acidification of rain and surface waters. A major impact in the ocean would also cause huge tsunamis. Radiation from the fireball would probably ignite global wildfires, and the impact might even generate earthquakes.

It seems clear that the global effects of a major meteorite impact could be devastating to life. What we must determine next is whether there is any direct evidence for the connection between major impacts and mass extinctions in geologic history.

The Geologic Record of Mass Extinctions

Most people know that dinosaurs became extinct about 65 million years ago, at the boundary between the Cretaceous and Tertiary Periods (commonly referred to as the *K–T boundary*). But many are not aware that other animal and plant species were also affected. Approximately one quarter of all known animal families then alive (including marine and land-dwelling species) became extinct at the end of the Cretaceous Period. This mass disappearance of species is clearly evident in the fossil record; it is the reason that early paleontologists selected this particular stratigraphic horizon to represent a major boundary in the geologic time scale.

The great Cretaceous–Tertiary extinction is not unique in Earth history, nor was it the most devastating of such occurrences; there have been at least five and possibly as many as 12 recognizable episodes of mass extinction during the

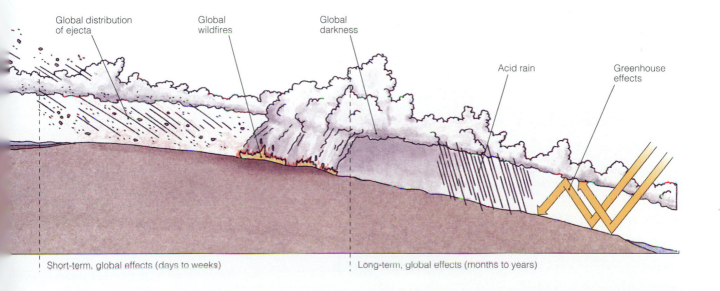

Global distribution of ejecta

Global wildfires

Global darkness

Acid rain

Greenhouse effects

Short-term, global effects (days to weeks)

Long-term, global effects (months to years)

past 250 million years (Fig. 10.20). The most devastating of these occurred 245 million years ago, at the end of the Permian Period. Perhaps as many as 96 percent of all species then alive died out at that time. Another great extinction occurred at the end of the Triassic Period, approximately coinciding with the meteorite impact that created Manicouagan Crater in Québec.

The Iridium Anomaly

The possibility that mass extinctions might be caused by devastating impacts was suggested at least as early as 1970.

In 1978, however, the connection between meteorite impacts and mass extinctions was dramatically substantiated by a piece of evidence uncovered by scientists from the University of California at Berkeley. They found anomalously high concentrations of a rare element, iridium, in clay-rich sediments marking the K–T boundary (Fig. 10.21).

This *iridium (Ir) anomaly* has been found in K–T boundary sediments throughout the world. What is its significance? Only two reasonable explanations have been proposed so far: Either the iridium was brought to the sur-

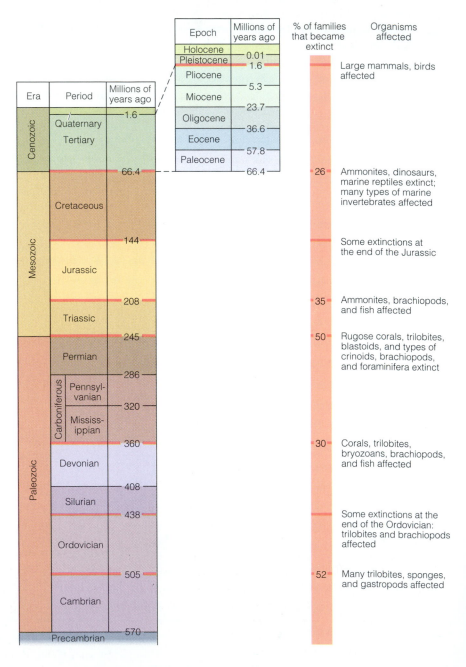

▲ **F I G U R E   10.20**
Mass extinctions in geologic history.

▲ F I G U R E 10.21

Geologists climbing a face of eroded limestone strata to observe a thin, dark layer of rock rich in the chemical element iridium. The iridium-rich layer, seen here in New Mexico, has been identified at many places around the world and is believed to have formed as a result of a world-encircling dust cloud formed by a great meteorite impact about 65 million years ago.

by the extraordinarily high pressures characteristic of major explosions. Grains of the mineral *stishovite,* which are formed only under conditions of intense pressure, have also been found. Also collected from some of the K–T boundary layers are tiny glass spherules, which are thought to be the remnants of melt droplets generated by the impact.

Periodicity of Extinctions

The unfolding story of the link between meteorite impacts and mass extinctions became more complicated when scientists from the University of Chicago proposed that extinctions may actually occur on a regular schedule (Fig. 10.22). Through computer modelling they demonstrated that extinctions seem to occur at regular intervals of approximately 26 million years. In the absence of an obvious terrestrial cause for this periodicity, the researchers suggested that it might be caused by the effects of periodic swarms of comets passing through the inner part of the solar system. The likelihood of a comet striking the Earth would be dramatically increased during these encounters. If this model turns out to be valid, the concept of comet impacts causing mass extinctions on a regular, cyclical basis would revolutionize our theories of how life on Earth evolved into its present state.

face of the Earth by a period of exceptionally intense volcanic activity, or it came from a sudden influx of extraterrestrial material. As mentioned earlier, the dust ejected into the atmosphere by a major impact would eventually settle out, forming a global layer of fine sediment like the iridium-rich clay layer that marks the K–T boundary. The fact that iridium is much more abundant in meteorites than in the crust of the Earth lends support to the hypothesis that the Ir anomaly resulted from a major meteorite impact.

Soot, Shock, and Spherules

On the basis of evidence presented so far, we cannot rule out the possibility that the iridium was ejected from the Earth's interior through volcanism. Nor can we definitively connect the event that generated the Ir anomaly with the mass extinction of species at the K–T boundary.

However, other lines of evidence appear to connect the K–T extinction to an extraterrestrial (i.e., impact) cause. For example, soot has been discovered in the K–T boundary clay layer at a number of locations. This has been interpreted as evidence of global wildfires ignited by the fireball. Grains of quartz that have undergone *shock metamorphism* have also been discovered. Shock metamorphism is a type of microscopic mineral fracturing that can be caused only

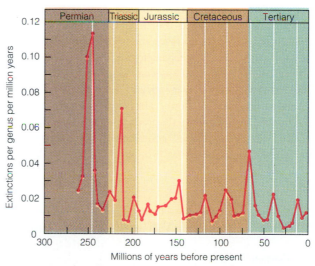

▲ F I G U R E 10.22

Periodicity in the fossil record of extinctions suggests that a regular, recurrent event has disrupted the Earth's biosphere every 26 million years. This diagram is based on the work of David Raup and John Sepkoski of the University of Chicago, who analyzed extinctions in 11,000 genera of marine organisms. The vertical lines represent intervals of 26 million years, which seem to be well aligned with the extinction peaks over the past 150 million years; for earlier times the correlation is not as close.

The Controversy Continues

To summarize the story so far, it is now generally accepted that:

- Major meteorite impacts can cause devastating world-wide effects.

- Such effects would leave characteristic signatures in the geologic record, such as Ir anomalies, sooty layers, shocked minerals, and glassy spherules.

- Many such pieces of evidence have been located, not only in the clay-rich layer that marks the Cretaceous–Tertiary boundary but in other layers marking major transitions in the geologic time scale.

The suggestion of a 26-million-year periodic cycle of mass extinctions and the significance of this observation are less widely accepted. And the direct cause and effect connection between a major meteorite impact and the extinction of the dinosaurs (and other species) at the end of the Cretaceous Period is still a subject of controversy. Let's take a closer look at this ongoing debate.

Volcanism Versus Impact Theory

A number of experts argue that many of the effects associated with a major impact could have resulted from an extended period of particularly intense volcanic activity. These effects include the Ir anomaly and global climatic changes caused by material ejected into the stratosphere. It has even been suggested that impact-generated volcanic activity could have caused the mass extinctions. Observational evidence seems to favor the impact theory over volcanism as the causal mechanism for the K–T extinction (Table 10.3). However, there is a remarkable correlation between known extinction events and the occurrence of very extensive basaltic lava flows. The largest known mass extinctions (those marking the Permian–Triassic and Cretaceous–Tertiary boundaries) coincide with the formation of the world's largest flood basalt provinces, which are located

in Siberia and India, respectively. These are no ordinary volcanic events; the Deccan Plateau flood basalts in India, for example, represent 1.5 million km^3 of lava extruded over a period of 100,000 to 1 million years. The consequences of such an extended period of volcanic activity must have been dramatic and widespread.

The Time Scale of Extinctions

Another major unsolved problem for the impact theory of mass extinctions is related to the timing of the extinctions themselves. The geologic record of sediments, and therefore the fossil record, is incomplete; in any given locality thousands or even millions of years of the record may be missing altogether. Even in the best preserved fossil strata, with extremely tiny and abundant organisms, there will be gaps of thousands of years in the fossil record. The situation becomes even more tenuous when one is dealing with larger, much less common fossils such as dinosaurs; in such cases, there are likely to be gaps of 1 million years or more. This means that what looks like an abrupt extinction on the geologic time scale may turn out to be quite gradual when examined on a finer time scale. Furthermore, it is often difficult to prove that changes observed in a particular species occurred at the same time throughout the world.

Many experts argue that this is the case with the K–T and other mass extinctions—that the decline of many species actually began many thousands of years before the actual transition point and continued for tens of thousands or even millions of years (Fig. 10.23). Mechanisms such as extensive, episodic volcanic eruptions, gradual climatic change, or changes in sea level may offer better explanations for the extended duration of these extinction episodes than an impact scenario.

The "Smoking Gun"

A perennial problem for those who favor the impact theory of mass extinctions has been the lack of a "smoking gun." That is, there was no large impact crater that could be

T A B L E 10.3 • Comparison of Observational Evidence from the K–T Boundary with a Meteorite Impact or Volcanism as the Causal Mechanism

Observational Evidence	Meteorite Impact	Volcanic Eruption	
		Quiet Basaltic	Violent Rhyolitic
Iridium	Yes: all types of meteorites are high in iridium	Possibly: high Ir found in Kilauea gas; no other known case of high Ir	No: high Ir not yet found in rhyolitic volcanics
Spherules	Yes: droplets of impact melt	Yes: drops of liquid basalt	No: violent eruptions produce angular fragments, not droplets
Shocked minerals	Yes: shocked quartz common at known impact sites	No: large explosions do not occur in this setting	Possibly: violent rhyolitic eruptions might produce sufficient pressure for shock metamorphism to occur
Soot (global)	Yes: global wildfires set off by fireball	No: fires only very local	No: fires only very local

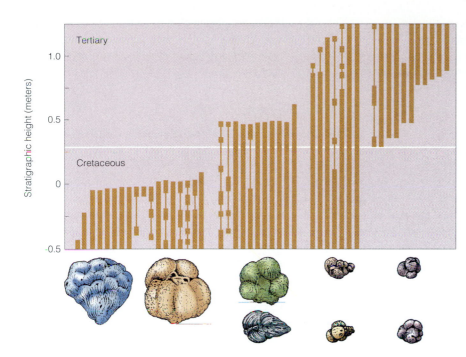

◀ F I G U R E 10.23
This graph, based on the work of Gerta Keller of Princeton University, shows a stairstep pattern of extinctions of fossil foraminifera, tiny shelled marine organisms. The important feature of the diagram is that the extinctions began before the Cretaceous–Tertiary boundary (marked by the horizontal line) and continued in stepwise fashion across the boundary. The entire sequence represents roughly a million years of Earth history. This suggests that the K–T extinction may not have been as sudden as the fossil record of larger, rarer fossils (like dinosaurs) seems to indicate.

shown to have formed exactly 65 million years ago, at the time of the K–T extinction. Recently this problem may have been resolved by the discovery of Chicxulub Crater, a 180-km-diameter crater in the northern Yucatán Peninsula of Mexico that appears to have been formed about 65 million years ago (Fig. 10.24). The presence of shocked min-

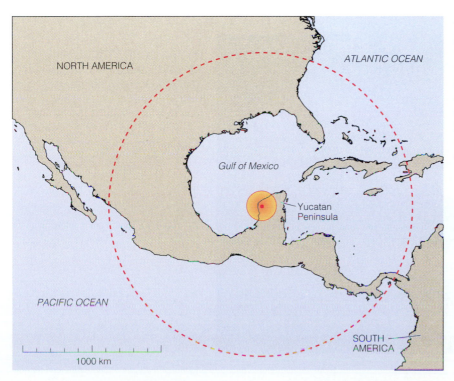

◀ F I G U R E 10.24
Recently discovered Chicxulub Crater near the Yucatán Peninsula may be the "smoking gun" sought by impact theorists—the impact made by the meteorite that caused the Cretaceous–Tertiary extinction. Effects of this impact were recorded in the rocks as far away as the red dashed circle.

eral grains in associated ejecta deposits supports the interpretation of the structure as an impact crater. The very large size of the crater is consistent with the production of the global effects discussed earlier—worldwide distribution of iridium dust in the stratosphere, and soot from wildfires. In addition, materials thought to be deposited by impact-generated tsunamis have been found in coastal areas as far away as Texas.

After more than 15 years of debate, the questions linger: What killed the dinosaurs? Was the Earth devastated by the effects of a major impact, or by an extended period of intense volcanism? Could events like these have caused a mass extinction of species, or were they just the final blow in a long, slow decline that had begun thousands of years earlier? Is the apparent cyclicity of mass extinctions related to periodic swarms of extraterrestrial objects? Whether or not these issues are resolved in the near future, the research surrounding these questions has contributed much valuable information about the evolution of life on the Earth. It has also fueled a fascinating and lively scientific debate.

WORLDS COLLIDING

Given the prevalence of impact cratering and the mass of material known to enter the atmosphere each year, what are the chances of a large impact occurring in the foreseeable future? It has been calculated that in a single human lifetime there is about a 1 in 10,000 chance that the Earth will be hit by a meteorite large enough to cause worldwide climatic changes leading to widespread crop failures and, perhaps, mass extinctions of species. These are about the same as the odds of dying from anesthesia during surgery, dying in a car crash during any 6-month period, or dying of cancer as a result of breathing automobile exhaust on a Los Angeles freeway every day.

Near Misses

Some rather unnerving near-collisions with large extraterrestrial objects have been recorded in the not too distant past. For example, on June 14, 1968, an Apollo object named 1566 Icarus passed within 6 million kilometers of the Earth. If the object had struck the Earth, it would have created a crater 10 km in diameter. The approach of 1566 Icarus had been predicted by astronomers, who assured the public that there was no cause for alarm.

An astonishing near-collision with a slightly smaller body occurred on April 10, 1972, when a rocky object somewhere between a few meters and 80 m in diameter approached the Earth at a velocity of 10 km/s. Impelled by gravitational attraction, it accelerated to 15 km/s and entered the Earth's atmosphere. It traveled over North America from south to north, dipping as low as 50 to 60 km above the surface and becoming visible as a brilliantly glowing object over Utah, Idaho, and Montana. If the object had fallen to the surface it would have struck somewhere in Alberta, causing an explosion about the same size as that of the bomb that destroyed Hiroshima in World War II. Instead, the object glanced off the atmosphere and returned to outer space (Fig. 10.25).

Observers of Apollo objects are identifying Earth approachers at the remarkable rate of 35 new objects each year. In all, there may be as many as 300,000 Earth-cross-

◀ **F I G U R E 10.25**
On August 10, 1972, a small asteroid passed at an altitude of just 58 km above the western United States. The asteroid is seen as a bright streak in this photo, taken at Grand Teton National Park in Wyoming.

BOX 10.3
•

THE HUMAN PERSPECTIVE

STRATEGIC PLANNING FOR A COSMIC STRIKE

Most of the asteroids in our solar system exist in an orbit between Mars and Jupiter; however, many have orbits dangerously close to the Earth's. Collisions and near-collisions are well documented. If the Earth were hit by a large asteroid or comet, as it has been many times in the past, the results would be catastrophic. The doomsday scenario for a major collision includes a blast large enough to annihilate a significant portion of the Earth's human, plant, and animal population; spew enough Sun-blocking dust into the atmosphere to change the global climate and destroy agriculture; and cause monstrous firestorms and tsunamis. What can we do to prepare for such an event or, better yet, to forestall it?

In 1991, at the request of the U.S. Congress, the National Aeronautics and Space Administration (NASA) presented a proposal to protect the Earth and its inhabitants from such impacts by establishing the Spaceguard Survey. The plan called for a global network of at least six automated wide-field telescopes to map the skies and track the orbits of asteroids and other planetary bodies that could strike the Earth. This plan was designed by experts from Australia, Finland, France, India, Russia, and the United States. NASA's early-warning system would locate 95 percent of potentially deadly large asteroids and predict their paths. In the unlikely case that one of them was on a trajectory for a direct hit on the Earth, the scientists would have a couple of decades to prepare a defense.

Once the trajectory of an Earth-approaching asteroid is known, it may be possible to plan a mission to intercept it. The most realistic proposal is to redirect the asteroid by exploding a nuclear bomb alongside it. The NASA proposal states that very powerful explosive devices would be required to provide enough energy to skew the path of a typical asteroid, an object the size of a mountain traveling toward the Earth at 25 km/s.

The problems associated with the Spaceguard Survey and intercept plans involve cost, technology, and human nature. The costs of the Spaceguard Survey alone are estimated at $50 million initially and $10 million annually for 25 years. Moreover, even after an intense search extending over 25 years, some potentially dangerous Earth-approaching objects might not have been identified. A final argument voiced by critics is that once the technology is in place to modify an asteroid's orbit, the likelihood of its misuse is greater than the probability of an asteroid hitting the Earth.

ing objects with diameters greater than 100 m. As many as 50 objects between 5 and 50 m in diameter regularly pass closer to the Earth than the Moon. Most of the newly identified Apollo objects are asteroids. However, with respect to possible future impacts, the greatest cause for concern comes not from an asteroid, but from a comet.

Comet Swift–Tuttle

Comet Swift–Tuttle is approximately 10 km in diameter, about the same size as the body that could have caused the extinction of the dinosaurs. The comet was named for the American astronomers who observed its passage in 1862. Small fragments of Swift–Tuttle are responsible for the Perseid meteor showers described earlier. The fragments are sprayed out of the comet's nucleus every time it passes the Sun and eventually become dispersed along the comet's orbit. Whenever the Earth crosses the path of the fragments, a meteor shower results.

About 200 comets regularly pass by the Earth as they follow their very elongated orbits from the outer edge of the solar system. Although it is often possible to predict the return of a comet with great accuracy, new (i.e., never before observed) comets have a disconcerting habit of appearing "out of nowhere." As noted earlier, comets can also change their trajectories, split apart, or even disappear without notice. Gas jets streaming out from the core of the comet can change the comet's course or weaken the already loosely bound material in the nucleus, causing it to fragment.

Objects as large as Swift–Tuttle strike the Earth only about once in every 10 to 30 million years. In 2126 Swift–Tuttle will pass within 60 lunar distances (about 23 million kilometers) of the Earth. This is a reasonably safe distance, and the comet probably will not prove to be of concern to our descendants . . . unless the gas jets from its nucleus happen to nudge it slightly off its course. No doubt, astronomers of the twenty-second century will be tracking Swift–Tuttle's course with interest.

SUMMARY

1. A meteorite is an extraterrestrial object that strikes the Earth. Glowing fragments of extraterrestrial material passing through the atmosphere are called meteors. Large meteorite impacts are rare, but such impacts may have had catastrophic effects on the Earth at various times throughout geologic history.

2. Meteorites can be classified into three general categories: stones, irons, and stony-irons. One group of stony meteorites, called chondrites, is thought to represent primitive material that has not been altered since early in the history of the solar system. Another group, the achondrites, resemble terrestrial igneous rocks. The irons are made of an alloy of nickel and iron, while the stony-irons are complex mixtures of stony (silicate) material and metal.

3. The asteroid belt, which contains at least 100,000 asteroids, is located between the orbits of Mars and Jupiter. The asteroids are probably either remnants of a planet that was broken up or rocky fragments that failed to gather together to form a planetary mass.

4. Many of the characteristics of asteroids suggest that they are the most likely parent bodies for most meteorites. Large asteroids have undergone differentiation into metallic cores and rocky outer layers, which would account for the compositional differences among meteorites. Collision and mutual fragmentation are common processes in the asteroid belt; such collisions would produce smaller fragments with erratic trajectories. Asteroids whose orbits cross within the Earth's orbit are called Earth-crossing asteroids or Apollo objects.

5. Comets may be the parent bodies for some types of meteorites. Comets are loosely packed aggregations of ice and silicate dust, resembling "dirty snowballs." When they pass close to the Sun they begin to glow as a result of the vaporization of volatile material, and they take on the characteristic appearance of a comet with a glowing head and long, wispy tails.

6. Impact cratering has played an important part in the modification of planetary surfaces throughout the history of the solar system. The size of the crater that forms, and the nature of the events associated with the impact, depend on the size and velocity of the incoming object. The Earth receives hundreds of tons of meteoritic material per day, but dust-sized particles are far more common than large objects. Meteorites with diameters greater than 100 m, which would cause significant cratering, strike the Earth about once every million years.

7. Impact craters are not as common on Earth as on the Moon because on Earth the processes of weathering and erosion continually wipe out or cover the evidence. The two main types of craters are the simple, small, bowl-shaped crater and the large, complex impact structure. When an impact occurs, material is thrown out of the crater, forming a blanket of ejecta. Fracturing and melting of rocks may occur at the bottom of the crater, and rock layers around the edge of the crater are overturned.

8. A very large impact would have regional or even global effects, including global darkness, extreme cold followed by enhanced warming, acidic rain, tsunamis, and global wildfires.

9. It has been postulated that major impacts might be the cause of mass extinctions of species at different times in geologic history, notably the extinction of the dinosaurs at the boundary between the Cretaceous and Triassic Periods. The presence of a layer of clay marking this boundary, which is anomalously high in the rare element iridium, seems to support this theory. However, an extensive period of widespread volcanism might also have contributed to the extinction of the dinosaurs.

10. During a single human lifetime there is a 1 in 10,000 chance that the Earth will be hit by a meteorite large enough to cause global climatic changes, crop failures, and perhaps mass extinctions. Several near misses have been observed, and a program is under way to identify potentially dangerous Earth-approaching objects. The most significant identifiable danger is associated with Comet Swift–Tuttle, which will pass within 60 lunar distances of the Earth in 2126.

IMPORTANT TERMS TO REMEMBER

Apollo object (p. 256)
asteroid (p. 254)
asteroid belt (p. 254)
chondrite (p. 254)
comet (p. 258)

crater (p. 260)
Earth-crossing asteroid (p. 256)
ejecta (p. 263)
iron (p. 254)
meteor (p. 252)

meteorite (p. 251)
stone (p. 254)
stony-iron (p. 254)

QUESTIONS AND ACTIVITIES

1. It has been suggested that it may one day be possible to land a spaceship on an asteroid and mine it for iron and nickel. Investigate this possibility; do you think it might be feasible? What are the main points in favor of such an expedition? What stumbling blocks might there be?

2. How do you think the inhabitants of the Earth should prepare for the eventuality of a collision with a large extraterrestrial object?

3. Scientists have estimated that the probability of civilization being wiped out by the effects of a large impact is somewhere between 1 in 6000 and 1 in 10,000. Write a fictional account of such an event from the point of view of a survivor. What would the first few days after the event be like? What kind of civilization do you think might emerge afterward?

4. Why are carbonaceous chondrites especially important in terms of our understanding of the origin of the solar system? Investigate the meteorite Allende. When and how was it found? What is so special about it?

5. Do you think a major meteorite impact caused the extinction of the dinosaurs? Do some research to support your opinion. Write a short summary of the theories about the cause of the extinction of the dinosaurs and the evidence in support of each theory.

6. Discuss the pros and cons of the impact theory of mass extinctions versus the volcanic eruption theory. Which of the types of evidence presented by experts do you find most convincing? Which do you find unconvincing? Can you think of a type of evidence that has not yet been uncovered that might settle the controversy one way or another?

7. How do scientists identify a circular structure as having been caused by a meteorite impact? This is not always a straightforward process because impact craters can be partially eroded, covered by sediment, or deformed (folded or faulted). Moreover, circular and ringlike geologic structures (such as volcanic calderas) are common. Do some research to find out how scientists find impact craters and what kinds of evidence they look for to confirm the impact origin of a circular structure.

UNITS AND THEIR CONVERSIONS

ABOUT SI UNITS

Regardless of the field of specialization, all scientists use the same units and scales of measurement. They do so to avoid confusion and the possibility that mistakes can creep in when data are converted from one system of units, or one scale, to another. By international agreement the SI units are used by all, and they are the units used in this text. SI is the abbreviation of Système International d'Unités (in English, the International System of Units).

Some of the units are likely to be familiar, some unfamiliar. The SI unit of length is the meter (m), of area the square meter (m^2), and of volume the cubic meter (m^3). The SI unit of mass is the kilogram (kg), and of time the second (s). The other SI units used in this book can be defined in terms of these basic units. Three important ones are:

- The newton (N), a unit of force defined as that force needed to accelerate a mass of 1 kg by 1 m/s^2; hence 1 N = 1 kg·m/s^2. (The period between kg and m indicates multiplication.)

- The joule (J), a unit of energy or work, defined as the work done when a force of 1 newton is displaced a distance of 1 meter; hence 1 J = 1 N·m. One important form of energy so far as the Earth is concerned is heat. The outward flow of the Earth's internal heat is measured in terms of the number of joules flowing outward from each square centimeter each second; thus, the unit of heat flow is $J/cm^2/s$.

- The pascal (Pa), a unit of pressure defined as a force of 1 newton applied across an area of 1 square meter; hence 1 Pa = 1 N/m^2. The pascal is a numerically small unit. Atmospheric pressure, for example (15 lb/in^2), is 101,300 Pa. Pressure within the Earth reaches millions or billions of pascals. For convenience, earth scientists sometimes use 1 million pascals (megapascal, or MPa) as a unit.

Temperature is a measure of the internal kinetic energy (expressed as movement) of the atoms and molecules in a body. In the SI system, temperature is measured on the Kelvin scale (K). The temperature intervals on the Kelvin scale are arbitrary, and they are the same as the intervals on the more familiar Celsius scale (°C). The difference between the two scales is that the Celsius scale selects 100°C as the temperature at which water boils at sea level, and 0°C as the freezing temperature of water at sea level. Zero degrees Kelvin, on the other hand, is absolute zero, the temperature at which all atomic and molecular motions cease. Thus, 0°C is equal to 273.15 K, and 100°C is 373.15 K. The temperatures of processes on and within the Earth tend to be at or above 273.15 K. Despite the inconsistency, earth scientists still use the Celsius scale when geological processes are discussed.

Appendix A provides a table of conversion from older units to Standard International (SI) units.

PREFIXES FOR MULTIPLES AND SUBMULTIPLES

When very large or very small numbers have to be expressed, a standard set of prefixes is used in conjunction with the SI units. Some prefixes are probably already familiar; an example is the centimeter (which is one hundredth of a meter, or 10^{-2} m). The standard prefixes are

tera	1,000,000,000,000	=	10^{12}
giga	1,000,000,000	=	10^{9}
mega	1,000,000	=	10^{6}
kilo	1,000	=	10^{3}
hecto	100	=	10^{2}
deka	10	=	10
deci	0.1	=	10^{-1}
centi	0.01	=	10^{-2}
milli	0.001	=	10^{-3}
micro	0.000001	=	10^{-6}
nano	0.000000001	=	10^{-9}
pico	0.000000000001	=	10^{-12}

One measure used commonly in geology is the nanometer (nm), a unit by which the sizes of atoms are measured; 1 nanometer is equal to 10^{-9} meter.

COMMONLY USED UNITS OF MEASURE

Length

Metric Measure

1 kilometer (km)	= 1000 meters (m)
1 meter (m)	= 100 centimeters (cm)
1 centimeter (cm)	= 10 millimeters (mm)
1 millimeter (mm)	= 1000 micrometers (μm) (formerly called microns)
1 micrometer (μm)	= 0.001 millimeter (mm)
1 angstrom (Å)	= 10^{-8} centimeters (cm)

Nonmetric Measure

1 mile (mi)	= 5280 feet (ft) = 1760 yards (yd)
1 yard (yd)	= 3 feet (ft)
1 fathom (fath)	= 6 feet (ft)

Conversions

1 kilometer (km)	= 0.6214 mile (mi)
1 meter (m)	= 1.094 yards (yd) = 3.281 feet (ft)
1 centimeter (cm)	= 0.3937 inch (in)
1 millimeter (mm)	= 0.0394 inch (in)
1 mile (mi)	= 1.609 kilometers (km)
1 yard (yd)	= 0.9144 meter (m)
1 foot (ft)	= 0.3048 meter (m)
1 inch (in)	= 2.54 centimeters (cm)
1 inch (in)	= 25.4 millimeters (mm)
1 fathom (fath)	= 1.8288 meters (m)

Area

Metric Measure

1 square kilometers (km²)	= 1,000,000 square meters (m²)
	= 100 hectares (ha)
1 square meter (m²)	= 10,000 square centimeters (cm²)
1 hectare (ha)	= 10,000 square meters (m²)

Nonmetric Measure

1 square mile (mi²)	= 640 acres (ac)
1 acre (ac)	= 4840 square yards (yd²)
1 square foot (ft²)	= 144 square inches (in²)

Conversions

1 square kilometer (km²)	= 0.386 square mile (mi²)
1 hectare (ha)	= 2.471 acres (ac)
1 square meter (m²)	= 1.196 square yards (yd²)
	= 10.764 square feet (ft²)
1 square centimeter (cm²)	= 0.155 square inch (in²)
1 square mile (mi²)	= 2.59 square kilometers (km²)
1 acre (ac)	= 0.4047 hectare (ha)
1 square yard (yd²)	= 0.836 square meter (m²)
1 square foot (ft²)	= 0.0929 square meter (m²)
1 square inch (in²)	= 6.4516 square centimeter (cm²)

Volume

Metric Measure

1 cubic meter (m³)	= 1,000,000 cubic centimeters (cm³)
1 liter (l)	= 1000 milliliters (ml)
	= 0.001 cubic meter (m³)
1 centiliter (cl)	= 10 milliliters (ml)
1 milliliter (ml)	= 1 cubic centimeter (cm³)

Nonmetric Measure

1 cubic yard (yd³)	= 27 cubic feet (ft³)
1 cubic foot (ft³)	= 1728 cubic inches (in³)
1 barrel (oil) (bbl)	= 42 gallons (U.S.) (gal)

Conversions

1 cubic kilometer (km³)	= 0.24 cubic miles (mi³)
1 cubic meter (m³)	= 264.2 gallons (U.S.) (gal)
	= 35.314 cubic feet (ft³)
1 liter (l)	= 1.057 quarts (U.S.) (qt)
	= 33.815 ounces (U.S. fluid) (fl. oz.)
1 cubic centimeter (cm³)	= 0.0610 cubic inch (in³)
1 cubic mile (mi³)	= 4.168 cubic kilometers (km³)

1 acre-foot (ac-ft)	= 1233.46 cubic meters (m³)
1 cubic yard (yd³)	= 0.7646 cubic meter (m³)
1 cubic foot (ft³)	= 0.0283 cubic meter (m³)
1 cubic inch (in³)	= 16.39 cubic centimeters (cm³)
1 gallon (gal)	= 3.784 liters (l)

Mass

Metric Measure

1000 kilograms (kg)	= 1 metric ton (also called a tonne) (m.t)
1 kilogram (kg)	= 1000 grams (g)

Nonmetric Measure

1 short ton (sh.t)	= 2000 pounds (lb)
1 long ton (l.t)	= 2240 pounds (lb)
1 pound (avoirdupois) (lb)	= 16 ounces (avoirdupois) (oz) = 7000 grains (gr)
1 ounce (avoirdupois) (oz)	= 437.5 grains (gr)
1 pound (Troy) (Tr. lb)	= 12 ounces (Troy) (Tr. oz)
1 ounce (Troy) (Tr. oz)	= 20 pennyweight (dwt)

Conversions

1 metric ton (m.t)	= 2205 pounds (avoirdupois) (lb)
1 kilogram (kg)	= 2.205 pounds (avoirdupois) (lb)
1 gram (g)	= 0.03527 ounce (avoirdupois) (oz) = 0.03215 ounce (Troy) (Tr. oz) = 15,432 grains (gr)
1 pound (lb)	= 0.4536 kilogram (kg)
1 ounce (avoirdupois) (oz)	= 28.35 grams (g)
1 ounce (avoirdupois) (oz)	= 1.097 ounces (Troy) (Tr. oz)

Pressure

1 pascal (Pa)	= 1 newton/square meter (N/m²)
1 kilogram/square centimeter (kg/cm²)	= 0.96784 atmosphere (atm) = 14.2233 pounds/square inch (lb/in²) = 0.98067 bar
1 bar	= 0.98692 atmosphere (atm) = 10⁵ pascals (Pa) = 1.02 kilograms/square centimeter (kg/cm²)

Energy and Power

Energy

1 joule (J)	= 1 newton meter (N.m)
	= 2.390×10^{-1} calorie (cal)
	= 9.47×10^{-4} British thermal unit (Btu)
	= 2.78×10^{-7} kilowatt-hour (kWh)
1 calorie (cal)	= 4.184 joule (J)
	= 3.968×10^{-3} British thermal unit (Btu)
	= 1.16×10^{-6} kilowatt-hour (kWh)
1 British thermal unit (Btu)	= 1055.87 joules (J)
	= 252.19 calories (cal)
	= 2.928×10^{-4} kilowatt-hour (kWh)
1 kilowatt hour	= 3.6×10^{6} joules (J)
	= 8.60×10^{5} calories (cal)
	= 3.41×10^{3} British thermal units (Btu)

Power (energy per unit time)

1 watt (W)	= 1 joule per second (J/s)
	= 3.4129 Btu/h
	= 1.341×10^{-3} horsepower (hp)
	= 14.34 calories per minute (cal/min)
1 horsepower (hp)	= 7.46×10^{2} watts (W)

Temperature

To change from Fahrenheit (F) to Celsius (C)

$$°C = \frac{(°F - 32°)}{1.8}$$

To change from Celsius (C) to Fahrenheit (F)

$$°F = (°C \times 1.8) + 32°$$

To change from Celsius (C) to Kelvin (K)

$$K = °C + 273.15$$

To change from Fahrenheit (F) to Kelvin (K)

$$K = \frac{(°F - 32°)}{1.8} + 273.15$$

Radioactivity

1 rad	= the unit of absorbed radiation, which corresponds to 0.01 J of energy per kg of material
1 rem	= the dose of ionizing radiation that gives the same biologic effect as 1 rad of 250 kvp x-rays

A rem is the product of the dose in rads and a quality factor that depends on the nature of the radiation; alpha particles and neutrons have factors of 10, whereas beta particles and gamma rays are weighted as 1; hence, 1 rad of gamma radiation = 1 rem, but 1 rad of alpha particles = 10 rem.

1 curie (Ci)	= a measure of the number of radioactive disintegrations per second; specifically, 37 billion disintegrations per second; (more commonly used in measuring radiation is the picocurie (pCi) = 10^{-12} Ci or .037 disintegrations per second).
1 sievert (Sv)	= 1 Sv = 1J/kg = 100 rem
1 gray (Gy)	= 1 J/kg = 100 rads (the SI unit of dose equivalent is the sievert; the gray is the SI unit of absorbed dose)
1 working level (WL)	= the concentration of radioactive gas with a potential alpha energy release of 1.3×10^5 MeV per liter of air (approximately equivalent to 200 pCi per liter of air).

TABLE OF THE CHEMICAL ELEMENTS

Element	Symbol	Atomic Number	Crustal Abundance, Weight Percent	Element	Symbol	Atomic Number	Crustal Abundance, Weight Percent
Actinium	Ac	89	Human-made	Iron	Fe	26	5.80
Aluminum	Al	13	8.00	Krypton	Kr	36	Not known
Americium	Am	95	Human-made	Lanthanum	La	57	0.0050
Antimony	Sb	51	0.00002	Lawrencium	Lw	103	Human-made
Argon	Ar	18	Not known	Lead	Pb	82	0.0010
Arsenic	As	33	0.00020	Lithium	Li	3	0.0020
Astatine	At	85	Human-made	Lutetium	Lu	71	0.000080
Barium	Ba	56	0.0380	Magnesium	Mg	12	2.77
Berkelium	Bk	97	Human-made	Manganese	Mn	25	0.100
Beryllium	Be	4	0.00020	Mendelevium	Md	101	Human-made
Bismuth	Bi	83	0.0000004	Mercury	Hg	80	0.000002
Boron	B	5	0.0007	Molybdenum	Mo	42	0.00012
Bromine	Br	35	0.00040	Neodymium	Nd	60	0.0044
Cadmium	Cd	48	0.000018	Neon	Ne	10	Not known
Calcium	Ca	20	5.06	Neptunium	Np	93	Human-made
Californium	Cf	98	Human-made	Nickel	Ni	28	0.0072
Carbon[a]	C	6	0.02	Niobium	Nb	41	0.0020
Cerium	Ce	58	0.0083	Nitrogen	N	7	0.0020
Cesium	Cs	55	0.00016	Nobelium	No	102	Human-made
Chlorine	Cl	17	0.0190	Osmium	Os	76	0.00000002
Chromium	Cr	24	0.0096	Oxygen[b]	O	8	45.2
Cobalt	Co	27	0.0028	Palladium	Pd	46	0.0000003
Copper	Cu	29	0.0058	Phosphorus	P	15	0.1010
Curium	Cm	96	Human-made	Platinum	Pt	78	0.0000005
Dysprosium	Dy	66	0.00085	Plutonium	Pu	94	Human-made
Einsteinium	Es	99	Human-made	Polonium	Po	84	Footnote[d]
Erbium	Er	68	0.00036	Potassium	K	19	1.68
Europium	Eu	63	0.00022	Praseodymium	Pr	59	0.0013
Fermium	Fm	100	Human-made	Promethium	Pm	61	Human-made
Fluorine	F	9	0.0460	Protactinium	Pa	91	Footnote[d]
Francium	Fr	87	Human-made	Radium	Ra	88	Footnote[d]
Gadolinium	Gd	64	0.00063	Radon	Rn	86	Footnote[d]
Gallium	Ga	31	0.0017	Rhenium	Re	75	0.00000004
Germanium	Ge	32	0.00013	Rhodium[c]	Rh	45	0.00000001
Gold	Au	79	0.0000002	Rubidium	Rb	37	0.0070
Hafnium	Hf	72	0.0004	Ruthenium[c]	Ru	44	0.00000001
Helium	He	2	Not known	Samarium	Sm	62	0.00077
Holmium	Ho	67	0.00016	Scandium	Sc	21	0.0022
Hydrogen[b]	H	1	0.14	Selenium	Se	34	0.000005
Indium	In	49	0.00002	Silicon	Si	14	27.20
Iodine	I	53	0.00005	Silver	Ag	47	0.000008
Iridium	Ir	77	0.00000002	Sodium	Na	11	2.32

Element	Symbol	Atomic Number	Crustal Abundance, Weight Percent	Element	Symbol	Atomic Number	Crustal Abundance, Weight Percent
Srontium	Sr	38	0.0450	Unnilhexium	Unh	106	Human-made
Sulfur	S	16	0.030	Unniloctium	Uno	108	Human-made
Tantalum	Ta	73	0.00024	Unnilpentium	Unp	105	Human-made
Technetium	Tc	43	Human-made	Unnilquadium	Unq	104	Human-made
Tellurium[c]	Te	52	0.000001	Unnilseptium	Uns	107	Human-made
Terbium	Tb	65	0.00010	Uranium	U	92	0.00016
Thallium	Tl	81	0.000047	Vanadium	V	23	0.0170
Thorium	Th	90	0.00058	Xenon	Xe	54	Not known
Thulium	Tm	69	0.000052	Ytterbium	Yb	70	0.00034
Tin	Sn	50	0.00015	Yttrium	Y	39	0.0035
Titanium	Ti	22	0.86	Zinc	Zn	30	0.0082
Tungsten	W	74	0.00010	Zirconium	Zr	40	0.0140
Unnilennium	Une	109	Human-made				

Source: After K. K. Turekian, 1969.

[a]Estimate from S. R. Taylor (1964).

[b]Analyses of crustal rocks do not usually include separate determinations for hydrogen and oxygen. Both combine in essentially constant proportions with other elements, so abundances can be calculated.

[c]Estimates are uncertain and have a very low reliability.

[d]Elements formed by decay of uranium and thorium. The daughter products are radioactive with such short half-lives that crustal accumulations are too low to be measured accurately.

THE GEOLOGIC TIME SCALE

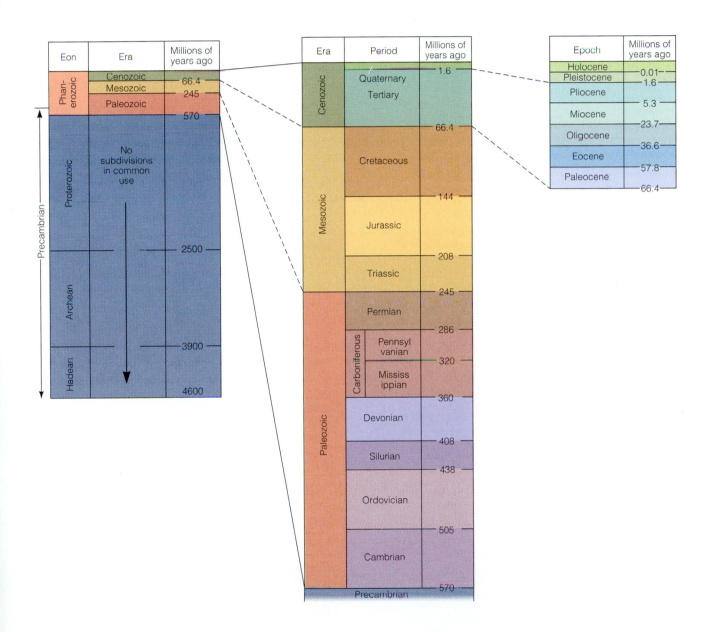

The geologic time scale. Absolute ages obtained from radiometric dates. Note that the Pennsylvanian and Mississippian Periods are equivalent to the Carboniferous Period of Europe. The time boundary between the Archean and Hadean is uncertain as no rocks of the Hadean Eon are known on the Earth. Hadean rocks are known to exist on other planets in the solar system.

PHOTO CREDITS

Introduction
Opener: ©Earth Satellite Corp./SPL/Photo Researchers. Fig. In.-1: © Steve O'Meara/Nature Stock. Fig. In.-3: ©Southern Illinoisian/AP/Wide World Photos. Fig. In.-4: © Panos Pictures. Fig. In.-5: Courtesy National Oceanic & Atmospheric Administration.

Part 1
Opener: ©Tsado/Tom Stack & Associates. Figure I-1: NRSC Ltd./Science Photo Library/Photo Researchers. Figure I-3: COMSTOCK, Inc. Figure I-5: E. R. Degginger/Bruce Coleman, Inc. Figure I-6: ©John S. Shelton. Figure I-7: ©Tom Van Sant/The Geosphere Project. Figure I-8: Earth Satellite Corporation. Figure I-9: Spencer Swanger/Tom Stack & Associates.

Chapter 1
Opener: ©David Robert Austen. Figure 1-9: Claus Meyer/Black Star. Figure 1-15: Bruce Heezen and Marie Tharp. Figure 1-16: Courtesy NASA/Lyndon B. Johnson Space Center. Figure 1-20a: ©John S. Shelton. Figure 1-20b: Francois Gohier/Ardea London.

Chapter 2
Opener: ©Robert Chasin/Black Star. Figure 2-3: William E. Ferguson. Figure 2-7: Michael Hochella. Figure 2-10a: Brian J. Skinner. Figure 2-10b: Brian Parker/Tom Stack & Associates. Figure 2-11: Brian J. Skinner. Box 2-2-1a: Barry L. Runk/Grant Heilman Photography. Box 2-2-1b: Runk/Schoenberger/Grant Heilman Photography. Box 2-2-1c: Brian J. Skinner. Figure 2-12: Brian J. Skinner.

Part 2
Opener: Les Stone/Sygma.

Chapter 3
Opener: Iwasa/Sipa Press. Figure 3-1: ©John S. Shelton. Figure 3-4a: William C. Gaustein. Figure 3-4b: Brian J. Skinner. Figure 3-14: ©Steve Lehman/SABA. Figure 3-15: Paul X. Scott/Sygma Photo News. Figure 3-16: A. Nogues/Sygma Photo News. Figure 3-21: ©Mark Downey/Gamma Liaison. Figure 3-22: Kevin Schafer/Tom Stack & Associates. Figure 3-23: Lysaght/Gamma Liaison. Figure 3-24: Steve McCutcheon. Figure 3-26: Courtesy USGS Photo Library, Denver, Colorado. Figure 3-33: Centre National d'Etudes Spatiales.

Chapter 4
Opener: ©Alberto Garcia/SABA. Figure 4-1 & 4-2: ©Krafft Explorer/Photo Researchers. Figure 4-3 & 4-4: J.D. Griggs/USGS. Figure 4-5: Brian J. Skinner. Figure 4-6a: William E. Ferguson. Figure 4-6b: J. D. Griggs/USGS. Figure 4-6c: Schofield/Gamma Liaison. Figure 4-7: G. J. Orme, Dept. of Army, Fort Bragg, N.C.. Figure 4-9: Roger Werth/Woodfin Camp & Associates. Figure 4-10: S.C. Porter. Figure 4-11: Steve Vidler/Leo de Wys, Inc. Figure 4-12a: ©Krafft Explorer/Photo Researchers. Figure 4-12b: Tom Bean/DRK Photo. Figure 4-13: Brian J. Skinner. Figure 4-14: S.C. Porter. Figure 4-15: Greg Vaughn/Tom Stack & Associates. Figure 4-16: Brian J. Skinner. Figure 4-18: Lyn Topinka/USGS. Figure 4-19: J. D. Griggs/USGS. Figure 4-23: G. Brad Lewis/Gamma Liaison. Figure 4-24: David Hiser/The Image Bank. Box 4-1-3a: ©T. Orban/Sygma. Box 4-1-3b: ©Peter Turnley/Black Star. Figure 4-27: David Robert Austen. Figure 4-29: NRSC Ltd./Science Photo Library/Photo Researchers.

Chapter 5
Opener: ©Sankei Shimbun/Symga Photo News. Figure 5-2: Bishop Museum. Figure 5-6: ©Steve McCutcheon.

Chapter 6
Opener: El Tiempo/©Sipa Press. Figure 6-2: Jacques Jangoux/Tony Stone Images/New York, Inc. Figure 6-5: S.C. Porter. Figure 6-6: Alex Soto/ Health Sciences Center for Educational Resources, University of Washington. Figure 6-7: USGS/ Vancouver. Figure 6-10: A. Lincoln Washburn. Figure 6-12: © George Herben Photography. Figure 6-19: Steve McCutcheon/Tony Stone Images/ Seattle. Figure 6-20: George

Plafker, U.S Geological Survey. Figure 6-23: F. O. Jones/U.S. Geological Survey. Figure 6-24: Hulton Deutsch. Figure 6-25: S.C. Porter. Box 6-2-1a: Los Angeles County Department of Public Works. Box 6-2-1b: Perry Ehlig. Figure 6-26: ©John Elk III. Figure 6-28: ©Bill Brooks/Masterfile. Figure 6-29: S.C. Porter.

Chapter 7
Opener: ©Alex S. MacLean/Landslides. Figure 7-1: Michael Nichols/Magnum Photos, Inc. Figure 7-2: Courtesy NASA. Figure 7-3: Hiroji Kubota/Magnum Photos, Inc. Figure 7-4: Sam C. Pierson, Jr./Photo Researchers. Figure 7-6: ©Robert Baumgardner. Figure 7-10: C.R. Dunrud/Courtesy USGS. Figure 7-14: Courtesy City of Long Beach. Figure 7-15: ©Jim Pickerell/Gamma Liaison. Figure 7-16: ©Stevan Stefanovic/OKAPIA/Photo Researchers. Figure 7-17: Mirco Toniolo/Gamma Liaison. Figure 7-18: ©James Fraser/Impact Photos.

Chapter 8
Opener: ©Frank Oberle/Tony Stone Images/New York, Inc. Figure 8-1: G. Hading/Photo Researchers. Figure 8-2: Chip Hires/Gamma Liaison. Figure 8-3: ©Wim Ruigrok/Sygma Photo News. Figure 8-4 & 8-7: S.C. Porter. Figure 8-8: Courtesy NASA. Figure 8-17: Hiroji Kubota/Magnum Photos, Inc. Figure 8-18: A. Holbrooke/Gamma Liaison. Figure 8-21: Earth Satellite Corporation.

Chapter 9
Opener: ©David R. Frazier Photolibrary/Photo Researchers. Figure 9-6: Greg Scott/Masterfile. Figure 9-10: Nicholas DeVore III/ Photographers Aspen. Figure 9-12: Vince Streano/Tony Stone Images/New York, Inc. Figure 9-13: G. R. Roberts. Figure 9-19: ©Mark Wexler/Woodfin Camp & Associates. Figure 9-20: © Spaceshots, Inc. 1-800-272-2779. Figure 9-21: Valerie Taylor/Ardea London. Figure 9-22b: James Stanfield/National Geographic Society. Figure 9-23b: Dick Davis/Photo Researchers. Figure 9-24: ©Tom Sobolink/Black Star. Figure 9-25: National Oceanic and Atmospheric Administration/ DLR— FRG/Roger Ressmeyer/©CORBIS. Figure 9-26: Otto/ Visions/Impact Photos. Figure 9-27: Courtesy NASA Goddard Space Flight Center/ A. F. Hasler, H. Pierce, K. Palaniappan, and M. Manyin. Figure 9-28: Stephen Krasemann/DRK Photo. Figure 9-30: Victor Englebert/Photo Researchers. Box 9-2-1: Dr. Richard Iegeckis/ Science Photo Library/Photo Researchers.

Chapter 10
Opener: Courtesy Canada Centre for Remote Sensing, Department of Natural Resources Canada. Figure 10-1: Pekka Parviainen/ Science Photo Library/Photo Researchers. Figure 10-2: Dr. Ursula Marvin, with permission of the Keeper of Manuscripts at the Universitatsbibliothek/Smithsonian Astro Observatory. Figure 10-3: Nabeel Turner/Tony Stone Images/New York, Inc. Figure 10-4: Dr. Ursula Marvin/Smithsonian Astro Observatory. Figure 10-5: Wards Science/ Science Source/Photo Researchers. Figure 10-7: ©World Perspectives/ Explorer/Photo Researchers. Box 10-1-1: Dr. Ursula Marvin/Smithsonian Astro Observatory. Figure 10-8: ©William K. Hartmann. Figure 10-11: ©Ronald E. Royer/JPL/Photo Researchers. Figure 10-13: ©1986 Max-Planck-Institut fur Aeronomie, Courtesy Harold Reitsema & Alan Delamere, Ball Aerospace/Roger Ressmeyer ©CORBIS. Figure 10-14: Novosti Press Agency/Science PhotoLibrary /Photo Researchers. Box 10-2-1: AP/Wide World Photos. Figure 10-18a: John Sanford/Science Photo Library/Photo Researchers. Figure 10-18b: ©Tom Till. Figure 10-21: Jonathan Blair/National Geographic Image Collection. Figure 10-25: J. Baker/ D. Milon/Photo Researchers.

GLOSSARY

A

Abrasion   Mechanical erosion that results from the impact of wind-driven grains of sediment. (Ch. 14)

Acid mine drainage   The acidified runoff from a mine, waste rock pile, or tailings pile. (Ch. 17)

Acid precipitation   Any precipitation with a pH lower than 5.0. (Ch. 18)

Active remediation   An approach to the cleanup of contaminated groundwater that involves active intervention. (Ch. 17)

Active volcano   A volcano that has erupted within recorded history. (Ch. 4)

Aerosols   Extremely fine particles or droplets that are carried in suspension; volcanic aerosols result from the reaction of volcanic gases with water vapor in the atmosphere. (Ch. 18)

Aftershock   An earthquake that occurs shortly after a major quake. (Ch. 3)

Amplitude   Half the height from the peak to the trough of a wave; or, the height of a water wave from the still-water line. (Chs. 3, 5)

Andesite   A fine-grained igneous rock with intermediate silica content. (Ch. 4)

Anthropogenic hazards   Human-generated hazards that arise from pollution and degradation of the natural environment (e.g., contamination of surface and underground water bodies, depletion of the ozone layer, and global climatic warming). (Essay II)

Apollo objects   Extraterrestrial bodies with earth-approaching orbits. (Ch. 10)

Aquiclude   A body of impermeable or distinctly less permeable rock adjacent to a permeable one. (Ch. 15)

Aquifer   A water-bearing rock or sediment unit. (Chs. 7, 15)

Artesian aquifer   An aquifer in which water is under hydraulic pressure; a confined aquifer. (Ch. 15)

Asbestos   An industrial term that refers to a group of minerals that grow with a fibrous crystal habit. (Ch. 18)

Asteroid   A small, irregularly shaped rocky body that orbits the sun. (Ch. 10)

Asteroid belt   A swarm of at least 100,000 asteroids nestled between the orbits of Mars and Jupiter, at the boundary between the inner (terrestrial) and outer (Jovian) planets. (Ch. 10)

Atmosphere   The mixture of gases, predominantly nitrogen, oxygen, argon, carbon dioxide, and water vapor, that surrounds the Earth. (Essay I)

Atom   The smallest individual particle that retains all the properties of a given chemical element. (Ch. 2)

Attenuation layer   An impermeable layer that underlies a waste-disposal site and slows the migration of leachate. (Ch. 16)

Aven   An open, chimneylike passageway that connects a cavern to near-surface rocks and unconsolidated overburden. (Ch. 7)

B

Bankfull discharge   The discharge measured when a river is at bankfull stage. (Ch. 8)

Bankfull stage   A condition in which a river's channel fills completely, so that any further increase in discharge results in water overflowing the banks. (Ch. 8)

Barrier island   A long, narrow, sandy island close to and parallel to the coast. (Ch. 9)

Basalt   A fine-grained igneous rock with low silica content. (Ch. 4)

Base metals   Metallic resources that include any of the common metals, such as copper or lead. (Ch. 13)

Beach   Wave-washed sediment along a coast. (Ch. 9)

Beach drift   The irregular movement of particles along a beach as they travel obliquely up the slope of a beach with the swash and directly down this slope with the backwash. (Ch. 9)

Biochemical oxygen demand (BOD)   The depletion of dissolved oxygen in a body of water that results from an excess of aerobic microorganisms. (Ch. 17)

Bioconcentration   Biologic processes leading to the buildup of contaminants in organisms. (Ch. 17)

Biodegradation   The decomposition of a substance through the action of microorganisms. (Ch. 17)

Biogas   Gas derived from decaying organic wastes, which is collected and used as a fuel. (Ch. 12)

Biogeochemical cycle   A natural cycle that describes the movements and interactions through the Earth's spheres of the chemicals essential to life. (Ch. 1)

Biomass energy   Any form of energy that is derived more or less directly from the Earth's plant life. (Ch. 12)

Biosphere   The totality of the Earth's living matter, including organic matter that has not been completely decomposed. (Essay I)

Body waves   Seismic waves that travel outward from an earthquake's point of origin and pass through the Earth. (Ch. 3)

Bore   A wall of water that results when an onrushing wave becomes concentrated in a long, narrow bay or river mouth. (Ch. 5)

C

Caldera A roughly circular, steep-walled volcanic collapse basin several kilometers or more in diameter. (Ch. 4)

Cap rock A rock with very low porosity and permeability that impedes migrating oil and gas and prevents them from going any farther. (Ch. 11)

Carrying capacity The maximum number of people an ecosystem can support, imposed by the limited resources of that ecosystem. (Essay III)

Catastrophism The concept that all of the Earth's major features, such as mountains, valleys, or oceans, have been produced by a few great catastrophic events. (Ch. 1)

Cave A natural underground open space, generally connected to the surface and large enough for a person to enter. (Ch. 7)

Cavern A large cave or system of interconnected cave chambers. (Ch. 7)

Channel The passageway in which a river flows. (Ch. 8)

Channelization Modifications to river channels, consisting of some combination of straightening, deepening, widening, clearing, or lining of the natural channel. (Ch. 8)

Chemical weathering The decomposition of rocks and minerals as chemical reactions transform them into new chemical combinations that are stable at or near the Earth's surface. (Ch. 14)

Chondrites Very old, stony meteorites that contain small, round, glassy-looking spheres called chondrules. (Ch. 10)

Climate The average weather conditions of a place or an area over a period of years. (Ch. 9)

Closed system A system whose boundaries allow the exchange of energy, but not matter, with the surrounding environment; a system in which the amount of matter is fixed and the contents cycle within the boundaries of the system. (Ch. 1)

Coal A black, combustible, sedimentary or metamorphic rock that consists chiefly of decomposed plant matter and contains more than 50 percent organic matter. (Ch. 11)

Coalification The process of coal formation, which involves the loss of volatile materials, including water, carbon dioxide, and methane, from peat. (Ch. 11)

Comet An icy extraterrestrial object that glows when it approaches the sun, producing a long, wispy tail that points away from the sun. (Ch. 10)

Compaction A decrease in the thickness and porosity of a sediment layer, due to the weight of overlying material and, sometimes, to withdrawal of fluids. (Ch. 7)

Composting Facilitating the breakdown of organic refuse such as food waste, leaves, or yard clippings by the action of bacteria and other naturally occurring microorganisms. (Ch. 16)

Compressional waves (*P waves*) Seismic body waves that consist of alternating pulses of compression and expansion acting in the direction in which the wave is traveling. (Ch. 3)

Cone of depression A conical depression in the water table immediately surrounding a pumped well. (Chs. 7, 15)

Confined aquifer An aquifer that is bounded by aquicludes. (Ch. 15)

Contaminants Materials that have harmful impacts and degrade the environment; see also *Pollutants*. (Essay IV)

Convergent margin A type of plate boundary where plates meet as they move toward each other. (Ch. 1)

Core The spherical mass, largely metallic iron, at the center of the Earth. (Ch. 2)

Coriolis effect An effect which causes any body that moves freely with respect to the rotating Earth to veer to the right in the Northern Hemisphere and to the left in the Southern Hemisphere. (Ch. 9)

Crater (volcanic) A funnel-shaped depression opening upward near the summit of a volcano, from which gases, tephra, and lava are ejected. (Ch. 4)

Crust The outermost and thinnest of the Earth's compositional layers, which consists of rocky matter that is less dense than the rocks of the mantle below. (Ch. 2)

Crystalline Any solid with a crystal structure; that is, an orderly, repeated geometric pattern of atoms. (Ch. 2)

Current A broad, slow drift of water in a particular direction. (Ch. 9)

Cyclone An atmospheric low-pressure system that gives rise to roughly circular, inward-spiraling wind motion, called vorticity. (Ch. 9)

D

Decommissioning The cleaning up and rehabilitation of a contaminated site after industrial use. (Ch. 13)

Deep-well disposal A waste-disposal procedure that involves injecting liquid wastes deep into the ground in a specially designed well. (Ch. 16)

Deflation Erosion that occurs when the wind picks up and removes loose rock fragments, sand, and dust. (Ch. 14)

Desert Land where annual rainfall is less than 250 mm or in which the rate of evaporation exceeds the rate of precipitation. (Ch. 9)

Desertification The spread of desert into nondesert areas. (Ch. 9)

Discharge The quantity of water that passes a given point on the bank of a river within a given interval of time; or, in the context of groundwater, the process whereby water leaves the subsurface to join a surface water body or to become surface runoff. (Chs. 8, 15)

Dissolution The chemical weathering process whereby minerals and rock materials pass directly into solution. (Ch. 7)

Divergent margin A type of plate boundary along which plates are moving apart from one another. (Ch. 1)

Divide The topographic high point that separates adjacent drainage basins. (Ch. 8)

Dormant volcano A volcano that has not erupted in recent memory and shows no signs of current activity, but is not deeply eroded. (Ch. 4)

Downstream flooding Flooding that continues for a long time and extends over an entire region. (Ch. 8)

Drainage basin The total area that contributes water to a river. (Ch. 8)

Drought An extended period of exceptionally low precipitation. (Ch. 9)

Dry deposition Materials transported in suspension in the air that eventually settle out as dry matter. (Ch. 18)

E

Early warning A public declaration that the normal routines of life should be altered for a period of time to deal with the danger posed by an imminent hazardous event. (Essay II)

Earth-crossing asteroids Asteroids whose orbits pass within the orbit of the Earth. (Ch. 10)

Earth resources   Materials of value to humans that are extracted (or extractable) from the solid Earth. (Essay III)

Economic geology   The field of study devoted to how and where the different kinds of valuable mineral deposits form. (Ch. 13)

Ejecta   A jet of rock and dust that results as a high-speed meteorite strikes and penetrates the surface of a planet. (Ch. 9)

Elastic rebound theory   The theory that earthquakes result from the release of stored elastic strain energy by slippage on faults. (Ch. 3)

Element   The most fundamental substance into which matter can be separated by chemical means. (Ch. 2)

El Niño   An anomalous warming of surface waters in the eastern equatorial Pacific Ocean. (Ch. 9)

Energy cycle   A natural cycle describing the movement of energy (solar, geothermal, and tidal) through the Earth system and its interactions with materials and processes in the biosphere, hydrosphere, atmosphere, and solid Earth. (Ch. 1)

Energy intensity   Energy consumption per unit of economic activity. (Ch. 11)

Enhanced recovery   The introduction of fluids into an oil well to facilitate the flow of oil. (Ch. 11)

Enrichment factor   The extent to which a material must be concentrated over and above its average abundance in the Earth's crust, in order to be economically recoverable. (Ch. 13)

Environment   The outer biophysical system, in which people and other organisms exist; anything—living or nonliving—that surrounds and influences living organisms. (Intro)

Environmental geology   The study of the functioning of Earth systems and how they affect and are affected by human activities. (Introduction)

Environmental radiation   The dose of radiation each of us receives in our normal daily lives. (Ch. 16)

Eolian erosion   Wind erosion. (Ch. 14)

Epicenter   The point on the Earth's surface directly above the focus of an earthquake. (Ch. 3)

Eruption column   A hot, turbulent mixture of gas and tephra that rises rapidly in the cooler air above the vent of an erupting volcano. (Ch. 4)

Eustatic   Global (refers to sea-level change). (Ch. 18)

Eutrophication   A process whereby the oxygen in water becomes depleted as a result of the uncontrolled growth of plankton and algae. (Ch. 17)

Evaporite deposits   Layers of salts that precipitate as a consequence of the evaporation of seawater. (Ch. 13)

Expansive soils   Soils that expand greatly when saturated with water. (Ch. 6)

Exploration geology   The branch of geology concerned with discovering new supplies of usable minerals. (Ch. 13)

Exponential growth   The increase in a population by a constant percentage per unit of time. (Essay III)

Extinct volcano   A volcano that has not erupted within recorded history, is deeply eroded, and shows no signs of future activity. (Ch. 4)

F

Fall   A sudden, vertical movement of Earth material—for example, from an overhanging cliff. (Ch. 6)

Fault   A fracture in a rock along which movement occurs. (Ch. 3)

Fission   The splitting of an atom into two smaller atoms, with an attendant release of heat energy. (Ch. 12)

Fissure   A long, linear or arc-shaped fracture in the Earth's crust. (Ch. 4)

Fissure eruption   A volcanic eruption that occurs when lava reaches the surface via elongated fissures. (Ch. 4)

Flash flood   A flood in which the lag time is exceptionally short—hours or minutes. (Ch. 8)

Flood   A discharge great enough to cause a body of water to overflow its natural or artificial confines and submerge surrounding land. (Ch. 8)

Flood-frequency curve   Flood magnitudes that are plotted with respect to the recurrence interval calculated for a flood of that magnitude at a given location. (Ch. 8)

Floodplain   The part of any stream valley that is inundated during floods. (Ch. 8)

Flood stage   A condition in which a river's channel fills completely, so that any further increase in discharge results in water overflowing the banks. (Ch. 8)

Flow   A mass-wasting process that involves the chaotic movement of mixtures of sediment and water (or sediment, water, and air). (Ch. 6)

Focus   The place where energy is first released to cause an earthquake. (Ch. 3)

Forecast   The short- or long-term prediction of the magnitude and time of occurrence of a hazardous event. (Essay II)

Foreshock   A swarm of small earthquakes that precedes a big quake. (Ch. 3)

Fossil fuels   Hydrocarbons, formed from the remains of plants or animals trapped in sediment, that may be used as fuel; primarily oil, natural gas, and coal. (Ch. 11)

Frequency   The time interval between crests of adjacent waves. (Ch. 5)

Frost heaving   The lifting of regolith by the freezing of water contained within it. (Ch. 6)

Fumarole   A vent or fracture from which volcanic gases are emitted. (Ch. 4)

Fusion   The joining together, or fusing, of two small atoms to create a single, larger atom, with an attendant release of heat energy. (Ch. 12)

G

Gangue   The nonvaluable minerals of an ore. (Ch. 13)

General circulation models (GCMs)   Climate models that portray interconnected processes in the atmosphere, hydrosphere, and biosphere. (Ch. 18)

Geologic hazards   The wide range of geologic circumstances, materials, processes, and occurrences that are hazardous to humans, such as earthquakes, volcanic eruptions, floods, or landslides. (Essay II)

Geologic isolation   The underground disposal of hazardous wastes in a geologic repository. (Ch. 16)

Geologists   Scientists who study the Earth. (Introduction)

Geology   The scientific study of the Earth. (Introduction)

Geopressurized zone   A high-pressure, high-temperature environment that holds large quantities of natural gas at great depths in the pore spaces of basinlike sedimentary rock sequences thousands of meters thick. (Ch. 12)

Geothermal energy Energy derived from the Earth's internal energy. (Ch. 1, 12)

Geyser A thermal spring equipped with a natural system of plumbing and heating that causes intermittent eruptions of water and steam. (Ch. 4)

Glaciations Periods during which the average surface temperature dropped by several degrees and stayed cool long enough for the polar ice sheets to grow larger; also referred to as *glacial periods* or *ice ages*. (Ch. 18)

Glowing avalanche A pyroclastic flow that is dense, gas-charged, and so hot that it incandesces. (Ch. 4)

Grade A term for the level of concentration of a metal or a mineral in an ore. (Ch. 13)

Gradient A measure of the vertical drop over a given horizontal distance; in the cases of streams, the vertical distance that a channel falls between two points along its course. (Chs. 6, 8)

Greenhouse effect The property of the Earth's atmosphere by which long-wavelength heat rays from the Earth's surface are trapped or reflected back by the atmosphere. (Ch. 18)

Ground-level ozone Ozone that occurs in the lower part of the atmosphere; tropospheric ozone. (Ch. 18)

Groundwater The water contained in spaces within bedrock and regolith. (Ch. 15)

Gullies Distinct, narrow stream channels that are larger and deeper than rills and that result as erosion from running water progresses. (Ch. 14)

H

Half-life The time it takes for half of the atoms in a given sample to decay. (Ch. 16)

Hazard assessment The process of determining when and where hazardous events have occurred in the past, the severity of the physical effects of past events of a given magnitude, the frequency of events that are strong enough to generate physical effects, and what a particular event would be like if it were to occur now; and portraying all this information in a form that can be used by planners and decision makers. (Essay II)

Hazardous wastes Wastes that pose a present or potential threat to humans or wildlife. (Ch. 16)

Historical geology The study of the chronology of events, both physical and biologic, that have occurred in the past. (Introduction)

Humus Partially decomposed organic matter derived from the decay of dead plants and animals in soils. (Ch. 14)

Hurricane A tropical cyclonic storm having winds that exceed 120km/h. See *Typhoon*. (Ch. 9)

Hydraulic gradient The slope of the water table. (Ch. 15)

Hydrocarbon An organic compound (gaseous, liquid, or solid) that consists primarily of carbon and hydrogen. (Ch. 11)

Hydrocarbon trap The combination of a source rock, reservoir rock, and cap rock that provides a barrier to the upward migration of liquid or gaseous hydrocarbons. (Ch. 12)

Hydroelectric power Power that is generated from the potential energy of stream water as it flows to the sea. (Ch. 12)

Hydrograph A graph in which river discharge is plotted against time. (Ch. 8)

Hydrologic cycle A natural cycle that describes the movement of water through the Earth system. (Ch. 1)

Hydrosphere The totality of the Earth's water, comprising oceans, lakes, streams, underground water, and all snow and ice, including glaciers. (Essay I)

Hydrothermal solutions Hot, water-rich solutions that can concentrate metallic ore minerals. (Ch. 13)

I

Igneous rock Rock formed by the cooling and crystallization of magma. (Ch. 1)

Impact crater A bowl-shaped depression that occurs as the result of the transfer of kinetic energy from an object to the surface of the Earth upon impact. (Ch. 10)

Incineration The burning of refuse, often in a specially designed facility. (Ch. 16)

Inexhaustible resource A resource that can never be depleted or used up. (Essay III)

Infiltration The process by which water from precipitation soaks into the soil. (Ch. 15)

Ion An atom that has excess positive or negative charges caused by electron transfer. (Ch. 2)

Irons Meteorites composed mainly of alloys of metallic nickel and iron. (Ch. 10)

Isohyetal map A map that depicts the distribution of rainfall over the area affected by a storm. (Ch. 8)

J

Jovian planets The giant planets (Jupiter, Saturn, Uranus, and Neptune) in the outer regions of the solar system, which are characterized by great masses, low densities, and thick atmospheres consisting primarily of hydrogen and helium. (Essay I)

K

Karst topography Topography characterized by many small, closed basins and a disrupted drainage pattern whereby streams disappear into the ground and eventually appear elsewhere in large springs. (Ch. 7)

L

Lag time The time that elapses between the onset of precipitation and the peak flood stage. (Ch. 8)

Lahar A volcanic mudflow. (Ch. 4)

Landslide Any perceptible downslope movement of bedrock, regolith, or a mixture of the two. (Ch. 6)

Lateral blast A sideways volcanic eruption of pulverized rock and hot gases. (Ch. 4)

Lava Magma that reaches the Earth's surface through a volcanic vent and pours out over the landscape. (Ch. 4)

Lava dome A formation that results from the squeezing out of sticky and viscous lava after a major volcanic eruption. (Ch. 4)

Leachate Contaminated water that results when water percolates through a pile of solid waste, picking up soluble materials. (Ch. 16)

Liquefaction The rapid fluidization of sediment as the result of a disturbance or an abrupt shock, such as an earthquake. (Chs. 3, 6)

Lithosphere The outermost rocky layer of the Earth; more precisely, the crust and the outermost portion of the mantle. (Essay I)

Load   The particles of sediment and dissolved matter that are carried along by a river. (Ch. 8)

Longshore current   A current within the surf zone that flows parallel to the coast. (Ch. 9)

M

Magma   Molten rock, sometimes containing suspended mineral grains and dissolved gases, that forms when temperatures rise sufficiently for melting to occur in the Earth's crust or mantle. (Ch. 4)

Mantle   The thick shell of dense, rocky matter that surrounds the Earth's core. (Ch. 2)

Mass-wasting   The movement of Earth materials downslope as a result of the pull of gravity. (Ch. 6)

Maturation   A series of complex physical and chemical changes that break down organic material into simpler liquid and gaseous hydrocarbons. (Ch. 11)

Mechanical weathering   The disintegration or physical breakup of rocks with no change in their chemistry or mineralogy. (Ch. 14)

Medical geology   The subfield of geology that studies the effects of chemical elements in the environment on human and animal health. (Ch. 17)

Metallic mineral resources   Minerals mined specifically for the metals that can be extracted from them. (Ch. 13)

Metamorphic rock   Rock whose original compounds or textures, or both, have been transformed into new compounds and new textures by reactions in the solid state as a result of high temperature, high pressure, or both. (Ch. 1)

Meteor   A glowing fragment of extraterrestrial material passing through the atmosphere. (Ch. 10)

Meteorite   A fragment of extraterrestrial material that strikes the surface of the Earth. (Ch. 10)

Meteorology   The study of the Earth's atmosphere and weather processes. (Ch. 9)

Mineral   Any naturally formed, crystalline solid with a definite chemical composition and a characteristic crystal structure. (Ch. 2)

Mineral deposits   Volumes of rock that contain enrichments of one or more minerals. (Ch. 13)

Mineral resource   Any element, compound, mineral, or rock that can be extracted from the ground and is of potential value as a commodity. (Ch. 13)

Modified Mercalli Intensity Scale   A scale used to compare earthquakes based on the amount of vibration people feel during low-magnitude quakes and the extent of damage to buildings during high-magnitude quakes. (Ch. 3)

Molecule   The smallest unit that retains all the properties of a compound. (Ch. 2)

N

Natural gas   Naturally occurring hydrocarbon compounds that are gaseous at ordinary temperatures and pressures. (Ch. 11)

Natural hazards   The wide range of natural circumstances, materials, processes, and events that are hazardous to humans, such as locust infestations, wildfires, or tornadoes, in addition to strictly geologic hazards. (Essay II)

NIMBY   "Not in my backyard"; a social attitude toward waste disposal that reflects the fact that no one wants to live next door to a waste-disposal facility. (Essay IV)

Nonmetallic mineral resource   A mineral resource, such as salt, gypsum, or clay, that is not used for the metals it contains but for its other physical or chemical properties. (Ch. 13)

Nonpoint sources   Broad areas where pollutants originate and enter the natural environment. (Ch. 17)

Nonrenewable resource   A resource that cannot be replenished or regenerated on a human time scale. (Essay III)

Nor'easter   A cyclonic storm that originates in regions of atmospheric instability in middle latitudes over land or coastal regions; named for its wind, which blows from the northeast. (Ch. 9)

Normal fault   A fault that is caused by tension—movement that tends to pull crustal blocks apart. (Ch. 3)

Nuclear energy   The heat produced during the controlled transformation of suitable radioactive isotopes. (Ch. 12)

O

Oceanography   The study of the physical, chemical, biologic, and geologic aspects of the Earth's oceans. (Ch. 9)

Ocean thermal energy conversion (OTEC)   The use of heat pumps to withdraw energy from ocean water. (Ch. 12)

Oil   The liquid form of petroleum. (Ch. 11)

Oil shale   A shale containing waxlike substances that break down to liquid and gaseous hydrocarbons when heated. (Ch. 12)

Ore   An aggregate from which one or more minerals can be extracted profitably. (Ch. 13)

Ozone hole   The phenomenon of stratospheric ozone depletion centered over the south polar region. (Ch. 18)

Ozone layer   A zone in the stratosphere where ozone occurs in unusually high concentrations. (Ch. 18)

P

Paleoflood hydrology   The study of ancient floods. (Ch. 8)

Paleoseismology   The study of prehistoric earthquakes. (Ch. 3)

Parent material   The rock and mineral regolith from which soil develops. (Ch. 14)

Particulates   Pollutants that are carried in suspension in the air as extremely fine, solid particles. (Ch. 18)

Passive remediation   An approach to the cleanup of contaminated groundwater which says, "Let nature take its course." (Ch. 17)

Peak discharge   The point at which the maximum discharge for a particular flood is reached. (Ch. 8)

Peat   An unconsolidated deposit of plant remains with a carbon content of about 60 percent. (Ch. 11)

Percolation   The movement of groundwater in a saturated zone. (Ch. 15)

Permeability   A measure of how easily a solid allows fluids to pass through it. (Ch. 15)

Persistant contaminants   Contaminants that take a very long time to decompose or to dissipate their harmful properties. (Ch. 17)

Petroleum   Gaseous, liquid, and semi-solid naturally occurring substances that consist chiefly of hydrocarbon compounds. (Ch. 11)

Photosynthesis   A process whereby light energy is used to cause carbon dioxide to react with water, synthesizing organic substances (carbohydrates) and releasing oxygen in the process. (Ch. 18)

Photovoltaic cells   Devices used to convert solar energy directly into electricity. (Ch. 12)

Physical geology   The study of the processes that operate at or beneath the surface of the Earth and the materials on which those processes operate. (Introduction)

Placer   A deposit of heavy minerals that have been concentrated by mechanical processes. (Ch. 13)

Plate tectonics   The special branch of tectonics that deals with the processes by which the lithosphere is moved laterally over the aesthenosphere. (Essay I, Ch. 1)

Point sources   Discrete sources of contaminants that can be represented by single points on a map. (Ch. 17)

Pollutants   Materials that have harmful impacts and degrade the environment; see also *Contaminants*. (Essay IV)

Pollution   Materials with harmful impacts on the natural environment, or the act of releasing such materials. (Essay IV)

Porosity   The percentage of the total volume of a body of regolith or bedrock that consists of open spaces or pores. (Ch. 15)

Precious metals   Gold, silver, and platinum group metals. (Ch. 13)

Precursor   A small physical change that precedes a catastrophic event. (Essay II)

Prediction   A statement of probability based on scientific observation. (Essay II)

Primary recovery   When oil flows from a well under its own pressure or flows easily with pumping. (Ch. 11)

Prior appropriation   A doctrine that grants water-use rights on the basis of historical precedent. (Ch. 15)

Pyroclast   A fragment of rock ejected during a volcanic eruption. (Ch. 4)

Pyroclastic flow   A hot, highly mobile flow of tephra that rushes down the flank of a volcano during a major eruption. (Ch. 4)

Pyroclastic rock   A rock formed from pyroclasts. (Ch. 4)

P waves   Compressional, or primary, waves. (Ch. 3)

R

Rad   A unit used to quantify radiation in terms of the amount of energy absorbed by biologic tissue. (Ch. 16)

Radiation   Alpha particles, beta particles, and gamma rays that are emitted from disintegrating nuclides during radioactive decay. (Ch. 16)

Radiatively active gases   Atmospheric gases that are efficient absorbers of infrared radiation; also known as *greenhouse gases*. (Ch. 18)

Radioactive waste   Leftover radioactive material or equipment for which there is no further economic use; also called *radwaste*. (Ch. 16)

Radon   A colorless, odorless radioactive gas produced by the radioactive decay of uranium in soils derived from uranium-bearing rocks. (Ch. 18)

Rapid onset hazards   Events that strike quickly and with little warning, such as earthquakes, flash floods, or sudden windstorms. (Essay II)

Raveling   The process of sinkhole formation initiated by the gradual downward movement of unconsolidated material into the aven. (Ch. 7)

Recharge   Replenishment of groundwater that occurs when rainfall and snowmelt enter the ground in recharge areas, where precipitation infiltrates and percolates downward through soil layers to reach the saturated zone. (Ch. 15)

Recurrence interval   The average interval between occurrences of two events, such as floods, of equal magnitude. (Ch. 8)

Recycling   The process of taking apart an old product and using the material it contains to make a new product. (Essay IV)

Reef   A generally ridgelike structure composed chiefly of the calcareous remains of sedentary marine organisms (e.g., corals, algae). (Ch. 9)

Regolith   The irregular blanket of loose rock debris that covers the Earth. (Essay I, Ch. 14)

Rem   A unit that measures the damage done to organisms by alpha particles, beta particles, and gamma rays. (Ch. 16)

Renewable resource   A resource that can be renewed, replenished, or regenerated. (Essay III)

Reserve   That portion of a resource for which there is detailed information concerning the quality and quantity of material available. (Ch. 13)

Reservoir   A natural storage place for materials or energy in Earth cycles. (Ch. 1)

Reservoir rock   A rock with a high proportion of interconnected pore spaces in which petroleum accumulates. (Ch. 11)

Residence time   The average length of time a substance remains in a particular reservoir. (Chs. 1, 18)

Resource   Any potentially valuable mineral compound, including hypothetical or speculative deposits of the material. (Ch. 13)

Resource recovery   Any means of extracting economically valuable components, whether materials or energy, from wastes. (Essay IV)

Resurgent dome   A volcanic dome formed when magma reenters the chamber and causes uplifting of the collapsed floor of a caldera. (Ch. 4)

Retardation   A measure of how much a contaminant has been slowed down relative to normal groundwater flow. (Ch. 17)

Reuse   Using a product repeatedly instead of throwing it away. (Essay IV)

Reverse fault   A fault that arises from compression; movement that tends to push crustal blocks together. (Ch. 3)

Rhyolite   A fine-grained igneous rock with high silica content. (Ch. 4)

Richter magnitude scale   A scale, based on the recorded amplitudes of seismic body waves, for comparing the amounts of energy released by earthquakes. (Ch. 3)

Rills   Small, narrow channels that first begin to cut into the regolith during erosion due to running water. (Ch. 14)

Riparian rights   A principle that views bodies of water as a common property resource, owned by all who have access to it. (Ch. 15)

Risk assessment   The process of establishing the probability that a hazardous event of a particular magnitude will occur within a given period and estimating its impact, taking into account the locations of buildings, facilities, and emergency systems in the community, the potential exposure to the physical effects of the hazardous situation or event, and the community's vulnerability when subjected to those physical effects. (Essay II)

River   A stream of considerable volume with a well-defined passageway. (Ch. 8)

Rock Any naturally formed, nonliving, firm, and coherent aggregate mass of mineral matter that constitutes part of a planet. (Ch. 2)

Rock cycle The cyclic movement of rock material, in the course of which rock is created, destroyed, and altered through the operation of internal and external Earth processes. (Ch. 1)

Runoff The portion of precipitation that flows over the surface of the land. (Ch. 8)

Run-up The water level achieved by a tsunami once it hits the shore; expressed as height in meters above normal high tide. (Ch. 5)

S

Safety factor The ratio of shear strength to shear stress, which determines the tendency of a slope to fail. (Ch. 6)

Salinization Contamination of groundwater by salt water. (Ch. 17)

Sanitary landfill A disposal site for waste in which a layer of soil isolates the waste from birds, insects, and rodents and minimizes the amount of precipitation that can infiltrate the refuse pile. (Ch. 16)

Saturated zone The groundwater zone in which all pore spaces are filled with water. (Ch. 15)

Scarp A cliff that is formed where displaced material has moved away from undisturbed ground upslope. (Ch. 6)

Secure landfill A landfill that is specially designed and engineered to contain hazardous materials. (Ch. 16)

Sediment Regolith that has been transported by any of the external Earth processes. (Ch. 1)

Sedimentary rock Any rock formed by chemical precipitation or by sedimentation and cementation of mineral grains transported to a site of deposition by water, wind, ice, or gravity. (Ch. 1)

Sediment pollution The contamination of sediment by hazardous substances; or situations in which sediment acts as a pollutant. (Ch. 17)

Sediment yield The amount of sediment eroded from the land by runoff and transported by streams. (Ch. 14)

Seiche A periodic standing-wave oscillation of the water surface in an enclosed water body such as a lake. (Ch. 5)

Seismic belts Large tracts of the Earth's surface that are subject to frequent earthquakes. (Ch. 3)

Seismic gaps Places along a fault where earthquakes have not occurred for a long time even though tectonic stresses are still active and elastic energy is steadily building up. (Ch. 3)

Seismic sea waves (also called *tsunamis*) Long-wavelength ocean waves produced by sudden movement of the seafloor. (Ch. 3)

Seismic waves Elastic disturbances that spread out in all directions from an earthquake's point of origin. (Ch. 3)

Seismograms The characteristic signatures and arrival times of each seismic wave of an earthquake, as recorded by seismographs. (Ch. 3)

Seismograph A device used to record the shocks and vibrations caused by earthquakes. (Ch. 3)

Seismology The study of earthquakes. (Ch. 3)

Sensitive soils Materials that lose shear strength as a result of remolding. (Ch. 6)

Septic tank A holding tank designed to receive domestic sewage from a single household. (Ch. 16)

Shear strength The internal resistance of a body to movement. (Ch. 6)

Shear stress The force acting on a body that causes movement of the body parallel to a slope. (Ch. 6)

Shear waves (*S* waves) Seismic body waves that consist of alternating series of sideways movements, each particle in the deformed solid being displaced in a direction perpendicular to the direction of wave travel. (Ch. 3)

Sheet wash Erosion by water through overland flow during heavy rains; also called *sheet erosion*. (Ch. 14)

Shield volcano A volcano that emits fluid lava and builds a broad, dome-shaped edifice. (Ch. 4)

Silicate mineral A mineral that contains the silicate anion, $(SiO_4)^{-4}$. (Ch. 2)

Sinkhole A large dissolution cavity that is open to the sky. (Ch. 7)

Sinkhole karst A landscape dotted with closely spaced, circular basins of various sizes and shapes. (Ch. 7)

Slide The rapid displacement of rock or sediment in one direction, with no rotation. (Ch. 6)

Slump A type of slope failure involving the downward and outward movement of rock or regolith along a curved, concave-up surface. (Ch. 6)

Smog A yellowish-brownish haze that results when airborne contaminants interact with sunlight and undergo a variety of complex photochemical reactions. (Ch. 18)

Soil Earth material that has been broken down and altered in such a manner that it can support rooted plant life. (Ch. 14)

Soil fertility The ability of a soil to provide the nutrients needed for plant growth. (Ch. 14)

Soil horizons The subhorizontal weathered zones formed as a soil develops. (Ch. 14)

Soil profile The succession of soil horizons between the surface and the underlying parent material. (Ch. 14)

Solar energy Energy that reaches the Earth from the sun. (Ch. 12)

Solar system The sun and the group of objects in orbit around it. (Essay I)

Source reduction The process of cutting back on waste generation by reducing the volume of products, increasing the useful lifetime of products, reducing the amount of packaging, and decreasing consumption. (Essay IV)

Source rock Rock containing organic material that is eventually converted into oil and natural gas. (Ch. 11)

Splash erosion The spattering of small particles of loose soils dislodged as raindrops strike bare ground. (Ch. 14)

Spring A flow of groundwater that emerges naturally at the ground surface. (Ch. 15)

Stage The height of a body of water above a locally defined reference surface. (Ch. 8)

Stones Meteorites made primarily of silicate minerals, including many of the same minerals that are abundant in the Earth's crust, which resemble terrestrial igneous rocks. (Ch. 10)

Stony-iron Meteorites composed of mixtures of stony (silicate) and metallic material. (Ch. 10)

Stoping The process of sinkhole formation initiated by the sudden wholesale collapse of the bedrock "roof" into the underground void; the process of roof collapse can also occur in magma chambers and in underground mines. (Ch. 7)

Storm surge   An abnormal, temporary rise in water levels in oceans or lakes due to low atmospheric pressure associated with storms. (Ch. 9)

Stratosphere   The layer of the atmosphere that extends from the top of the troposphere (the tropopause) to an altitude of 50 km. (Ch. 18)

Stratovolcano   A volcano that emits both tephra and viscous lava and that builds up steep, conical mounds of interlayered lava and pyroclastic deposits. (Ch. 4)

Stream   A body of water that flows downslope along a clearly defined natural passageway, transporting particles and dissolved substances. (Ch. 8)

Streamflow   The flow of surface water in a well-defined channel. (Ch. 8)

Strike-slip fault   A fault that arises from stresses that lead to horizontal, or translational, motions of the fault blocks. (Ch. 3)

Subsidence   The sinking or collapse of a portion of the land surface. (Ch. 7)

Surf   Wave activity between the line of breakers and the shore. (Ch. 9)

Surface waves   Seismic waves that are guided by and restricted to the Earth's surface. (Ch. 3)

S **waves**   Shear, or secondary, waves. (Ch. 3)

System   A portion of the universe that can be isolated from the rest of the universe for the purpose of observing changes. (Ch. 1)

T

Tar   An oil that is viscous and does not flow. (Ch. 11)

Tar sands   Deposits of dense, viscous, asphaltlike oil that are found in a variety of sedimentary rocks and unconsolidated sediments. (Ch. 12)

Technological hazards   Hazards associated with everyday exposure to hazardous substances, such as radon, mercury, asbestos fibers, or coal dust, usually through some aspect of the use of these substances in our built environment. (Essay II)

Tephra   A loose assemblage of pyroclasts. (Ch. 4)

Tephra cone   A buildup of tephra around the vent of a volcano. (Ch. 4)

Terrestrial planets   The innermost planets of the solar system (Mercury, Venus, Earth, and Mars), which have high densities and rocky compositions. (Essay I)

Thermal inversion   A phenomenon whereby a warm air mass moves into an area and traps a pocket of cool air underneath it. (Ch. 18)

Thermal pollution   Excess heat that can harm or kill plants or animals. (Ch. 17)

Thermal spring   Descending groundwater that comes into contact with the hot rock near the magma chamber of a volcano, is heated, and rises to the surface along fractures. (Ch. 4)

Tidal bore   A large, turbulent, wall-like wave of water caused by the meeting of two tides or by the rush of tide up a narrow inlet, river, estuary, or bay. (Ch. 9)

Tidal energy   Energy derived from gravitational interactions among the Earth, the moon, and the sun. (Ch. 12)

Tide   The twice-daily rise and fall of the ocean's surface that results from the gravitational attraction of the moon and sun. (Ch. 9)

Topsoil   A layer of fertile soil near or at the surface of the uppermost horizon in a soil profile. (Ch. 14)

Tornado   A cyclonic storm with a very intense low-pressure center. (Ch. 9)

Toxicity   The capacity of a substance to cause adverse effects in a living organism. (Ch. 17)

Trade winds   A globe-encircling belt of winds in the low latitudes, which blow from the northeast in the Northern Hemisphere and the southeast in the Southern Hemisphere. (Ch. 9)

Transform fault margin   A type of plate margin along which two plates slide past one another. (Ch. 1)

Translational slide   The rapid displacement of masses of rock or sediment involving the movement of relatively coherent blocks of material along well-defined, inclined surfaces such as faults, foliation planes, or layering. (Ch. 6)

Trap   A geologic situation that includes a source rock (to contribute organic material), a reservoir rock (to allow for the migration of oil), and a cap rock (to stop the migration). (Ch. 11)

Troposphere   One of the four thermal layers of the atmosphere, which extends from the surface of the Earth to an altitude of 10 to 16 km. (Chs. 9, 18)

Tsunami   A very long wavelength ocean wave that is generated by a sudden displacement of the sea floor; also called *seismic sea wave*. (Ch. 5)

Tsunami earthquake   An earthquake which generates a tsunami that is anomalously large with respect to the magnitude of the quake. (Ch. 5)

Tsunamigenic earthquake   An earthquake that generates a tsunami. (Ch. 5)

T-value   The tolerable annual rate of soil loss; losses that are balanced or offset by natural regeneration of the soil. (Ch. 14)

Typhoon   A term used to describe a tropical cyclonic storm that originates in the western Pacific Ocean. See *Hurricane*. (Ch. 9)

U

Unconfined aquifer   An aquifer whose upper surface coincides with the water table and therefore is in contact with the atmosphere (through overlying soil layers). (Ch. 15)

Unconventional hydrocarbon   Any hydrocarbon fuel that occurs in an atypical reservoir, such as a tight sandstone or a geopressurized zone, or in an unconventional form. (Ch. 12)

Uniformitarianism   The principle that the same external and internal processes we recognize in action today have been operating unchanged, though at different rates, throughout most of the Earth's history. (Ch. 1)

Upstream flooding   Severe but local flooding caused by intense, infrequent storms of short duration. (Ch. 8)

V

Viscosity   The internal property of a substance that offers resistance to flow. (Ch. 4)

Volcano   The vent from which magma, solid rock debris, and gases erupt. (Ch. 4)

Vulnerability   A concept that encompasses the physical effects of a natural hazard as well as the status of people and property in the area. (Introduction)

W

Waste   The residual materials and by-products that are generated by human use of the Earth's resources and wind up unwanted and unused. (Essay IV)

Water table   The upper surface of the saturated zone of groundwater. (Ch. 15)

Wave base   The effective lower limit of wave motion, which is half of the wavelength. (Ch. 9)

Wave-cut cliff   A cliff cut by wave action at the base of a rocky coast. (Ch. 9)

Wave energy   A secondary expression of solar energy; used to ring bells and blow whistles that serve as navigational aids. (Ch. 12)

Wavelength   The distance between the crests of adjacent waves. (Ch. 5)

Wave trap   A situation in which the energy of a long section of a wave is concentrated on a particular stretch of coastline as a result of seafloor topography or shoreline configuration. (Ch. 5)

Weather   The state of the atmosphere at a given time and place. (Ch. 9)

Weathering   The chemical alteration and physical breakdown of rock and sediment when exposed to air, moisture, and organic matter. (Ch. 14)

Wet deposition   Materials transported in suspension in the air, which eventually settle out as precipitation. (Ch. 18)

Wind energy   A secondary expression of solar energy used to power ships and windmills. (Ch. 12)

Z

Zone of aeration   A zone in which open spaces in regolith or bedrock are partially filled with air; also called the *vadose zone* or the *unsaturated zone*. (Ch. 15)